WHAT WILL HAPPEN
DURING POLE SHIFT?

The ultimate disaster! Enormous tidal waves will roll across the continents as oceans become displaced from their basins. Hurricane winds of hundreds of miles per hour will scour the planet. Earthquakes greater than any ever measured will change the shape of the continents. Volcanoes will pour out huge lava flows, along with poisonous gases and choking ash. Climates will change instantly, and the geography of the globe will be radically altered. If the pole shift is less than a full 180°, the polar icecaps will melt rapidly, raising sea levels, while new icecaps will begin to build. And large numbers of organisms, including the human race, will be decimated or even become extinct, with signs of their existence hidden under thick layers of sediment and debris or at the bottom of newly established seas . . .

"Thoughtful and disturbing . . ."

—BRISTOL PRESS

**PREDICTIONS
AND PROPHECIES OF THE
ULTIMATE DISASTER**

POLE SHIFT

JOHN WHITE

**ARE
PRESS**

**ASSOCIATION FOR
RESEARCH AND
ENLIGHTENMENT**

A.R.E. Press, Virginia Beach, Virginia

Grateful acknowledgment is made for permission to reprint the following: Excerpts from *Readings on Earth Changes*, by Aron and Doris Abrahamsen, copyright © 1973 by Aron and Doris Abrahamsen. Reprinted by permission of the authors. Excerpts from *The Secret Doctrine*, by H. P. Blavatsky, copyright © 1938 by The Theosophical Publishing House. Reprinted by permission of The Theosophical Publishing House. Excerpts from *Cataclysms of the Earth*, by Hugh Auchincloss Brown, copyright © 1967 by Hugh Auchincloss Brown. Reprinted by permission of Steiner Books. Excerpts from the Edgar Cayce readings, copyright © 1971 by The Edgar Cayce Foundation. Reprinted by permission. Excerpts from "Frozen Mammoths and Modern Geology," by W. R. Farrand, *Science*, Vol. 133, pp. 729–35, 17 March 1961. Copyright © 1961 by the American Association for the Advancement of Science. Reprinted by permission. Excerpts from *Season of Changes, Ways of Response*. Copyright © 1974 by Associations of the Light Morning. Reprinted by permission of the publisher. Excerpts from *The Ultimate Frontier*, by Richard Kieninger, copyright © 1963 by The Stelle Group. Reprinted by permission of The Stelle Group. Excerpts reprinted by permission of Coward, McCann & Geoghegan, Inc., from *A World Beyond*, by Ruth Montgomery. Copyright © 1971 by Ruth Montgomery. Excerpts from "The Riddle of the Frozen Mammoths," by Ivan Sanderson. Reprinted from *The Saturday Evening Post*, copyright © 1960 The Curtis Publishing Company. Reprinted by permission of Sabina Sanderson. Excerpts from Frank C. Hibben's *The Lost Americans* are copyright © 1961 by Frank C. Hibben. Excerpts from *Life & Death of Planet Earth*, by Tom Valentine, copyright © 1977 by Tom Valentine. Reprinted by permission of the author. Excerpts from *Worlds in Collision*, by Immanuel Velikovsky, copyright © 1950 by Immanuel Velikovsky. Reprinted by permission of Doubleday & Company, Inc. Excerpts from *Earth in Upheaval*, by Immanuel Velikovsky, copyright © 1955 by Immanuel Velikovsky. Reprinted by permission of Doubleday & Company, Inc. Excerpts from "When the Sky Rained Fire: The Velikovsky Phenomenon," by Fred Warshofsky, reprinted with permission from the December 1975 *Reader's Digest*. Copyright © 1975 by The Reader's Digest Association, Inc. Excerpts from *Portents and Promises* by Joey Jochmans, copyright © 1979 by Joey Jochmans. Used by permission of the author.

This Berkley book contains the complete text of the original hardcover edition. It has been completely reset in a type face designed for easy reading, and was printed from new film.

POLE SHIFT

Printing History
Doubleday edition published 1980

A.R.E. Press,   13th Printing, July 1996

ISBN: 87604-162-4

Acknowledgments

The research and writing that make up this book were supported by many people who generously gave their time and information. Without their help the project would never have been completed, and it is with deep gratitude that I acknowledge their contributions.

Piero and Francesca Lattanzi, visiting geologists at Yale, provided technical assistance in dealing with data from the earth sciences.

Renee Kra, editor of the journal *Radiocarbon*, aided my research on mammoths in Chapter 2, as did Dr. Ellen M. Prager at the University of California, Berkeley. Sabina Sanderson kindly allowed me to use quotations from her husband, Ivan T. Sanderson's, article in *The Saturday Evening Post*. Dr. William R. Farrand and *Science* magazine did the same for Farrand's article.

Hugh Auchincloss Brown, Jr., with whom I met and corresponded, was most helpful on the chapter about his father. And without the unstinting support of Mrs. Dorothy Starr, publisher-editor of the *Pole Watchers' Newsletter*, the chapter could not have been written. Her service extended into many other areas of the book. I am greatly indebted to her for constant and generous assistance.

At the direction of Dr. Immanuel Velikovsky, Leroy Ellenberger and Jan Sammer reviewed Chapter 6, making useful comments and correcting some errors.

In Chapter 7, Peter Warlow reviewed my use of his work, offering helpful corrections and suggestions.

Charles Hapgood graciously provided research background, constructive criticism and permission to quote from his works and use his illustrations (in the photograph section) in Chapter 8. William Markowitz, of Nova University, also gave me useful data.

Chapter 9 would not have been possible without the thoughtfulness of Riley T. Crabb, director of the Borderland Science Research Foundation, in Vista, California, who provided a copy of Adam Barber's book.

Kenneth M. Skidmore, editor of A.R.E. Publications, verified the quotations from Edgar Cayce in Chapter 11 and gave me permission to use them. Paul and Walene James offered guidance and critical comments on the chapter.

Aron and Doris Abrahamsen were most helpful to me in Chapter 12. They provided biographical data and background information, along with permission to quote from the Abrahamsen readings. Dr. William Kautz and Dr. Jeffrey Goodman were also helpful in reviewing the chapter.

Without the biographical information given by Paul Solomon, along with access to the Solomon readings, Chapter 13 could not have been completed. I am grateful for his aid and permission to quote from the readings.

In "Other Visions, Other Voices," the Heritage Store kindly gave me permission to quote from *Season of Changes, Ways of Response*. So did Lenora Huett for her book *The Path to Illumination*. Baird Wallace, author of *The Space Story and the Inner Light*, and Bella Karish, channel for the reading from Eternal Cosmos quoted here, likewise graciously consented to allow use of their work.

The chapter on Native American prophecies was assisted by many people, to all of whom I offer heartfelt thanks: Sun Bear and Wabun, Alex Hladky, Thomas Banyacya, Frank Waters, Ted

Krueger, Nathan Koenig, Twylah Nitsch and Friends of the Hopi.

Chapter 16 is indebted primarily to Dr. Joey Jochmans, who gave unstintingly of his research material and unpublished writings. I am most grateful for allowing prepublication use of his material.

Tom Valentine gave me strong support in researching and writing this book, especially Chapter 17. His permission to quote from his publications and to use illustrations from *The Life & Death of Planet Earth* is most generous. I am also grateful to Malcolm Carnahan, president of The Stelle Group, for permission to quote from *The Ultimate Frontier*. Astrologers Robert Hand and Capel McCutcheon graciously provided me with data to verify The Stelle Group's prediction. I was also assisted in the matter by astrologers Brenda Weiner and Dane Rudhyar. Two graduate students of astronomy at Yale, Paul Clark and Horace Smith, likewise assisted by performing astronomical calculations.

Chapter 19 would not have been possible without the help of Rosemarie Stewart, editor of the Theosophical Publishing House, who guided me through the massive volumes of *The Secret Doctrine*.

Many others contributed to this book in various ways. I received research help from Ralph Amelan, Larry E. Arnold, Tom Bearden, Dr. Doris Cecarelli, Trevor J. Constable, William R. Corliss, Susanne Crayson, Sophie and Charles Dutton, C. Leroy Ellenberger, William J. H. Foster, Al Fry, Harry Gaetano, Lewis M. Greenberg, Johann Kloosterman, Rita Livingston, William Markowitz, Dr. Irving Michelson, Wally and Jenny Richardson, W. G. Roll, Ed Rosenfeld, Wes Thomas, Frank Tribbe, Dr. Thomas C. Van Flandern, Alan Vaughan, Roger Wescott, Edward M. Weyer, and my father, Robert P. White. After the manuscript was in rough draft, valuable criticism was offered by Larry E. Arnold, Jerome Ellison, Dr. Jeffrey Goodman, Dr. Gerrit L. Verschuur, Dr. Roger W. Wescott, and Dr. David D. Zink.

To my illustrator, Sue Tilberry, I offer a grateful "Well done."

To my typist, Bobbi West, I offer thanks for patiently turning my rough copy into smooth form.

The gratitude I have for these people, and to anyone I may have neglected to mention, should not be construed as meaning they share any of the responsibility for the data, interpretations and opinions that follow. That responsibility is mine alone.

To the higher humanity, *Homo noeticus*

And to those sages and seers, ancient and modern,
whose selfless service has prepared the birthplace.

Contents

Introduction

By ALAN VAUGHAN

If an uneasy feeling grips you as you read this book, as it did me, then John White is making his point. As farfetched as it may seem by current scientific orthodoxies, an occult tradition and a scientific underground merge to suggest that a cataclysmic shift of our planet's poles may be coming soon.

White's thorough scholarship in unearthing both pro and con pronouncements from scientists and unnerving prophecies from psychics, ancient and modern, puts the thesis of polar shift under careful scrutiny. Has it happened before? Will it happen again? If so, when? And what can we do about it?

If White's thesis has only a thread of truth, it could profoundly change geology's concept of how epochs of life—and death—change on the face of the earth. If the psychic predictions have any validity, then we may learn firsthand by the turn of the century what happens when the poles of the earth shift.

Evaluating psychic predictions in advance is not easy, I have learned in a ten-year study of prophecy (*Patterns of Prophecy*, Dell, 1976). Because psychics are right about some things does not mean they are right about others. Indeed, I suspect that by the very nature of the universe it is impossible for psychics to be correct in their predictions most of the time. Possibilities change and shift according to the free will of human consciousness; blueprints of life are continually being revised.

On the other hand, there are cases of startling prophetic accuracy over a period of many years.

The symbolic metaphors in which prophecy is so often couched indicate an origin from the deepest unconscious, much like dreams. It may be that predicted dates for prophesied events are mere embroidery from the conscious mind. In examining thousands of predictions, I have learned not to take dates very seriously. Yet, what if a number of psychics agree on taking dates very seriously? Further, what if a number of psychics agree more or less on the time for some predicted event—as they do in predicting the shift of the poles between 1998 and 2001? Are they elaborating on some previous prediction of which they have conscious knowledge? Or are they independently tuning in on some extraordinarily cataclysmic event?

Parapsychology cannot yet supply the answers to those questions. However, in my opinion, it would be foolish to take extreme actions in either believing or disbelieving. It would be foolish to invest your life savings in an ark; it would be equally foolish to reject this book out of hand.

Read it carefully; catalog your objections and reexamine them. Does orthodox science *really* have the answers? Keep in mind that its leading theories dramatically shift from time to time; there are no unassailable truths, only the latest theorizing.

I can recall as a fourth-grade geography student, when admiring a relief map of the globe made by our class, telling the teacher that it looked as if Africa and South America were once joined. That's only a coincidence, she told me, for we know it could not have been that way. But thirty years later the theory of continental drift has been acclaimed as orthodoxy, and Africa and South America *were* once joined.

So keep an open mind as you read this provocative book. It could be only a coincidence that "crackpot" scientists and modern-day seers foresee a similar fate for the earth. Or it could be that they are tapping a cosmic pattern of action that will be repeated within our children's lifetimes. It could be enormously exaggerated. It might even be influenced by the beliefs of human consciousness.

I, for one, will keep a careful eye out for any fulfilled predictions that are on the psychic timetable to precede the prophesied pole shift. In the meanwhile, I'm grateful to John White for bringing together this material in a balanced and thoughtful way. If there's something to it, we should be finding out sooner than we may like to think.

San Francisco
October 1978

What will the future bring? From time immemorial this question has occupied men's minds, though not always to the same degree. Historically, it is chiefly in times of physical, political, economic and spiritual distress that men's eyes turn with anxious hope to the future, and when anticipations, utopias and apocalyptic visions multiply. One thinks, for instance, of the chiliastic expectations of the Augustan age at the beginnings of the Christian era, or of the changes in the spirit of the West which accompanied the end of the first milennium. Today, as the end of the second millenium draws near, we are again living in an age filled with apocalyptic images of universal destruction.

CARL JUNG, *The Undiscovered Self*

Predicting the future has always been an uncertain business. But we can confidently predict an ever-increasing volume of end-of-the-world predictions between now and the year 2000. Perhaps the year 1000 meant little to men of the Middle Ages, but the year 2000 means a great deal to modern numerologists, who believe that there is an overpowering significance to certain numbers and dates. Already there have been numerous predictions that "something," a great war, a great religious revival, perhaps something as catastrophic as the end of the world will take place in that year.

DANIEL COHEN, *How the World Will End*

Preface

The Doomsday question—When and how will the human race die out?—has assumed a new perspective as scientific knowledge has advanced in several areas.

The subject is a matter of heated controversy on campuses and in laboratories across the nation. Interviews with astronomers, geophysicists, biologists, and health experts disclose that they believe total human extinction is not necessarily as distant a possibility as many of us would choose to think.

While most scientists regard as remote the likelihood of human extinction in the near future, it is real enough, some assert, that governments should start seeking ways to limit the risks. As the earth hurtles through space at 1.3 million miles an hour, there is the chance of catastrophe from both cosmic and terrestrial causes, but the damage might be reduced by timely precautions.

Doomsday predictions are almost commonplace nowadays. Hardly a week goes by, it seems, without some new warning of worldwide destruction being headlined in the media, as in the *Chicago Tribune* (18 November 1978, p. 14) article quoted above. Visions of nuclear Armageddon and ecological suicide abound. Lately we have added drought, carbon dioxide buildup, new diseases, and the onset of another ice age to the list of impending catastrophes. Beyond these are the more distant but nevertheless real possibilities of annihilation through contact with supernovae or black holes in space, and impact with celestial bodies such as a comet or an asteroid. As astronomer Gerrit L. Verschuur says in his recent *Cosmic Catastrophes,* "Prognostications about our future life on earth . . . in view of all the cosmic dramas that can play themselves out . . . are likely to be less than optimistic" (p.202).

The number of works dealing with the variety of ways in which global disaster might strike grows almost daily. To name only a few that came to my attention in the seven years I have been working on this project, there is Daniel Cohen's 1973 *How the World Will End* and that important 1974 "distant early warning" book by Alfred L. Webre and Phillip H. Liss, *The Age of Cataclysm.* Fred Warshofsky's *Doomsday: The Science ot Catastrophe* and Martin Ebon's *Doomsday: How the World Will End—and When,* both published in 1977, elaborated on the theme of "world's end"—a theme that reached bestseller status in Hal Lindsay's *The Late Great Planet Earth.* In 1978, Verschuur's *Cosmic Catastrophes* and Joseph Goodavage's *Our Threatened Planet* appeared, followed last year by Isaac Asimov's *A Choice of Catastrophes.* Fiction writers have also made their contribution to the image of planet-wide calamity. Larry Niven and Jerry Pournelle's recent *Lucifer's Hammer,* for example, dramatizes the death of civilization when a comet strikes the earth, while in Fred Hoyle's earlier science fiction novel *The Black Cloud* interstellar hydrogen gas nearly ends all terrestrial life. And let us not forget movies, especially *Star Wars,* in which Darth Vader's ultimate weapon, the Death Star, disintegrates an entire planet. Nearly all these were overlooked by *Time* (5 March 1979) in a roundup article, "The Deluge of Disastermania."

At the risk of being ignored by a public already numbed by dire predictions, I have written a book about a particular "apocalyptic image of universal destruction." I could not do otherwise. And if it finally proves that I was chasing a chimera, I will welcome the news. At the moment, however, I must attempt to draw the attention of science and society to still another apparent threat to the community of life.

The most common term for this new danger is "pole shift" or "polar shift." Also called "axis shift," "axis flip" or "a shifting of the poles," the concept underlying these terms is that of a sudden and radical displacement of the planet's axis of rotation or—an alternate view among pole shift theorists—a slippage of the planet's solid crust over the molten interior so that the polar locations change. This, "pole shift" means an event in which the North and South poles move—as much as 180°, according to some sources—either because the planet tumbles in space, changing its angular position relative to the sun, or because the geographical points on the surface of the planet marking its spin axis are shifted due to crustal slippage. The time involved is said by most sources to vary from a few days to as little as a few hours.

What would result from a pole shift? We can describe it as "the ultimate disaster." Enormous tidal waves would roll across the continents as the oceans became displaced from their basins. Electrical storms with hurricane winds of hundreds of miles per hour would sweep the planet. Tremendous earthquakes and lava flows would wrack the land. Poisonous gases and ash would fill the skies. Geography would be altered as seabeds rose and land masses submerged. Climates would change instantly. And if the shift were less than a full end-over-end, the polar ice caps, exposed to strong sunlight by having moved out of the frigid zones, would melt rapidly—within a few thousand years at most—while new ice caps would begin to build at the new polar locations. Last of all, huge numbers of organisms would be destroyed, including people, with signs of their existence hidden under thick layers of debris, sediment and ice or at the bottom of newly established seas.

Has this ever happened? Might it happen in the future? That is what this book is all about. I have attempted to be comprehensive in gathering data bearing on the possibility of a pole shift. The result is a wide-ranging report that presents descriptions of the ultimate disaster and, in a brief afterword, prescriptions for preparing to survive it.

If you ask in the scientific community, "Has it happened? Might it happen?" the vast majority will probably respond, "No, it hasn't happened and it will not happen. You're wasting our time and raising a false alarm." I must, in truth, admit at the outset that there is a substantial scientific case against the concept of sudden polar shifting. It will be presented throughout the book.

On the other hand, a number of independent researchers, claiming the scientific method as the basis of their information, maintain that such a worldwide catastrophe has indeed happened before—many times—and that it will happen again soon. In addition, some contemporary psychics are adding their voices to the chorus of warnings. These psychics have established a record as genuine channels for paranormal information in other fields, and therefore have some degree of authority for their predictions of impending planetary cataclysm at the end of this century. Last of all, certain ancient prophetic traditions have described signs of events that will precede an "end time," when the earth is shaken to its very foundations. According to various "keepers of the prophecy"—those who maintain and interpret their prophetic traditions—that time is now upon us.

Three lines of evidence, then, seem to be converging on a single idea: a shifting of the poles only two decades from now. That idea is what will be examined here. To anticipate the findings, I will say that the examination will prove inconclusive. The questions raised above cannot be answered definitively—yet. That is because the magnitude of the topic requires much vaster study than one man—a layman at that—can provide. Nevertheless, I feel I have drawn together for the first time all the key data bearing on the subject and have identified the various issues to be considered in further investigations.

My lack of expertise for addressing these questions was clear to me from the start. It was also clear, however, that if there was the slightest validity to the various predictions and prophecies, the human race and all other life on earth would be endangered on an apparently unprecedented scale. Nothing in modern history except the threat of nuclear annihilation comes close to equaling the lethal dimensions inherent in the possibility of a pole shift.

Therefore, as a thinking and concerned member of the community of life, I *had* to make this investigation and write this book. Because I am inexpert, I sought competent help whenever I could. Admittedly, my research has only skimmed the surface. Much deeper study of many critical issues is needed. This book is *not* an attempt to convince anyone that a pole shift is coming—only that there is significant indication that one *might* be coming and that the possibility therefore ought to be rigorously examined. A premature closure of the question could be fatal. The stakes may be no less than the life or death of civilization.

To defuse some of the criticism I think likely to arise, I will raise some questions for you, the reader, no matter what your educational or scientific background. Consider this: If you felt you had knowledge bearing on the immediate safety and welfare of the entire human race, what would you do? How would you behave? What would you see as the proper way to handle that information?

This philosophic-yet-pragmatic concern was foremost in my mind as I proceeded with my research. I therefore ask potential critics to consider the work, not the man, as the important focus of attention. If some parts of it are flawed and incomplete, even glaringly so, take what is useful in the spirit with which it is offered. The purpose of this book is to raise a question—to cause the subject of pole shift to be considered thoughtfully and to stimulate further investigations into this awesome subject. Outreach, not outrage, is what I hope for. The world needs it, even if the ultimate disaster never occurs.

In order to communicate widely, I have tried to make this book simple and readable. Many quotations are used in order to do justice to the people involved. Except for that, I have minimized the scholarly apparatus of footnotes in favor of a more streamlined design. Primary references are noted at the end of each chapter and quotation page numbers at the end of each quote. The book is simplified but not, I trust, oversimplified. Its structure is based on the three lines of evidence mentioned above. Part I sets the perspective—a perspective based on geological and archaeological mysteries that contain tantalizing hints about the need for a radical revision of our notions of planetary and human history. Part II examines the predictions of a pole shift by those theorists who say their evidence is drawn from scientific research. The third part looks at the modern psychic predictions, the fourth part at ancient prophecies. In each of these sections, I have tried to present the information in an interesting, reportorial fashion. The format for the chapters in Parts II–IV is basically the same: introductory material and historical background followed by the essential facts, with occasional critical comments. A discussion and evaluation of the data is made in Part V, where I try to get to "the big picture." Some possibilities are considered, some potentially fruitful areas for new research are identified, and some tentative conclusions are drawn.

In the closing pages, I offer a few suggestions on preparing for the ultimate disaster. If it should come to pass, those attuned in consciousness to the cosmic processes by which our civilization might draw to a close are those most likely to be part of the succeeding one. For the end of an age is also the beginning of another. Thus evolution proceeds. Thus the mighty drama of creation goes on, with higher forms of life upon the stage. This could be the outcome at the end of the century, but it is not inevitable. The choice, as you will see, may be ours.

Preface to the A.R.E.® Edition

As I write this in early 1985, enough time has passed to judge the validity of some of the predictions reported in *Pole Shift*. I feel it is important to comment briefly on them and on new evidence bearing on the pole shift concept.

Edgar Cayce's Predictions. I point out in Chapter 11, which deals with the Cayce predictions, that by the most liberal interpretation 1982 would be the outer limit for the earth change activity, "in the physical aspect of the west coast of America," to begin in the "early portion" of the 1958–'98 period he mentioned. Cayce had stated, "The earth will be broken up in the western portion of America." (3976-15) I also point out that the presence or absence of a "hit" for that prediction will certainly bear on our assessment of Cayce's pole shift prediction.

How do things stand in 1985? Clearly, California and the rest of the west coast of America are still there. So Cayce was wrong —in the timing. He may, nevertheless, be right with regard to seismic processes at work in that area. Eminent scientists have declared that a great earthquake is imminent, and state officials are girding for "the Big One." It is not a matter of "if" but of "when." One seismologist put the odds at 1 in 4 of a "killer quake" bringing death to 30,000 people and serious injury to another 100,000 by the year 2000. So, just as clearly, something potentially catastrophic is abuilding in the area that Cayce indicated in the 1930s.

As you will see in the book, Cayce's predictions, as well as those of others, indicate that the activity of our minds can have a direct influence on physical nature. This is termed biorelativity. Prayer, meditation, positive thought, and loving service are, from that point of view, instruments for pacifying the planet and bringing stability to unstable geological areas. You will also see in the book that the true prophet does not want to see his dire predictions come true; rather, he seeks to be contradicted. His prophetic words of foreseen disaster are intended to warn people in time to avert that disaster. Through a change of mind and behavior, people can to some extent defuse the circumstances leading up to catastrophe, as well as get out of the way in time.

True prophecy stems from deep cosmic attunement. It is not the product of logical extension—that is, merely rational projection. Nor is it precognition in the ordinary sense. Prophecy is distinguished from precognition by its moral dimension, the spiritual framework in which it is issued. All psychic abilities have some biological survival value, but prophecy goes beyond the biological realm by pointing to the divine, transcendent Source of Life, whose loving concern for human welfare is being ignored by people.

Thus, Edgar Cayce, "the sleeping prophet," cannot be said to have missed entirely in his earth changes predictions. It may strike some as a naive rationalization to consider the possibility that Cayce's warnings, amplified by others who likewise attuned to the planet's seismic processes after he brought attention to the situation, helped to reduce the severity of what would otherwise have occurred. Strictly speaking, we will not know until the next century whether his pole shift prediction is correct. But even if it does not happen—which we should fervently want to be the case—we must acknowledge that many of the signs and precursors to which Cayce pointed can be seen in the news every day, and we must acknowledge that Cayce pointed to them long before anyone else.

Two A.R.E. publications are recommended for those interested in Edgar Cayce's statements about earth changes and a pole shift. They are *Earth Changes Update* by Hugh Lynn Cayce and Mark Thurston's *Visions and Prophecies for a New Age*. Both appeared after *Pole Shift* was published.

The Jupiter Effect. This 1974 book predicted a great earthquake in California in 1982 due to an unusual alignment of the major planets. The authors, John Gribben and Stephen Plagemann, later withdrew their prediction, feeling that scientific debate had shown it was based on an untenable hypothesis. That didn't stop many other people from picking up this idea and running quite wild with it, however. Public attention was riveted on March 10, 1982, as the time when the predicted alignment would bring catastrophe through its effect on Earth. March 11 dawned quite routinely, leaving many so-called psychics and others who foresaw disaster looking rather silly, including some notable fundamentalist Christian ministers.

They had two reasons for embarrassment. First, their predictions had proved totally false. Second, they had not based their predictions on psychic means or even mere straightforward astronomical or astrological calculations. In reality, the much-feared alignment of planets never happened in March, 1982; it occurred instead on November 11, 1982. Even that was not, strictly speaking, an alignment. It was only an unusual bunching of the planets within a 72-degree arc on the far side of the sun, with Earth by itself on the other side. That was quite different from the March gathering of planets, in which Earth was included within the group on the same side of the sun across 100 degrees of arc.

In any case, nothing happened—either in March or November. Neither gravitation nor unusual solar activity triggered earthquakes and volcanic eruptions. This outcome was entirely in line with the pronouncements of the scientific community. I mention this because Chapter 18, which deals with the Stelle Group, includes a discussion of a pole shift being triggered by a near-grand alignment of planets scheduled for May 5, 2000. In view of the Jupiter non-Effect, one has to look rather skeptically at the trigger mechanism identified by the Stelle Group. In fairness to them, however, I must note that their sources say there will be various other factors bringing Earth to a critically unstable condition then, and the effect of the near-grand alignment will be much more powerful than the 1982 event.

Paul Solomon's Predictions. Some of the most specific and dire predictions are mentioned in Chapter 13, which deals with Paul Solomon. Since the predictions are supposed to have occured between 1982 and 1984, it is clear that they are incorrect.

Again, however, as with Cayce's predictions, there may nevertheless be some truth to them in terms of having identified earth changes that have simply been modified in intensity and timing. For example, in 1983 the U.S. Geological Survey reported that a 100-mile-long fault zone extended through the Midwest and that it is one of the most dangerous in the nation, with the potential to trigger a natural disaster unequaled in American history. Compare that with the statement through the Solomon source that the North American continent will be "split in half, down its very center." Disaster prediction is hardly the place to discuss poetic license, but that may be somewhat appropriate in discussing this particular prediction. It may also be so, as I noted above for Cayce, that caring people who direct their attention inwardly and outwardly to the planet have ameliorated the situation through biorelativity.

Another prediction—that Japan will almost totally disappear beneath the sea—may be considered in a similar manner. Although the Japanese islands are still there, newspaper headlines early this year read "Japan on Alert for Earthquakes." Seismologists there have alerted the government and populace that the geological stress level in the vicinity of Tokyo is so high that a large earthquake could occur at any time. Japan is bracing for it through nationwide earthquake drills to train people in survival skills. It seems to me, then, that some credit might be due to the Source, given the relatively long range at which the predictions were made.

POLE SHIFT

I

PREHISTORY
A Guide to the Future

Chapter One

Planet in Motion
— and Mystery

Whirling like a gyroscope, our planet moves through space. Its motion is extremely complex. It spins, revolves, wobbles, sinuates, nods. At the same time, its internal systems swirl, heave, subside, drift. The result is an intricate, multidimensional web of cycles within cycles within cycles. Some are regular and dependable guides to the future. Others are erratic and surprising, often disastrous. Sunrise, the tides, lunar phases, the seasons—these are accurately known. Hurricanes, earthquakes, volcanoes, droughts—these are not.

Our planet completes one rotation on its axis in 23 hours, 56 minutes and 4.091 seconds, producing day and night. It orbits the sun at an average distance of 93 million miles in 365 days, 5 hours, 48 minutes and 45.51 seconds, producing the year. Its equator is inclined to the ecliptic, or plane of the earth's orbit, at an angle of 23° 45′, producing the seasons.

These facts are commonly known—in an approximate way, of course. Now consider some additional data about our moving home in space.

A point on the equator moves at almost 1,040 miles per hour due to the planet's rotation. Simultaneously, that same point is also moving at about 1,100 miles a minute due to orbital revolution around the sun. And as the galaxy of which we are a part wheels through space around the galactic center, the motion of that same point is measured at more than 150 miles a second.

As complex as the motion of our hypothetical point may seem, moving in so many directions at so many rates of speed at once, there is still more to it. Superimposed on the generally smooth rotation and revolution of the earth are a variety of small but important perturbations of its otherwise clockworklike travels.

The orbit of the earth around the sun is slightly irregular because the gravitational pull of the sun on the earth-moon system turns the orbit into a serpentine path that departs from its idealized line of travel about fifteen hundred miles to either side.

The motion of the axis is also irregular. Over a fourteen-month period, it zigs and zags counterclockwise around the geographical North Pole in what is called the Chandler wobble, shifting as much as six inches a day, for a maximum of seventy-two feet. This movement has two components. One is annual and is associated with seasonally varying planetary conditions such as atmospheric pressure and snow loading. The other, called the Chandler component (after its discoverer) and having a fourteen-month period, is a free oscillation. When the annual component and the Chandler component are out of phase, the two tend to dampen or cancel each other. Polar motion is very small then. But when the two are in phase, the path of the axis may wander as much as the six inches a day just mentioned. This in-phase portion of the cycle repeats approximately every six years. Theories of the wobble's origin vary. Some scientists believe it originates within the earth; others attribute it to changes in surface conditions; still others think it has an astronomical cause. Presently, however, the data are scanty, so the source of the wobble continues to be a mystery.

In addition to wobbling, the axis nutates, or nods, as if bowing its head. Every 18.6 years, due to the earth's gravitational relationship with the moon, it completes a nutation of 9.2 seconds of arc—about 1/360 of a degree.

These combined motions have sometimes been termed "the

dance of the pole." Their periods are relatively short, however, compared to some others that affect the planet. One is the 25,800-year precession of the equinoxes, or points in the sky marking the beginning of spring and fall. The equinoxes only appear to precess, due to their slow retrograde (westward) motion. Actually, the earth's axis moves counterclockwise in a circular path among the stars. Like a top slowing down, the North and South poles each trace out the base of a cone in space, with the apexes point to point at the earth's center. Presently, the North Pole points at Polaris, in the Little Bear constellation. But, three thousand years ago, Alpha Draconis was the pole star, and by A.D. 14,000 the axis will have precessed enough to make Vega the new north-marking star.

Our aforementioned point moves in still longer cycles than the precession of the equinoxes. Over a 41,000-year period, the earth's angle of tilt to the ecliptic varies between 21° 39′ and 24° 36′. Presently 23° 26′, this angle of obliquity changes at the rate of .013° per century.

Another cycle through which our point moves is a result of changes in the average distance to the sun as the earth's orbit is transformed from elliptical to circular and back again. This cycle takes about ninety-three thousand years to complete, and brings the earth about 3 million miles nearer to or farther from the sun than the typical distance.

If we turn now from astronomy to geology, we see that in addition to all this motion, our point on the equator is undergoing both slow and rapid displacements, vertically and horizontally, as the surface of the planet moves. For the earth is not a solid unit. Rather, it consists of a thin rigid crust floating atop a series of concentric layers of molten or semisolid material; with a solid metallic core. The crust is constantly rising and falling in a long-term motion due to isostasy, the tendency of the solid surface to seek hydrostatic balance in its red-hot, viscous underlayment. As changes in the distribution of surface mass occur due to erosion, evaporation, damming, mountain building, etc., the crust slowly rides higher in some places and sinks lower in others. The weight of a large glacier, for example, presses the crust inward, often many hundreds of feet. When it disappears, the crust isostatically rebounds over thousands of years. At present, parts of the United States are being uplifted while other parts are subsiding. The Great Plains and the Rocky Mountains are rising at the rate of one to five millimeters per year. The central-southern California coastline is moving downward at twice that rate, although the Palmdale Bulge, along the San Andreas Fault east of Los Angeles, is rising at a much faster and surprising rate. Most of the Atlantic and Gulf coasts are also subsiding, and no significant area of the country is really stable. In fact, to a lesser degree the same can be said of the entire planet, for there are diurnal tides in the crust—daily displacements of up to nine inches due to the effect of sun and moon, just as in the oceans.

Amazing as these facts may seem, there are still enormous gaps in our knowledge of the earth. These gaps amount to mysteries— major unanswered questions about the earth's structure, physical processes, past history and future state. We will investigate some of those mysteries in the following pages. To set the scene, let's examine our planet in motion more closely. In doing so, we will see that science is still far from having a unified and comprehensive theory of the earth—a point central to our theme.

Several thousand years ago, the Greeks had a rather close estimate of the diameter of our planet, but with the advent of satellite technology we have gained precise knowledge. We know now that the earth, though more nearly spherical than a billiard ball, is nevertheless not perfectly round. Its equatorial diameter is 7,926.41 miles, giving a circumference of 24,901.55 miles. But its polar diameter is about 27 miles shorter. Consequently, the earth is not a true sphere but an oblate spheroid—the shape you'd get if you pressed down lightly on a basketball.

And a good thing it is, too, that the earth bulges in the middle. If it did not, it would have no stability whatsoever and could spin about its axis pointing in any direction. Equatorial bulge is what imposes gyroscopic stability on the earth and keeps its axis of rotation pointed in the same direction. (As we have seen, of course, it is acted upon by various forces—precession, Chandler wobble, etc.—that slowly change both the direction in which the spin axis points and its location on the surface of the globe.)

The earth's bulge owes its existence to the plastic material below the crust, which yields to pressure and tends to flow. As the earth spins, centrifugal force throws the material outward. This produces the difference between polar and equatorial diameters, creating the equivalent of a flywheel rim that allows the earth to gyroscopically stabilize.

Why the earth spins, nobody knows, but spin it does, requiring an energy of 2.14×10^{36} ergs to move. Interestingly, the earth's spin seems to be slowing down, so the length of a day—the time it takes the earth to turn once on its axis—is increasing. The rate of increase is not great. It has been estimated at rates varying from one second every six hundred thousand years to two seconds every 100,000 years. While a precise figure is far from settled, the idea nevertheless has been used to explain a problem discovered through satellite measurement of the earth, namely, that the observed flattening at the poles is greater than expected by about .5 percent, giving an equatorial bulge about two hundred meters larger than calculated on the basis of assumed rheological conditions inside the earth. Some scientists think this extra, nonhydrostatic bulge is the relic of the earth's faster rotation millions of years ago.

The crust of our planet consists of crystallized rock, but it is not a seamless shell. The term "crust" originated when it was thought that the earth consisted of a liquid interior with a thin covering, much like the scum that collects atop a cooling pot of soup. Most earth scientists now think that the crust, although solid, consists of six large plates and several smaller ones that float on the deeper layer and slowly move, jostling one another like blocks of ice on a river in the spring thaw. These plates, made of both continents and surrounding ocean floor, move independently in various directions at rates varying from one to ten centimeters per year. Their motion is thought by geoscientists to originate from convection currents of magma—molten rock deep in the earth— which move the plates through friction. Some magma wells up to the surface at cracks in the sea floor called mid-ocean ridges. When a plate collides with another, the denser rock of the sea bottom slides under the lighter material of the continents. Should continental masses collide, their edges pile up and form mountains. This concept of the earth, which combines the recent discovery of sea-floor spreading with the older notion of continental drift, is known as global plate tectonics, or tectonic plate theory.

There are technical terms for the various structures of the earth, but it is an unfortunate fact in geology that these terms are not clearly defined. The crust of the earth is often called the *lithosphere,* but the nature and dimensions of the lithosphere are not agreed upon. Some scientists use the term to denote only that part of the earth above the Mohorovicic discontinuity, or Moho, a boundary surface marking the beginning of a denser rock whose chemical composition is markedly different from the surface rock. The Moho is an average of fifteen kilometers below the ocean basins and about thirty kilometers below the continents.

Beneath the crust, according to the thinking of these scientists, is the *mantle,* a shell of red-hot magma hundreds of miles thick that extends down to the earth's pressure-solid magnetic core. Thus Robert B. Gordon, of Yale, in his 1972 textbook *Physics of the Earth,* writes, "Material above the Moho is called the *crust* . . . while everything between the Moho and the core is the *mantle"* (p. 134).

Others define lithosphere differently. They consider it to be a layer of crystallized rock extending downward an average of two hundred kilometers under the continents but less than half that beneath the oceans. According to this conception, the Moho is found in the upper part of the lithosphere. In this case, the term "crust," when used synonymously with "lithosphere," would denote a much thicker shell of rock.

This semantic confusion continues with regard to the lower structures of the earth. Some, such as Gordon, consider the mantle to extend upward to the Moho, so that the mantle is partially molten, partially solid. Others separate the crust from the lithosphere by an intermediate shell of molten, uncrystallized material called the *asthenosphere.* Still others consider the asthenosphere to be the topmost layer of the mantle but simply having a different density. Because of the extreme pressures existing in the asthenosphere and the mantle, their materials, despite high heat, are semisolid, plastic. If some of this material happens to flow to the surface, through volcanoes or mid-ocean ridges, it is called lava.

The geologist J. Tuzo Wilson, of the University of Toronto, in his Introduction to *Continents Adrift and Continents Aground,* comments usefully on this imprecision in geology. One geologist, he remarks, distinguishes both the crust and the mantle from the lithosphere and the asthenosphere. According to this view,

All are shells of the earth but the first pair are distinguished by different chemistries, the second pair by different strengths. The crust is thin and formed of silicious rocks resting upon more basic rocks of the mantle. Its thickness can be measured because earthquake waves have different velocities in the rocks of the crust and those of the mantle, and can be reflected and refracted at the boundary between them, called the Mohorovicic discontinuity. . . .

The lithosphere comprises the crust and the uppermost part of the mantle. This is the strong layer of which the plates are made. It rests upon the asthenosphere, a deeper layer of the mantle so hot that it has lost its brittleness and is capable of slow creep (p. 6).

As if these distinctions were not complicated enough, Wilson says there is a further source of confusion involving identification of the boundary between the lithosphere and the asthenosphere. This has long been thought to be the top of a low-velocity layer, a zone below the Moho where temperatures are assumed to be great enough to partially melt the grains of rock, thus slowing the seismic waves. The trouble is that this low-velocity layer has not been detected beneath continents, yet continents move with the same "slow creep" mentioned in the quote above.

To account for this puzzling situation, another view of the asthenosphere has been proposed, Wilson notes. According to this particular view, the asthenosphere does not begin at the low-velocity layer but at a deeper level, where pressure and temperature combine to allow the solid mineral grains to be sheared, enabling motion to take place. The depth of this hypothetical boundary can be calculated by laboratory experiments, but it has not been detected in nature.

Another mysterious aspect of our planet is its magnetic field. This field, generally attributed to the nickle-iron core of the earth acting as a dynamo due to convection currents in the mantle, fluctuates in strange ways. First, the strength of the magnetic field varies, sometimes dropping to zero. Second, the magnetic poles wander—as much as 80° over the past millions of years. (The magnetic poles should not be confused with the rotational poles, which are the geographical ends of the axis of spin. At present, the magnetic north pole is in northern Canada, about 11° away from the true North Pole.) Third, the magnetic field reverses polarity. More than 170 polarity changes or field reversals have occurred during the past 80 million years. All this has been learned only in the past three decades, through study of paleomagnetism, the magnetism of rocks, although scientists admittedly know very little about the field and its daily, yearly, and long-term changes.

It is known, however, that at this very moment the earth's magnetic field strength is decreasing. Over the past twenty-five hundred years it has weakened by 50 percent. If this trend continues, the field will disappear altogether—perhaps as soon as early in the next century—and then presumably reappear with reversed polarity. J. M. Harwood and S. C. R. Malin predicted in *Nature* (12 February 1976) that the next reversal of the planet's magnetic poles will occur about A.D. 2030. The time involved in a magnetic pole shift is unknown. Theoretically, the "flip" is instantaneous once the field strength has decayed to zero. But for practical purposes—i.e., at a level that can be geologically observed in rocks—it takes at least several decades, and more likely several centuries. The total event, from highest field intensity before the reversal to highest field intensity after, may take as long as twenty thousand years.

Since detailed measurements of the earth's magnetic field have been made only in the past two decades, no one is sure what the change in strength really means. But several effects of magnetic pole shifts are clear. Magnetic compasses, for example, would be useless, causing great difficulty in navigation. (Inertial gyrocompasses should continue to function, however, since they are not dependent on the magnetic field.) Also, with no magnetic field to shield the planet from solar and cosmic radiation, its inhabitants would be largely unprotected from severe sunburn, skin cancer and possibly other, unsuspected effects. It is a fact that entire species of simple creatures have become extinct in the past in correlation with magnetic field reversals. A few years ago it was suggested in *Science News* (2 February 1974) that the earth's magnetic field influences atmospheric pressure in the upper atmosphere and that the average pressure system seems to move westward with the magnetic drift. If this is so, a drop of magnetic field strength to zero might disrupt the world's weather.

This point brings us to still another mystery of the earth: ice ages. The phenomenon was given definitive recognition by the Swiss naturalist Louis Agassiz, father of glaciology, who published his theory of the ice ages in 1840. It is now known that in the past million years alone, massive ice sheets have covered much of North America and Europe at least nine times. Moreover, during the past billion years there have been at least four epochs of ice covering major portions of the globe. These epochs occurred every 250 million years or so. Agassiz did not have these figures, but he saw clearly the presence of gigantic ice mantles in the past. These catastrophic periods, he said, were—among other things— responsible for the frozen carcasses of mammoths found in Siberia, whose flesh was still edible upon discovery. (We will examine this mystery—yet another—in the next chapter.)

Agassiz at first conceived of the ice ages as times when polar ice caps* spread outward from their Arctic and Antarctic starting points. But this concept, still widely held in science, could not explain the fact, discovered later, that ice sheets of continental size have appeared in temperate and even tropical zones. Africa, Madagascar, India, Guyana, Argentina, southeastern Brazil— what could have caused these regions to be covered by ice several thousand feet thick? Moreover, why have some glaciers flowed from the tropics to higher latitudes? Dr. William Stokes points out in his 1960 *Essentials of Earth History:*

> In South Africa the glaciers moved principally from north to south—away from the Equator. In central Africa and Madagascar, other deposits show that the ice moved northward, well within the tropic zone. Most surprising has been the discovery of great beds of glacial debris in northern India, where the direction of movement was northward . . . in Australia and Tasmania, where the ice moved from south to north . . . movement in Brazil and Argentina was toward the west (p. 220).

On the other side of the question, why were some of the most frigid parts of the planet—northern Greenland, the interior of Alaska, northern Siberia—never glaciated?

The puzzling phenomenon of ice ages is compounded by the fact that cold alone will not produce glaciers. It also takes a lot of heat. True, glaciers grow from snow accumulating faster than it can melt. But in order for snowfall to increase, there must be an increase of evaporation in tropical areas, and that requires greater heat. The increased moisture is carried through atmospheric circulation patterns to the polar areas. There heat is more easily lost to space, allowing the moisture to fall as snow. A worldwide temperature drop, proposed by some theorists to explain ice ages, cannot account for the necessary increased evaporation.

This was dramatically illustrated in the recent findings announced by geochemists Samuel Epstein and Crayton J. Yapp at

---

*The proper term for the polar glacial coverings is "ice sheet." The term "ice cap" technically refers to the glacial covering of a mountain or local geologic feature, and therefore denotes something much smaller than an ice sheet. However, some of the people quoted here use the terms interchangeably. For the sake of style and readability, I have followed their example.

the California Institute of Technology. According to the *San Diego Union* (12 March 1978), the men analyzed heavy hydrogen in ancient wood as an indicator of past climate. Their surprising results: winter temperatures in North America twenty-two thousand years ago, during the height of the last ice age, may actually have been higher than they are today! In areas not covered by ice, they reported, the winters were generally warmer and summers cooler than they are today.

In addition, other enigmatic facts uncovered in the past century by geologists and archaeologists raised important questions bearing on the search for an explanation of the ice ages. Coral, for example, was found in Alaska, coal in Spitzbergen and Antarctica. Likewise, many other areas of the earth were found to have geological and climatological characteristics in striking contrast with their present condition.

Many theories have been offered since Agassiz to explain the ice ages. In fact, two Lamont-Doherty Geological Observatory oceanographers, David Ericson and Goesta Wollin, estimated a decade ago that there had been nearly one theory a year for more than a century. Some theories invoke astrophysical causes such as solar variability in heat production or interception of the sun's radiation by clouds of interstellar gas and dust. Others invoke atmospheric causes such as dust particles from volcanic eruptions, which reduce reception of the sun's warmth. And still others attribute the ice ages to geological factors, specifically a tilting of the earth's axis or a displacement of the crust.

This last possibility—polar shifts—is, of course, my theme—a theme that is in the news even today. Two recent newspaper reports highlight the situation. A 1978 United Press International article headlined "Alaska Was Tropical, Experts Say," states that tropical rain forests bordered the Gulf of Alaska 45 million years ago. The expert quoted is Jack Wolfe, a paleobotanist with the U.S. Geological Survey in Menlo Park, California. If his findings are supported by further studies, Wolfe told the UPI, "it would mean the Earth's axis of rotation was once less inclined toward the sun. This could help explain major changes in climate."

The second article, appearing in the *New Haven Register* (5 March 1978), reported the discovery of fossil amphibians and reptiles as big as deer in a mountainous region of Antarctica. The animals lived during the Triassic period, about 200 million years ago. Today the largest land animal in Antarctica is an insect (penguins and whales being sea creatures). The research team making the discovery was headed by a biologist, John W. Cosgriff, Jr., who noted that many of the species found there were also found in rock deposits of the same age in Africa, India, and Australia. This supports the theory that all present land masses were originally joined into one supercontinent that broke apart, he said. He also noted that Antarctica's climate apparently was temperate to subtropical when the fossil animals were alive, adding, "Clearly, the south pole was then elsewhere than on the Antarctic continent."

Less inclined toward the sun? Elsewhere than on Antarctica? We have seen that science recognizes a change in the earth's angle of inclination to the ecliptic. In fact, this is one of the critical factors in the latest candidate for explaining ice ages. The theory was announced by geologist Dr. James D. Hays, of Columbia University, head of a team of scientists who claimed in *Science* (10 December 1976) that the "fundamental cause" of the ice ages had been identified. Their work involved a reconstruction of the earth's climatic history for the past half million years, which was then compared with cycles of change in the shape, tilt and seasonal positions of the earth's orbit. The team found that climatic cycles correlated with three cycles of change in the earth's orbital geometry, described above. The first cycle, with a 93,000-year period, involves changes in the shape of the earth's orbit, ranging from nearly circular to elliptical and back. The second cycle, about 41,000 years long, involves changes in the tilt of the earth's spin axis with respect to the orbital plane, varying from 21° 39′ to 24° 36′, causing changes in the planet's climate. The third cycle is the precession of the equinoxes or, more properly, the poles.

In a press conference, the Hays team released a statement saying, "We are certain now that changes in the earth's orbital geometry caused the ice ages. The evidence is so strong that other modifications must now be discarded or modified."

Extraordinary claims require extraordinary proof. As we shall see, the Hays theory does not explain all the facts—facts such as the quarter-billion-year periods without ice.

Nor does tectonic plate theory, which also claims to be able to unravel the mystery of ice ages. By reassembling the now-drifted continents into an original supercontinent (named Pangaea by scientists), it appears that the glaciated areas occurred at times when they were in proper positions for a polar or near-polar ice cap to form. This is said to explain why, for example, the debris carried from Brazilian glaciers included rocks known in south-western Africa but unknown in neighboring areas of South America.

The concept of global plate tectonics is now so well accepted that many people do not realize there are still many questions that the theory has not answered. There are, in fact, some major dissenters in the scientific community who do not accept the concept, plausible as it seems in light of data such as that from the Brazilian glacial deposits. One geophysicist, Paul S. Wesson, of Cambridge University, published an article in the *Journal of Geology* in 1972 that lists seventy-four shortcomings of various versions of continental drift and plate tectonics.

Since the theory of plate tectonics is relatively new—having developed only since the early 1960s—it may be that many of its flaws and gray areas are the result of incomplete knowledge of the earth's structure, rather than from internal inconsistency of the theory. Nevertheless, the mysteries of our planet in motion must be considered by any theory that purports to be a unified and comprehensive theory of the earth. Among those mysteries are the origin of the earth itself, the origin of continents and ocean basins, the motive force behind plate tectonics, the origin, nature and release of the heat of the earth, and the cause of radical climatic changes that have been correlated with the termination of so many species of animals. The theory must also answer questions such as these: What caused the ice ages? What brought them to an end? What caused periods of extreme volcanism? What caused massive lava flows? What caused mid-ocean ridges and fissures? What caused mountains to be built? What caused sea levels to change? What caused the earth's magnetic field strength to fluctuate? What caused the earth's magnetic poles to wander and to reverse? What caused sudden changes in the direction of continental drift?

These questions have not been answered in a satisfactory manner so far, and sometimes the "answer" to one contradicts the "answer" to another. Science is often at odds with itself, as we shall see. It was for this reason that in the nineteenth century T. H. Huxley wrote regarding the sanctity of certain scientific pronouncements, ". . . theories do not alter facts, and the universe remains unaffected even though texts crumble."

To try to answer these questions, come with me now on a journey through time and space—a journey that will eventually pass beyond the physical world and enter the metaphysical. Our object is to examine what seems to be the most mysterious motion of our planet: pole shifts. Along the way, we will explore many fascinating topics. We will consider lost civilizations, occult traditions, the evolution of man, and life in the next century. We will also listen to ancient prophecies and contemporary psychics. And as we press to the frontiers of knowledge, our search will ultimately require us to understand humanity's relationship with the universe. We will find, paradoxically, that outer space and inner space merge, that the past and the future unite.

We begin our journey inside the Arctic Circle.

References and Suggested Reading

Catastrophist Geology. Caixa Postal 41.003, Santa Teresa, Rio de Janeiro, Brazil. A magazine dedicated to study of discontinuities in earth history.

DOTT, ROBERT H., and BATTEN, ROGER L. *The Evolution of the Earth.* New York: McGraw-Hill, 1971.

GORDON, ROBERT B. *Physics of the Earth.* New York: Holt, Rinehart and Winston, 1972.

SULLIVAN, WALTER. *Continents in Motion.* New York: McGraw-Hill, 1974.

UYEDA, SEIYA. *The New View of the Earth.* San Francisco: W. H. Freeman, 1978.

WILSON, J. TUZO, ed. *Continents Adrift and Continents Aground.* San Francisco: W. H. Freeman, 1976.

WYLLIE, P. *The Way the Earth Works.* New York: John Wylie, 1976.

Chapter Two

The Riddle of the Frozen Mammoths

In June 1977, a Russian bulldozer operator was removing the sunwarmed surface layer of mud from the permafrost—the permanently frozen ground—in a remote mountainous province of northeastern Siberia. He was part of a free-lance team of gold prospectors working the sides of a little stream that flowed into the Kolyma River and thence into the Arctic Ocean. And he was finding the work monotonous.

Suddenly he noticed a block of muddy ice containing a curious dark mass. He knew that no shovel or pick could wrest it from the rock-hard grip of the frozen ground, so, intrigued, he turned to the more primitive but attested local method of loosening the permafrost. Diverting the stream to thaw the ice, he was soon amazed to see the contours of a small elephantlike creature appearing through the frozen block. And thus was discovered the first perfectly preserved complete woolly mammoth.

The prospectors christened their historic find "Dima," and word of it spread quickly. The Soviet Academy of Sciences soon took charge, at first flying Dima three hundred miles southeast to a special refrigerated chamber and later transferring it to the Zoological Institute, in Leningrad, on the other side of the U.S.S.R., where scientists from many institutes came to examine it.

Dima, a male, had died at the age of six months. The cause: blood poisoning due to a leg injury. Measuring 45 inches long and 41 inches high and weighing 139 pounds, he had reddish, chestnut-colored hair, tiny ears, and a 22-inch trunk with two distinctive "fingers," or appendages, at its end, just as pictured in Stone Age paintings and sketches from France and Spain.

According to Tass, the official Soviet news agency, Dima was still alive when "something hit his leg in two places and during his last hours of life he stopped eating." A UPI report in November 1978 stated:

> Tass said an investigation indicated the mammoth was grazing with a larger herd of long-extinct species of elephant on the arid subarctic steppes of Siberia. Nearby were wild horses, goats and bison.
>
> The scientists were also surprised to find that the mammoth . . . had been buried by a mud flood shortly after death.
>
> Tass said that such floods were previously known about only in the south of Siberia. Scientists speculated that melting glaciers caused the flooding.

The Tass release raises some intriguing questions. How, for instance, could Dima be perfectly preserved in those conditions? First, he died a lingering death; that is the way blood poisoning works. Some days would pass before his expiration. How is it that Dima, in his weakened condition, didn't fall prey to scavengers? The same question applies after Dima's demise. And since the climatic condition was warm enough to melt glaciers and cause flooding, it must have been well into spring or summer. Decomposition of flesh would be rapid then, even when buried in mud. Weeks, even months, would have to pass before the cold could turn Dima's burial site into permafrost.

Under such conditions, Dima's preservation is nothing short of miraculous—if we accept the official explanation, that is. But there is an alternative explanation, and we will examine it shortly.

Dima is not the first frozen mammoth to be discovered—only

the first perfectly preserved one. Four other essentially complete specimens of *Mammuthus primigenius* were reported prior to the twentieth century, the first being recorded in 1692. According to Dr. William R. Farrand in his *Science* (17 March 1961) article "Frozen Mammoths and Modern Geology," there have been at least thirty-nine discoveries of frozen mammoth remains, some with soft parts preserved. A few even had their eyeballs intact. Other authorities say that as many as eighty-odd mammoths have been found. Except for two very fragmentary carcasses from Alaska, all the frozen cadavers have come from northern Siberia.

The most famous is the Berezovka mammoth, found by hunters in 1900 on the bank of the Berezovka River after a landslide exposed its head. It remained there—67° 32′ N., seventy miles north of the Arctic Circle—for a year while careful excavation plans were made by Russian scientists. Czar Nicholas II took a deep interest in the matter and gave orders to be kept informed. When the scientists of the Imperial Academy of Science began to remove the remains, the stench from rotting exposed flesh made work almost unbearable at first, although within a few days they became accustomed to it. The body was dismembered and transported across Russia to St. Petersburg (now Leningrad) in a sealed railroad car. At each station, crowds pressed forward eagerly to view the prehistoric beast. But the Czar had ordered it to be strictly guarded, so the public was disappointed. When the car was opened, the mammoth was taken to the Zoological Institute for examination, reassembly and reconstruction from damage by weather and foxes, which had destroyed most of the face. The skeleton was removed and preserved separately. An artificial head and ears were mounted on the body, and some missing hair and wool were replaced with strands from other specimens. The final result, according to a 1903 *Nature* article, was "a triumph of the taxidermist's art." It is still on display in the Zoological Institute's museum, where Dima will probably also become an exhibit.

Examination of the Berezovka mammoth revealed that it was a small male. Unchewed grass and buttercups in its mouth and undigested vegetation in its stomach indicated death in mid to late summer. The original report states that the animal died "during the second half of July or the beginning of August." Its position in the permafrost had been one of sitting on its haunches, with its pelvis bone, right foreleg and several ribs broken. The contents of its stomach consisted chiefly of field grasses; nine kinds were found, but no evergreens were present. A report by G. N. Kutomanov in the *Bulletin of the Academy of Sciences of St. Petersburg* (Vol. 8, No. 6, 1914) observes that "no evergreens have ever been found in the stomach of a mammoth" (pp. 377–88). The Berezovka bull evidently died of suffocation. Its genital organ was erect—a condition inexplicable in any other way than suffocation, according to I. P. Tolmachoff in a 1929 *American Society Transactions* report.

In addition to preserved specimens, hundreds of thousands of mammoth bones and tusks have been found in excellent condition. An estimated fifty thousand tusks were taken from Siberia alone between 1660 and 1915. (It was the lure of "white gold"—ivory mammoth tusks up to sixteen feet long and weighing nearly four hundred and fifty pounds—that brought collectors to the Arctic tundra. Traffic in fossil ivory has been going on since the time of Pliny the Elder.) One of the world's foremost authorities on Ice Age elephants, Professor Nikolai Vereshchagin, chairman of the Soviet Academy's Committee for the Study of Mammoths and the man who eventually took charge of Dima, estimates that even the fifty thousand tusks represent a small fraction of the mammoth population. According to John Massey Stewart's article "Frozen Mammoths from Siberia Bring the Ice Ages to Vivid Life" (*Smithsonian*, December 1978), Vereshchagin calculates that "the heavy erosion of the Arctic coast spills thousands of tusks and tens of thousands of buried bones each year into the sea and that along the 600-mile coastal shallows between the Yana and Kolyma lie more than half a million tons of mammoth tusks with another 150,000 tons in the bottom of the lakes of the coastal plain" (p. 68).

The number of mammoths found increases as one goes north, being most numerous in the New Siberian Islands, which lie between the Arctic coast of Siberia and the North Pole. Their bones are often found in what have come to be called "elephant graveyards" and "mammoth boneyards," huge collections of bones mixed indiscriminately with almost no skeletons remaining intact. Vereshchagin himself found such a site in Siberia in 1970. Protruding from a riverbank for more than two hundred yards were thirty-five hundred densely packed bones. Beneath them, through six to twelve feet of loam, which Vereshchagin's team hosed away by fire engine pump, another such deposit was discovered: thirty-five hundred more mammoth bones, also densely packed, stretching along the full two hundred yards of the surface bones.

At first Dima's death was thought to have occurred about nine thousand to twelve thousand years ago, but later studies have pushed the date back to forty-four thousand years. This is about the time the Berezovka mammoth died—forty-four thousand years ago, plus or minus thirty-five hundred years, according to radiocarbon (C^{14}) dating techniques. A few specimens are nearly fifty thousand years old. Yet other mammoths have been radiocarbon-dated as expiring at various later times up to the end of the last Ice Age, about nine thousand years ago. And although it is widely assumed by paleontologists that mammoths became extinct then, according to the journal *Radiocarbon* (Vol. 15, No. 1), one mammoth tusk from a site in Bavaria, Germany, was dated at the University of Kiel (on the basis of an average of three separate dates) at around 1900 B.C. Still more astounding, a mammoth bone found in Mexico and associated with stone implements was dated, again according to *Radiocarbon* (Vol. 2), at 2,640 ± 200 years old. This would be 690 B.C. ± 200!

There are even hints, Ronald J. Willis tells us in *Info Journal* (Spring 1967), that the mammoth persisted into modern times. Eskimos preserve memory of mammothlike creatures. Indian legends speak of mammoths living until shortly after the founding of the U.S.A. None other than Thomas Jefferson, who was extremely interested in the mammoth, recounted in his memoirs the story of a Mr. Stanley, captured by Indians in the West and passed from tribe to tribe, apparently as a curiosity, until he passed the mountains west of the Missouri River to a place where the rivers flowed westward. Stanley described seeing mammoth bones and was told by Indians that the beast still existed in the northern region of their territory.

From Siberia come other tantalizing tales of modern mammoths, where natives reported to scientists early this century that the great beasts lived on. Bernard Huevelmans, in his 1959 book *On the Track of Unknown Animals*, tells the story of a Russian hunter who in 1918 noticed the tracks of two huge animals. This occurred in the taiga, the vast forest that covers all the middle of Siberia. The tracks measured about two feet across and eighteen inches front to back. He followed them for days, noting a huge heap of dung made of vegetable matter, and tree branches pushed about as if a huge bulk had smashed through them. Then,

All of a sudden I saw one of the animals quite clearly, and now I must admit I really was afraid. It had stopped among some young saplings. It was a huge elephant with big white tusks, very curved; it was a dark chestnut color as far as I could see. It had fairly long hair on the hind-quarters, but it seemed shorter on the front. I must say that I had no idea that there were such big elephants. It had huge legs and moved very slowly. I've only seen elephants in pictures but I must say that even from this distance (300 yards) I could never have believed any beast could be so big. The second beast was around. I saw it only a few times among the trees; it seemed to be the same size (p. 351).

Mammoths stood up to fourteen feet tall, the largest of all elephants. These "lords of the tundra" roamed the ancient world from North Carolina to Alaska and throughout most of Russia and Europe into Britain and Ireland. Contrary to popular belief, however, mammoths were not Arctic animals. Studies have shown that their hairy coat and skin were not adapted to Arctic conditions. Thick fur in itself means little; tigers of the tropics have thick fur. And the layer of fat, about three inches thick, which is found under the skin of mammoths testifies primarily to ample food supply, rather than insulation against a frigid climate, just as a camel's hump does. The key piece of evidence is the lack of sebaceous glands in the skin of mammoths. Skin adapted to cold regions of the earth has sebaceous glands to secrete oil. This keeps the skin and fur from drying out and cracking. Without lubrication, flesh will soon become dehydrated as the cold draws out moisture, and the organism will die. The evidence shows, in short, that the common sheep is better adapted to Arctic conditions than mammoths were.

How did these giant herbivores become extinct? How did Dima die and end up frozen intact? What about the other frozen mammoths? This is the riddle we will explore in this chapter—a riddle that is one of the key pieces of data in theories of polar shifting.

Catastrophists, or members of a particular school of geological-archaeological thinking, argue that mammoths and other animal species were exterminated en masse by cataclysms of the earth. Pole shift proponents say specifically that the cataclysm was a lurching of the planet on its axis or a slippage of the crust around the inner mantle that produced a flood and near-instant freezing. Yet the enormous range of time over which mammoths are known to have died forces the conclusion, as Dwardu Cardona points out in "The Problem of the Frozen Mammoths" (*Kronos*, Winter 1976) that "the mammoths in question could not have been the victims of the same cataclysm." Interestingly, the most recent date for a mammoth's death, 690 B.C., is the time that the foremost pole shift theorist—whom we'll meet in Chapter 6—offers as the occasion of a pole shift.

Let's try to penetrate this mystery—as hard and murky as permafrost itself—by looking in detail at the arguments, pro and con, concerning the demise of the mammoths.

First, however, we must see their death in context. The context is not pleasant. It involves the wholesale termination of many species. The number of deceased mammoths—at least 117,000, on the basis of tusks and bones found to date—is only a very small part of the animal population involved. Paleontologist W. B. Scott, in his *History of Land Mammals in the Western Hemisphere*, estimates that mammals over three-fifths of the earth's land surface were decimated at the end of the Pleistocene Epoch. According to Professor Frank C. Hibben, in his colorful narrative of prehistoric North America, *The Lost Americans*, some forty million animals lost their lives in a violent cataclysm that encompassed nearly all of the northern hemisphere. Listen to an oft-quoted passage from Hibben's chapter "End of a Universe":

The Pleistocene period ended in death. This was no ordinary extinction of a vague geological period which fizzled to an uncertain end. This death was catastrophic and all-inclusive. . . . The large animals that had given the name to the period became extinct. Their death marked the end of an era.

But how did they die? What caused the extinction of forty million animals? . . . the [extinction] was of such colossal proportions as to be staggering to contemplate. . . .

The "corpus delicti" . . . may be found almost anywhere. . . . the animals of the period wandered into every corner of the New World not actually covered by the ice sheets. Their bones lie bleaching in the sands of Florida and in the gravels of New Jersey. They weather out of the dry terraces of Texas and protrude from the sticky ooze of the tar pits of Wilshire Boulevard in Los Angeles. Thousands of these remains have been encountered in Mexico and even in South America. The bodies lie as articulated skeletons revealed by dust storms, or as isolated bones and fragments in ditches or canals. The bodies of the victims are everywhere in evidence.

... in the great bone deposits of Nebraska, we find literally thousands of these remains together. The young lie with the old, foal with dam and calf with cow. Whole herds of animals were apparently killed together, overcome by some common power.

... the muck pits of Alaska are filled with evidences of many thousands of animals killed in their prime. The best evidence that we could have that this Pleistocene death was not simply a case of the bison and the mammoth dying after their normal span of years is found in the Alaskan muck. In this dark gray frozen stuff is preserved, quite commonly, fragments of ligaments, skin, hair, and even flesh. We have gained from the muck pits of the Yukon Valley a picture of quick extinction. . . .

Neither the Pleistocene animals nor their untimely end are phenomena peculiar to the American continents. Asia was deeply involved.

The uniformitarian school of geological-archaeological thought—which will be introduced in Part II—argues that there is no riddle, because the situation is well understood. It could have been due to epidemics or caveman "overkill." But, more likely, the argument goes, a fairly rapid (but not sudden) change of climate brought a change of vegetation, resulting in insufficient food for the giant beasts. The rare mammoth specimens that have been found frozen are due to freak accidents in which individuals fell into icy ravines or were buried under collapsing cliffs, riverbanks, etc., or simply plunged under their own weight through the ice of frozen lakes and rivers, sinking to their death. As Farrand puts it in his *Science* article, speaking directly against the notion of a pole shift, "All evidence now at hand supports the conclusions of previous workers that no catastrophic event was responsible for the death and preservation of the frozen woolly mammoths. The cadavers are unusual only in that they have been preserved by freezing; the demise of the animals, however, accords with uniformitarian concepts" (p. 733). He adds, "There is no direct evidence that any woolly mammoth froze to death."

In his book on continental drift, *Continents in Motion*, Walter Sullivan, science editor of *The New York Times*, takes an anti-pole shift stance regarding the death of mammoths and other exterminated animals in the frozen northlands. He asserts, "The chaotic jumbles of bones and trees uncovered by placer-mining can be explained without resort to oceanic invasions of the hinterland" (p. 35) and cites a researcher who argues that the elephant graveyards were formed by streams carrying animal remains to locations where debris collected. Sullivan also points out that the recently obtained datings of mammoths do not concur with the dates proposed by various pole shift advocates such as Hugh Auchincloss Brown, Charles Hapgood, and Immanuel Velikovsky—all of whom we will meet in the next section.

Daniel Cohen, in a rebuttal of the catastrophist position, puts it this way in *How the World Will End:* "Russian scientists have shown how the mammoths could have been preserved by falling into deep crevasses. . . . No overnight catastrophic deep freezes need be brought in to account for the frozen mammoths. Besides, both the number and state of preservation of the mammoths is usually exaggerated by those who like to see mysteries where there are none" (pp. 215-16).

No mystery? Or might it be a failure to confront *all* the evidence and thereby *evade* a mystery? Listen to what has been offered in support of catastrophic polar shifting as the cause of the frozen mammoths.

Ironically, some of the data come from a uniformitarian—Farrand himself. In his article, Farrand comments on the vegetation associated with the Berezovka and Mamontova mammoths (the latter being an incomplete carcass found in 1948 in Siberia): "In general, this floral assemblage is 'richer . . . somewhat warmer and probably also moister' than the present flora of the tundra in which frozen mammoth carcasses are now found. Quackenbush found 'large trees' associated with fossil mammoths in a now-treeless part of Alaska and also came to the conclusion that the climate was somewhat milder when the mammoths lived" (p. 730).

Continuing his discussion, Farrand points out that the Berezovka mammoth lived in a climate "slightly warmer than the present" and therefore, when the radiocarbon dating is considered, must be assigned to the last interglacial period. However, he said, the Mamontova mammoth presents a problem. Since it is younger in age, dating from the last glaciation, and since the associated flora is that of a warmer latitude, "an apparent paradox remains—that the climate in northern Siberia was warmer than at present at some period in late glacial time when climates elsewhere on the earth were cooler than at present" (p. 733).

Farrand does not attempt to resolve the climate paradox. Instead, he concludes that the healthy, robust condition of the cadavers and their full stomachs argue against death by slow freezing, and especially sudden freezing. "Histological examination of fat and flesh of the Berezovka mammoth," he writes, "showed 'deep penetrating chemical alteration as a result of the very slow decay,' and even the frozen ground surrounding a mammoth had the same putrid odor, implying decay *before* freezing" (p. 734).

Furthermore, he argues, the only direct evidence of the mode of death indicates that at least some of the frozen mammoths died of asphyxia, either by drowning or by being buried alive by a cave-in or mudflow. "Asphyxiation is indicated by the erection of the penis in the case of the Berezovka mammoth..." (p. 734).

He concludes that since no significantly smaller Eurasian fauna of that time have been found frozen and well preserved—large woolly rhinoceros have been found, but in far fewer numbers*—the near uniqueness of the mammoth situation "points to some peculiarity of their physique as a contributing factor." Most likely, Farrand argues, the late-summer or early-fall weather melted the ground enough to make locomotion difficult for the huge beasts, and occasionally trapped one by various means that slowly buried and suffocated it with mud and water. "There appears to be no need to assume the occurrence of catastrophe.... On the contrary, the frozen giants are indicative of a normal and expected (uniformitarian) circumstance of life on the tundra" (p. 734).

Apparently a sensible explanation, Farrand's view is not without loopholes. For example, in the same article he also states that the mammoths' "broad, four-toed feet...were advantageous *in marshy pastures*" (emphasis added)—the same terrain in which he presumes they would lose their footing. Moreover, as Cardona points out, enough mammoth cadavers have been found standing in an upright position to dispel all illusions of their having slipped—unless they happened to regain their legs after slipping, in which case, Cardona says, it would be more than obvious that they could not have been killed by the fall. Cardona continues:

---

*In 1979 gold miners in Alaska discovered the fully preserved carcass of a long-horned buffalo-like animal more than 20,000 years old. The carcass was found in permafrost sixty feet below the surface. Its well-preserved flesh was like beef jerky and its stomach contents, like that of the Berezovka mammoth, were intact, containing Ice Age plant life. Scientists at the University of Alaska's paleontological laboratory are studying the animal.

And how can one account for freezing, *sudden or otherwise,* in the warmth of the Arctic summer which, according to Farrand himself, is warm enough to carpet the tundra with a "relatively luxuriant vegetation." He states: "It is amazing what 24 hours of sunshine a day will do!" How more amazing that the same 24 hours of sunshine a day failed to decompose the dead mammoths which, if Farrand is right, had to await the return of winter before commencing to freeze (p. 80)!

Farrand's article was a reply to one written by the multifaceted Ivan T. Sanderson, biologist, author and founder of the Society for the Investigation of the Unexplained (SITU).* Sanderson's "Riddle of the Frozen Giants" had appeared in *The Saturday Evening Post* a year earlier (16 January 1960), provoking much comment. Sanderson, well known for his television appearances as a naturalist, declared in it nothing less than the proposition that the crust of the earth had shifted suddenly and radically, and that this catastrophe provided the answer to the riddle.

Sanderson argues thus: If the crust of the earth were to come "unstuck" from the central body of the spinning earth, it would start to move and new parts of it would drift in over the poles. The equatorial bulge would crack open the crust, triggering volcanic action, which, in addition to releasing surface flows of lava, would eject masses of dust particles, steam and other gases into the upper atmosphere. Winds beyond anything recorded would be whipped up by this turbulence, and vast cold fronts would build up with violent extremes of temperature on either side. But the most important element would probably be the gases which had been shot up highest of all. What would happen to them? Sanderson asks rhetorically. We will hear his answer in a moment, but first we must consider some critical data that are Sanderson's unique contribution to this controversial topic.

Sanderson went to what he describes as "frozen-food technicians" to learn what is necessary to preserve meat in the condition of the frozen mammoths when first discovered (a condition that changed rapidly through putrefaction upon exposure to air). He learned "two vital facts." The first is that temperatures of $-20°$ F. or lower are needed. The second and more important is that the

*SITU membership information can be obtained from: SITU, Membership Services, R.F.D. 5, Gales Ferry, CT 06335. SITU's journal *Pursuit* can be reached at P.O. Box 265, Little Silver, NJ 07739.

freezing must be very rapid, and the faster the better. As the meat freezes, crystals form in the water and other liquids contained in the cells. The size of the crystals depends on the speed of freezing; the faster the process, the smaller the crystals are. Above a certain size, the crystals burst the cells, allowing the meat to become dehydrated upon thawing and thus marring its flavor.

"Unless we have tremendous cold outside," Sanderson writes, "the center of the animal—and notably its stomach—will remain comparatively warm for some time, probably long enough for decomposition to start in its contents, while the actual chilling of the flesh will be slow enough for large crystals to form within its cells. Neither event occurred with the mammoths" (p. 82).

We can now return to Sanderson's question, which he answers this way:

> . . . the frozen-food experts have pointed out that to do this [quick-freeze a mammoth], starting with a healthy, live specimen, you would have to drop the temperature of the air surrounding it down to a point well below minus-150 degrees Fahrenheit. There are two ways of freezing rapidly—one is by the blast method, the other by the mist process; these terms explain themselves. Moreover, the colder air or any other gas becomes, the heavier it gets. If these volcanic gases went up far enough they would be violently chilled by the "cold of space," as it is called, and then as they spiraled toward the poles, as all the atmosphere in time does, they would begin to descend. When they came upon a warm layer of air, they would weigh down upon it and pull all the heat out of it and then would eventually fall through it, probably with increasing momentum and perhaps in great blobs, pouring down through the weakest spot. And if they did this, the blob would displace the air already there, outward in all directions and with the utmost violence. Such descending gases might well be cold enough to kill and then instantly freeze a mammoth (p. 83).

One of the major points of contention here is the condition of the flesh upon discovery. Sanderson claims that "the flesh of these animals was remarkably fresh and some was devoured by the explorers' sledge dogs" (p. 82). Farrand likewise notes the "healthy, robust condition of the cadavers" and many firsthand accounts attesting that the flesh was "fibrous and marbled with fat," looking "as fresh as well-frozen beef or horsemeat." However, his agreement with Sanderson on this point is only apparent, for he goes on to state, as given above, that the tissue was altered by slow decay before freezing. This contradiction in his argument is unresolved.

A second major disagreement concerns the nature of the Siberian terrain. Farrand concludes that the woolly mammoths lived in a tundra region similar to that in which they are found today, but the climate was slightly warmer and perhaps moister; they were nevertheless well adapted to the cold climate, as evidenced by their long hair, warm underwool, and thick subcutaneous fat layer.

On the other hand, there is Sanderson's portrait of the mammoth's environment. As he puts it:

> It now transpires, from several studies, that mammoths, though covered in a thick underwool and a long overcoat—and in some cases having quite a layer of fat—were not specially designed for arctic conditions; a little further consideration will make it plain that they did not live in such conditions.
>
> That they did not live perpetually or even all year round on the arctic tundra is really very obvious. First, the average Indian elephant, which is a close relative of the mammoth and just about the same size, has to have several hundred pounds of food daily just to survive. For more than six months of the year, there is nothing for any such creature to eat on the tundra, and yet there were tens of thousands of mammoths. Further, not one trace of pine needles or of the leaves of any other trees were in the stomach of the Berezovka mammoth; little flowering buttercups, tender sedges and grasses were found exclusively. Buttercups will not grow even at forty degrees, and they cannot flower in the absence of sunlight. A detailed analysis of the contents of the Berezovka mammoth's stomach brought to light a long list of plants, some of which still grow in the arctic, but are actually much more typical of southern Siberia today. Therefore, the mammoths either made annual migrations north for the short summer, or the part of the earth where their corpses are found today was somewhere else in warmer latitudes at the time of their death, or both (pp. 82–83).

In a later article published in *Pursuit* (October 1969), Sanderson revised some of these figures upward by saying, "A large elephantine needs some half a ton minimum of fresh green food a day to maintain itself. . . . For a minimum of eight months out of the year there is nothing for such large animals to eat north of the tree line in the Arctic. . . ."

One of the earliest scientists to address the question of mammoth extinctions, Henry H. Howorth, noted in the journal *Nature*

(26 January 1888), "We have not merely the mammoth carcasses to account for, but the trees found with these great beasts *still rooted*, and the land and freshwater shells showing a different climate when he lived." And while Howorth thought that the climate was "by no means a warm one," he noted "debris of trees— 'large stems', with their roots fast in the soil'—and found in places where no vegetation, save lichens, grows at present. . . ." Farrand himself mentions that L. S. Quackenbush, reporting in 1909, found "large trees" associated with fossil mammoths in a now-treeless part of Alaska.

How, then, one has to ask, can Farrand conclude that the terrain was "a tundra region similar to that in which they are found today"? Its present condition—treeless—is not what the geological record reveals at the time the mammoths roamed, millennia ago. Moreover, the presence of large trees implies that the surface was quite solid, not the "marshy pastures" assumed by Farrand.

Additionally, there is evidence implied by the condition of the many thousands of tusks taken from Siberia that sudden freezing was indeed involved. A traveler named Wrangell reported in the nineteenth century, according to a *Saturday Review* report (14 January 1888), that the ivory brought from northern districts of Siberia "is often as fresh and white as that from Africa." Obviously there would not have been a fossil ivory trade at all if the tusks had been decomposed or of unacceptable quality. Since the ivory thus retrieved was still in perfect and workable condition, Cardona reasons in his examination of the matter, it proves that the tusks themselves must have frozen suddenly. For as Professor Richard Lydekker reported to the Smithsonian Institution in 1899, Cardona notes, "Exposure in their ordinary condition would [have] speedily deteriorate[d] the quality of the ivory" (p. 78).

Evidence of both a sudden freezing condition and a warmer climate with firmer soil than Farrand allows is seen in this datum offered by Sanderson in his 1969 *Pursuit* article:

> In the New Siberian Islands . . . whole trees have turned up; and trees of the family that includes the plum; and with their leaves and fruits.* No such hardwood trees grow today any-

---

*The fruit was ripe, according to its discoverer, the Arctic explorer Baron Toll. Moreover, the roots were also well preserved in the permafrost. See Bassett Digby's *The Mammoth* (New York: Appleton, 1926).

where within two thousand miles of those islands. Therefore the climate must have been very much different when they got buried; and, please note, they could not have been buried in frozen muck which is rock-hard, nor could they have retained their foliage if they were washed far north by currents from warmer climes. They must have grown thereabouts, and the climate must have been not only warm enough but have had a long enough growing period of summer sunlight for them to have leafed and fruited.

He immediately adds, "Ergo, either what is now the Arctic was at the time as warm as Oregon, or the land that now lies therein was at that time elsewhere. Geophysicists don't go for an overall warming of this planet to allow such growth at 72 degrees north; otherwise everything in the tropics would have boiled! Thus, we are left with the notion that either the whole earth's crust shifted, or bits of it have drifted about." Since the time scale proposed for continental drift does not allow for such movement in so short a period, the logical conclusion to this piece of Sanderson's argument is that the crust shifted.

Evidence contrary to another of Farrand's points—the lack of frozen specimens of smaller Eurasian fauna contemporary with mammoths—is offered to some degree by Walter Sullivan in *Continents in Motion*. In an account of his work in 1935 for the American Museum of Natural History examining the mines around Fairbanks, Alaska, for fossil bones, Sullivan writes, "My trophy of the summer was the lower part of a super-bison leg with its fur, tendons, hoof and some flesh intact" (p. 36). While a super-bison may be comparable in size to a woolly rhino, if not a mammoth, a ground squirrel is not. Yet Sullivan reports that along the banks of prehistoric stream beds were circular patches, the size of a dinner plate, which, when cut around the edge with a hunting knife, could be lifted to expose a ground-squirrel nest. "Inside, almost always, we found a squirrel family curled up for its winter nap. . . . And in a number of cases there were still bits of fur, even flesh, clinging to the [Ice Age] skeletons." Although these animals were North American, not Eurasian, they nevertheless contradict the point Farrand tries to make.

Another important point overlooked by Farrand when he argues that the well-preserved mammoth specimens "must have died suddenly probably from asphyxia resulting from drowning in a lake or bog or from being buried alive by a mudflow or cave-in of a riverbank" is the statement of Dr. I. P. Tolmachoff, reported by A.S.W. (initials only given) in *Nature* (30 July 1903), that "the ice surrounding the [Berezovka] carcass was not that of a lake or river, but evidently formed from snow" (p. 298).

A.S.W. states immediately after the sentence above, "It is thus quite likely that the mammoth was quietly browsing on grassland which formed the thin covering of a glacier, and fell into a crevass which was obscured by the loose earth." We have already seen, however, that the animal died in summer (Farrand states that it was "probably late summer or early fall"). It would have decomposed long before enough snow could have fallen to entomb it in a permafrost grave. Moreover, it seems unlikely that the large trees noted in association with mammoths by several observers would or could grow on the "thin covering of a glacier" and certainly not without collapsing it long before a single mammoth crossed it. Last of all regarding this point is Sanderson's remark, "... there are not—and never were—any glaciers in Siberia except on the upper slopes of a few mountains, and ... the [mammoths] are never found in mountains, but always on the level plains and only a little above sea level" (p. 82).

To the argument that mammoths died from drowning, Sanderson replies that elephants are "the very best of swimmers" and, owing to the huge amounts of vegetable matter they have in their stomachs at all times, which develops much gas, it is almost impossible to sink them. They would have to be in an advanced state of decomposition or even to have burst. Yet clearly this is not the case.

Neither ice, water nor mud can be invoked as the agency that slew and preserved the frozen mammoths, according to Sanderson. Rather, he says, it probably happened this way:

> Consider now our poor mammoth placidly munching away in his meadow, perhaps even under a warm sun. The sky need not even cloud over, and there need not even be a dust haze where he is living, which would appear to have then been about where Central Asia is today. All of a sudden, in a matter of minutes, the air begins to move in that peculiar way one may experience today at the end of the arctic summer when the first cold front descends and the temperature may drop sixty degrees in an hour.
>
> All the mammoth feels is a sudden violent tingling all over his skin and a searing pain in his lungs; the air seems suddenly

to have turned to fire. He takes a few breaths and expires, his lungs, throat, eyeballs, ears and outer skin already crystallized. If he is near the center of the blob [of supercold descending gases], the terrible mist envelops him, and in a few hours he is a standing monument of what is virtually rock. Nor need there be any violence until the snow comes softly to pile up on him and bury him. And here we leave him for a moment and turn to his distant cousin chewing away in Alaska, just outside the area where the blob descends. What happens to him?

The sky here probably does cloud over, and it may even start to snow, something he has not before encountered in September, when he is in the north on his summer migration. He starts to pad off for cover. But then comes a wind that rapidly grows and grows in fury and explodes into something unimaginable. He is lifted off his feet and, along with bison, lion, beaver from ponds and fish from rivers, is hurled against trees and rocks, torn literally to bits and then bowled along to be finally flung into a seething caldron of water, mud, shattered trees, boulders, mangled grass and shrubbery and bits of his fellows and of other animals. Then comes the cold that freezes the whole lot, and finally when the holocaust is over, the snow to cover it all.

The notion of worldwide earthshaking volcanic eruptions ending the Pleistocene was suggested well before Sanderson, who simply adopted the concept into his larger one. Hibben, in his chapter quoted from above, notes that the idea has "considerable support, especially in the Alaskan and Siberian regions." Interspersed in the muck depths, he says, and sometimes through the very piles of bones and tusks themselves are layers of volcanic ash. He declares:

There is no doubt that coincidental with the end of the Pleistocene animals, at least in Alaska, there were volcanic eruptions of tremendous proportions. It stands to reason that animals whose flesh is still preserved must have been killed and buried quickly to be preserved at all. Bodies that die and lie on the surface soon disintegrate and the bones are shattered. A volcanic eruption would explain the end of the Alaskan animals all at one time, and in a manner that would satisfy the evidences there as we know them. The herds would be killed in their tracks either by the blanket of volcanic ash covering them and causing death by heat or suffocation or, indirectly, by the volcanic gases. Toxic clouds of gas from volcanic upheavals could well cause death on a gigantic scale. If every

individual, old and young, were killed, extinction would naturally follow.

Throughout the Alaskan mucks, too, there is evidence of
atmospheric disturbances of unparalleled violence. Mammoth
and bison alike were torn and twisted as though by a cosmic
hand in godly range. . . . Mixed with the piles of bones are trees,
also twisted and torn and piled in tangled groups; and the whole
is covered with the fine sifting muck, then frozen solid.

Storms, too, accompany volcanic disturbances of the proportions indicated here. Differences in temperature and the
influence of the cubic miles of ash and pumice thrown into the
air by eruptions of this sort might well produce winds and blasts
of inconceivable violence. If this is the explanation for the end
of all this animal life, the Pleistocene period was terminated
by a very exciting time, indeed (pp. 163–64).

In spite of the tempting ramifications of this concept, Hibben
cautions in conclusion that it cannot account for the loss of all the
Pleistocene life. Extinction of animals took place in Florida, Texas
and other southern regions where there is not evidence of a volcanic
eruption or any sweeping disturbance of the sort seen in Alaska.
"Any good solution to a consuming mystery must answer all of
the facts," he says (p. 164).

All of the facts. Contrary to Cohen's pronunciamento of "no
mystery," there is indeed a riddle to be solved, and no theory
offered so far—neither Sanderson's, Farrand's, nor any other—
has done it completely. We have seen that the C^{14} dates of various
mammoths indicate they were not victims of a single cataclysmic
event, as Sanderson assumes. On the other hand, the evidence is
clearly there for at least one such event, and this evidence is not
addressed at all by Farrand.

One of the principal theorists on the extinction of giant animals
in the Americas, Dr. Paul S. Martin, of the University of Arizona,
proposed in 1967 that *Homo sapiens* was the cause. Martin noted
that beginning in Africa about forty thousand years ago, the extinctions seemed to have spread into Europe, northern Asia, and,
about ten thousand years ago, across the Bering Strait to the
Americas. He postulates that this was due to the use of big-game
hunting technology by fierce and skillful hunters. The effect on
animals having no previous experience with man was, Martin
contends, catastrophic.

A major challenge to Martin's view has arisen in the past few
years, however, and it has the archaeological community in an

uproar. The controversy is due to the results of new dating techniques used first on several Indian skulls excavated from the San Diego area in the 1930s. In 1974 Dr. Jeffrey Bada, of the Scripps Institution of Oceanography, applied amino acid racemization, a laboratory process that measures the rate at which amino acids found in living things decay after death. Bada found the skulls to be forty-four thousand to forty-eight thousand years old. More recently, he dated a fully modern Indian skull as being seventy thousand years old—tens of thousands of years earlier than man is thought to have been in the New World!

All of the facts. This is the riddle of the frozen mammoths. It is a baffling maze of data tending to indicate that our planet underwent, a least once, a gigantic cataclysm. Yet so far the evidence has not been comprehensively explained, and there are pieces of the puzzle that do not fit the cataclysm concept. Do the frozen mammoths tell us a story of several pole shifts, or of a single one plus some random accidents? Or merely random local events? Or is there an alternative explanation such as a passing comet whose frozen gases, hitting the earth's atmosphere, could have produced the effects now seen in the Arctic regions—the sudden asphyxiation and freezing of animals, the violent windstorms, the heaps of battered and torn bodies? We will see other lines of evidence bearing on this in later chapters.

Perhaps Dima, now undergoing investigation by Soviet and American scientists, will tell us the final answer. And perhaps that answer will shed light on an even greater mystery: whether there is any truth to those persistent tales of resplendent civilizations that vanished in prehistory.

References and Suggested Readings

CARDONA, DWARDU. "The Problem of the Frozen Mammoths," *Kronos,* Winter 1976.

CORLISS, WILLIAM R. *Strange Planet.* Glen Arm, MD: The Sourcebook Project, 1975. (Reprints many inaccessible articles on strange phenomena, including nearly a dozen on frozen mammoths and mammoth graveyards.)

FARRAND, WILLIAM R. "Frozen Mammoths and Modern Geology," *Science,* 17 March 1961.

HIBBEN, FRANK C. *The Lost Americans.* New York: Thomas Y. Crowell, 1968.

HUEVELMANS, BERNARD. *On the Track of Unknown Animals.* New York: Hill & Wang, 1959.

Pursuit, 2008 Spencer Road, Newfield, NY 14867.

SANDERSON, IVAN T. "Riddle of the Frozen Giants," *The Saturday Evening Post,* 16 January 1960.

STEWART, JOHN MASSEY. "Frozen Mammoths from Siberia Bring Ice Ages to Vivid Life," *Smithsonian,* December 1977.

SULLIVAN, WALTER. *Continents in Motion.* New York: McGraw-Hill, 1974.

WILLIS, RONALD J. "Man and the Mammoth in the Americas," *Info Journal,* Vol. 1, No. 1 (Spring 1967). Available from the International Fortean Organization, 7317 Baltimore Avenue, College Park, MD 20704.

Chapter Three

The Death of Civilization
—Again!

It is a persistent idea of great antiquity that prehistoric civilizations existed before our own—vast in extent, great in strength, high in learning, sophisticated in science. The myth of Atlantis is only the best-known variation on this theme. Lemuria, Agartha, Shamballa, Ultima Thule, Yuga—there are many more.

These lost civilizations, legend has it, were as highly evolved as our own—and sometimes more so—but gigantic cataclysms destroyed them and obliterated nearly all traces of their existence. The disasters often involved global changes in the surface of the planet—"earth changes," as they are popularly called. Survivors from the lost cultures are said to have preserved a portion of the knowledge accumulated over many centuries and, after colonizing new areas, provided many of the seeds from which the succeeding civilization arose.

The Greek philosopher Plato, in his *Timaeus*, is one of the earliest to present this notion. There he records the words of an Egyptian priest to Solon, grandfather of Socrates:

> O Solon, Solon, you Hellenes are but children. . . . There is no old doctrine handed down among you by ancient tradition nor any science which is hoary with age, and I will tell you the reason behind this. There have been and will be again many destructions of mankind arising out of many causes, the greatest having been brought about by earth-fire and inundation. Whatever happened either in your country or ours or in any other country of which we are informed, any action which is noble and great or in any other way remarkable which has taken place, all that has been inscribed long ago in our temple records, whereas you and other nations did not keep imperishable records. And then, after a period of time, the usual inundation visits like a pestilence and leaves only those of you who are destitute of letters and education. And thus you have to begin over again as children and know nothing of what happened in ancient times either among us or among yourselves.
>
> As for those genealogies of yours which you have related to us, they are no better than tales of children; for in the first place, you remember one deluge only, whereas there were a number of them. And in the next place there dwelt in your land, which you do not know, the fairest and noblest race of men that ever lived of which you are but a seed or remnant. And this was not known to you because for many generations the survivors of that destruction made no records.

Fanciful thinking? Mere embroidery on imagination? We shall see that it may not be. . . .

In the Moslem year 919, which is 1513 in the Christian calendar, a map was painted on parchment and signed with the name of an admiral of the Turkish navy, Piri ibn-Haji Memmed, also known as Piri Re'is. Rendered in varying shades of brown, a large fragment of this map was discovered in 1929 in the old Imperial Palace in Constantinople, where it aroused great attention because, from the date, it appeared to be one of the earliest maps of America. Examination revealed that this map differed in a significant way from all other maps of America drawn in the sixteenth century: it showed South America and Africa in correct relative longitude. This was remarkable because navigators at the time had no means of finding longitude except by guesswork.

Charles Hapgood (whom we'll meet in Chapter 5) reports part of the story in his book *Maps of the Ancient Sea Kings:*

> Another detail of the map excited special attention. In one of the legends inscribed on the map by Piri Re'is, he stated that he had based the western part of it on a map that had been drawn by Columbus. This was indeed an exciting statement because for several centuries geographers had been trying without success to find a "lost map of Columbus" supposed to have been drawn by him in the West Indies. . . .
>
> Piri Re'is made other interesting statements about his source maps. He used about twenty, he said, and he stated that some of them had been drawn in the time of Alexander the Great, and some of them had been based on mathematics. The scholars who studied the map in the 1930s would credit neither statement. It appears now, however, that both statements were essentially correct.

After it was discovered, cartographers and the public heard nothing more of the map until 1953, when Captain Arlington H. Mallery, an engineer, navigator and author, happened to have it brought to his attention by a friend at the U.S. Navy Hydrographic Office, M. I. Walters. In 1951 Mallery had published a book, *Lost America,* that gave evidence that the prehistory of our continent was quite different from the conventional view. Among many pieces of evidence that Mallery pointed out was the fact—discovered by him—that ancient maps of Greenland showed landforms beneath the ice cap now covering them. How, he asked, could this have been known?

This question was especially strong in Mallery's mind when Walters brought him the map because, a short time before, he had seen an article in *Journal of Geography* showing seismic sounding maps of both Greenland and Antarctica that revealed features of the land beneath massive amounts of ice. In the case of Greenland, Mallery knew, the features compared almost exactly with what was shown on the ancient maps he had studied.

Therefore, when Mallery saw the Piri Re'is map, he was struck almost immediately by the correspondence between some of its features and those on the seismic map of Antarctica. After some study, Mallery stated his opinion that the southernmost part of the map showed bays and islands on the coast of Queen Maud Land in Antarctica. This was a shocking conclusion, for two reasons. First, when the map was drawn, in the sixteenth century, nobody was supposed to have known what Antarctica looked like. Historians officially date its discovery about 1818, and it was not fully mapped until after 1920. Second, and even more astounding, Antarctica had not been free of ice for thousands of years!

Mallery asked others to examine his findings. Among them were a man who had been to Antarctica, the Reverend Daniel L. Linehan, S.J., director of the Weston Observatory, of Boston College, and the Reverend Francis Heyden, S.J., director of the Georgetown University Observatory, in Washington, D.C. After much study, these astronomers, along with the cartographer Walters, felt convinced Mallery was correct.

Finding themselves in agreement, Mallery, Linehan and Walters took part in a radio panel discussion of the subject sponsored by Georgetown University. On 26 August 1956, the Georgetown University Forum began its 510th consecutive broadcast from the campus of Georgetown University. The topic: "New and Old Discoveries in Antarctica." I was able to obtain a transcript of the program, which I reproduce here at length for its historic value and extreme interest. Moderator Matthew Warren began:

WARREN: Mr. Walters, to open our discussion, would you give us a more detailed explanation of your studies?

WALTERS: Captain Mallery was introduced to me by the Head Engineer of the Hydrographic Office shortly following World War II, with the request that I check with him on an old map [of Greenland] which he had run across. I procured from our files a number of present-day charts and together we examined the map in question, which on first sight appeared to be like a hobgoblin and of no real value. However, we began to check various islands, capes and peaks, and this old map showed astounding accuracy in presenting the land and water areas in their exact locations. This was ten years ago. From that time on, Mr. Mallery has made a great many visits to the Hydrographic Office and each time he was brought to me for the purpose of rendering whatever assistance I could. The deductions and conclusions which he has reached I consider of the utmost importance in solving some of the secrets of our world history and its peoples, the extent of its water areas and the demarcation of its coastlines. In other words, where present-day charts are inaccurate or lacking in survey information, he has been able, with the use of these old maps, to supply missing data and correct the errors. I want to say that I am here to back Mr. Mallery and all he has to say on the subject and I stand squarely behind him in his deductions. Perhaps he can also shed some light on the mystery involved as to the various races that have inhabited this continent previous to the American Indian. I consider that he has contributed in a great measure to the scientific knowledge of the world, and that his work has been and will be of great value in the future.

WARREN: You say that these maps have been checked by the Hydrographic Office of the U. S. Navy?

WALTERS: Yes.

WARREN: As far as you are concerned, are they accurate?

WALTERS: Yes, they are.

WARREN: How old are these maps?

WALTERS: These maps go back 5,000 years and even earlier. But they contain data that go back many thousand years previous to that.

WARREN: Tell us a bit more of the maps, if you will, Mr. Mallery.

MALLERY: The important thing about these maps is that they bring home to us the fact that the oldest human records that we have, and which are absolutely authentic, are the navigation charts. A land map can be almost anything. If your map is wrong, you can go over to the nearest gasoline station and inquire the way. But if you are on a ship and the map is wrong, why, the captain usually goes down with the ship. That is the rule of the sea. That is what we have found with these ancient maps — they are maps that have survived for thousands of years and have been preserved only because they were accurate. The first captain under whom I sailed as First Officer in World War II was a Captain Myrdal. His family settled in Bergen, Norway, in 985 A.D. In other words, the records of that Captain and his ancestors — who had been skippers all those years — carried back for 1,000 years. This is just one fifth of the time to which we can date back these maps.

The maps of Greenland have this advantage — we were able to check, of course, through the Hydrographic Office, and afterwards by the seismic soundings of the French Polar Expeditions — that these maps actually recorded the subglacial topography of Greenland. About three years ago, the Chief Engineer of the Hydrographic Office handed me a copy of a map which had been sent to him by a Turkish naval officer. He suggested that I examine it in the light of the information we already had on the ancient maps. After making an analysis of it, I took it back to him and requested that the Officer check both the latitude and longitude and the projection. When they asked why, I said, "There is something in this map that no one is going to believe coming from me, and I don't know whether they will believe it coming from you." That was the fact that Columbus had with him a map that showed accurately the Palmer Peninsula in the Antarctic continent.

Here is a copy of two thirds of that map, which shows that Columbus had with him a map of Yucatan, Guatemala, all of South America to the Strait of Magellan, and a substantial part of the coast of Antarctica.

WARREN: It is difficult for us to understand today how they could have been so accurate so many thousands of years ago when we are just now—or just recently—come to know the modern scientific methods of mapping. How is it possible?

MALLERY: That is a problem we have puzzled over too. We cannot understand how they could have been so accurate. Of course, in the first place, it was evident that there was little ice then. But, second, they had a record—for example, of every mountain range in northern Canada and Alaska—which the Army Map Service did not have at the time that Mr. Walters and I checked. They have since found them. But to get back—just how they were able to do it—you will probably recall the tradition of the Greeks and the airplane—maybe they had the airplane. But the fact is that they did it. Not only that, they knew their longitude absolutely correctly. This was something that we did not know ourselves until about two centuries ago.

WARREN: Father Linehan, you have been in seismic expeditions in the Arctic and Antarctic. Do you share the enthusiasm over these new discoveries?

LINEHAN: I certainly do—yes. We are finding out things by seismic methods which seem to prove a lot of the drawings which were made on these old maps, the land masses, the projections of mountains, the seas, the islands are all being checked by these maps themselves. I think with more seismic work—to strip some of the ice off the lands that these maps have made—more will be proved in showing these maps to be correct than we are apt to believe today.

WARREN: What is a seismic exploration, Father?

LINEHAN: A seismic exploration is sort of using a man-made earthquake. We study earthquakes by studying the earthquake waves that come through the earth, and by looking at these waves in our records, we are able to tell what sort of material they went through. With a man-made earthquake, we set off a charge of dynamite, and in studying these waves, their velocity, the time they are reflected from various discontinuities below the surface of the earth, we are able to determine the depth from which these reflections come. It is like timing an echo. . . .

WARREN: Father Linehan, this gives you a map of the earth under the water's surface, does it not?

LINEHAN: That is right. Not only the water, but under the ice as well. Actually, what we do is strip the ice from the continent of Antarctica, or Greenland, wherever it may be, and try to see the land as it was mapped by these various charts that Mr. Mallery has here.

WARREN: Then how do you project one upon the other?

LINEHAN: We would determine the elevation of the land mass below the ice above the sea level, and check it against the maps which he had—which were apparently made when there was no icecap covering Antarctica—and then try to prove whether or not his maps were correct. What we have done to date shows his maps to look extremely correct.

WARREN: You, too, share this enthusiasm, do you not, Mr. Walters?

WALTERS: Yes, I do.

WARREN: You have examined the maps and to your satisfaction found they compare pretty favorably with our present maps?

WALTERS: Yes, they do, surprisingly favorably.

WARREN: How have you gone about it?

WALTERS: We have taken the old charts and the new charts that the Hydrographic Office produces today and made comparisons of the soundings of salient peaks and mountains. We have found them to be in astounding agreement. In this way we have checked the old work very closely. We put very much confidence in what Captain Mallery has disclosed.

WARREN: Mr. Mallery, this must then lead to the conclusion that there were competent explorers and map makers along the coasts of the Atlantic long before Columbus.

MALLERY: Several thousand years before. Not only explorers, but they must also have had a very competent and far-flung hydrographic organization, because you cannot map as large a continent as Antarctica, as they apparently have—half of it—or as extensive an area as Greenland, or half the continent of North America, as we know they did—probably 5,000 years ago. It can't be done by any single individual or small group of explorers. It means an aggregation of skilled scientists who are familiar with astronomy as well as the methods required for topographic surveying.

LINEHAN: I have a question, Mr. Mallery, that I would like to ask you. Looking at these maps that you are peeling off here, the type of projection is very odd. It looks nothing like our modern-day maps. I would find myself hard put to locate myself on some of these maps. Could you give us a brief explanation—to an unseen audience—as to this type of map today?

MALLERY: Nordenskjöld, the great Swedish explorer and cartographer, spent 18 years trying to solve this projection. I would not have been able to solve it if it had not been for the fact that we found the map of North America that showed so many points in latitude and longitude correctly, as we could judge by our present maps, that I was able to compute the method which they used to draw their grid on which their chart was plotted. A grid is formed by the latitude and longitude lines together—and on which a map can be read. These lines alone make it possible.

This projection, as we call it, is entirely different from anything used in modern cartography. It was probably the forerunner of our present Mercator charts, a straight-line drawing, and like the Mercator was used exclusively for navigational purposes.

WARREN: Father Linehan, have you found any seismic studies—or by these studies—any areas in Antarctica which agree with these maps?

LINEHAN: Actually, my own seismic studies there were not on any long, extended area. Most of my work was of an engineering aspect for the U. S. Navy in locating on various bases as to whether the ice was thick enough to establish a camp. Jan Viers of the Norwegian-Swedish-British Expedition ran quite a seismic line in Queen Maud Land and that seismic line agrees very well, extremely well with the maps that Captain Mallery is speaking about. He shows that even though the ice surface itself may rise as much as 3,000 feet above sea level, below that ice you have land masses, and we believe rock — almost bare rock; the glacier would have scraped off any other debris from it — and this rock goes below sea level. He even mapped the nunataks — an Eskimo word for mountains rising above the snow — as an extension of the seismic line. They agree extraordinarily well with the position of the islands found on these old maps. We feel that the seismic data done most extensively by Jan Viers proves these ancient maps without any doubt for this one area alone around the Weddell Sea and the Queen Maud Land area.

WARREN: Mr. Mallery, other than the startling fact that there were map makers 5,000 years ago, what does this mean to scientists today? Does it change their conception of the movement of ice?

MALLERY: I think it means very definitely that the ice age or an intermediate ice age took place at a much later time than at present we think. At the time I first took up this question of the ice glacier that was shown on the map of North America, Dr. Capps, the specialist of the Geological Survey, estimated that the great glacier shown on the map in the range of Alaska was at least 15,000 years old — rather, was in existence 15,000 years ago and had disappeared. Yet this map — which we are very sure dates back about 5,000 years ago — showed that glacier. These other maps indicate the glacier had disappeared or else had not arrived at the time these maps were made or had only begun to arrive.

In the map of Queen Maud Land, the map indicates that the glacier had just begun to appear at the middle of Queen Maud Land, but the bay on the map that Columbus had was still entirely uncovered. Now only the peaks of the mountains that were on the island show above the ice. The ice has added about a mile at least since that map was made.

WARREN: Where were these maps when you discovered them?

MALLERY: In the Library of Congress.

WARREN: This is interesting. Why is it that no one had discovered these before, I wonder?

MALLERY: They had been discovered, but no one could read them. It just happened that Mr. Walters and I, in our search for the old maps, uncovered that ancient map of North America which was brought to Iceland in 1568 by Father Thorsden, who was a parish priest in Staden, Iceland. He brought it to Iceland in 1568 and it was deposited, I presume, in the Cathedral in Iceland and somehow it got over to this country. No one paid any attention to it, however, until we finally, after gradually checking back over these other ancient maps, happened to notice one day that it had marked upon it the old Icelandic name for glacier, and also another Icelandic name for lava fields. So I took it over to Dr. Capps of the Geological Survey and from there we were able to go to town on the maps because we had the projection.

WARREN: Father Linehan, are there any other good seismic proofs of these maps?

LINEHAN: I think the greatest proof that has come, to date, and which has given Captain Mallery the greatest satisfaction, is that of Paul-Emile Victor and his studies of Greenland. The maps that Captain Mallery had showed Greenland to consist of at least three islands, and Victor's seismic data has proved this conclusively— and they worked independently. Victor has shown that these islands exist, that there is a great sound or fjord which runs through the middle of Greenland. This was a marvelous discovery by Victor and a great work on his part, but Captain Mallery knew it all the time. . . .

WARREN: Mr. Mallery, aside from the discoveries in Antarctica, you have made a number of studies from an archaeological standpoint, have you not, in North America?

MALLERY: In metallurgy—with the assistance of the National Bureau of Standards, and Batelle [Institute] and some other laboratories—we have excavated a number of iron furnaces in Ohio and Virginia. The iron that we obtained has been checked by these laboratories and was found to be made by the same processes of the metallurgists of Europe. The British Museum, you might say, upset the applecart of the archaeologists who have always claimed that the Egyptians did not make iron. But the British Museum set some of the tools from Egypt to a metallurgist and were astounded to find that they were using powdered metallurgy, the process which has made our atom bomb possible. So, 5,000 years ago, the Egyptians were using the same process we thought we had just discovered today to make the atom bomb, and the timing of the process agrees with the timing of the ancient maps.

* * *

Ancient maps and sophisticated metallurgy suggest that, as the title of Andrew Tomas's best seller about lost civilizations and mysteries of antiquity proclaims, we are not the first. Tomas is only one of a substantial number of researchers, historians and scholars who are challenging, on various grounds and to various degrees, the prevailing view of history. It does not appear, these people say, that current theories of man's gradual ascent from primitive beginnings are correct. Ascent there may have been*— but it was far earlier than supposed and it may have happened not once but *many* times. Too many things are out of place in the historical record to justify the notion that civilization had its genesis six or seven thousand years ago after humanity finally emerged from the Stone Age. Geological anomalies and archaeological "erratics" abound, misplaced in time and space. How, for example, can we explain the Greenland and Antarctic mappings that show land free of ice? And how can we explain the following:

• In June 1851 a metallic vessel was blown out of an immense mass of rock by workmen who were blasting on Meeting House Hill, in Dorchester, Massachusetts. The bell-shaped vessel, about $4.5'' \times 6.5''$, was described like this: ". . . resembles zinc in color, or a composition metal, in which there is a considerable portion of silver. On the sides there are six figures of a flower, or bouquet, beautifully inlaid with pure silver, and around the lower part of the vessel is a vine, or wreath, inlaid also with silver. The chasing, carving, and inlaying are exquisitely done by the art of some cunning workman. This curious and unknown vessel was blown out of solid pudding stone, fifteen feet below the surface. . . ."

• In June 1891 a woman broke a lump of coal in half preparatory to putting it in the scuttle. A chain dropped out of the lump, but not completely. The ends of the chain remained stuck in the coal. Theoretically, the coal was formed in the Carboniferous era and was several hundred million years old.

---

*There may also have been *descent* in the form of "gods from outer space" arriving on earth in UFOs to interact with primitive man. At least there are intriguing hints in mythology and history of extraterrestrial visitations. But the question is extremely complicated and beyond the scope of this book to examine in detail.

• In 1961 in California, an object was found inside a solid geode. After being sliced in two, the geode revealed an object that looks amazingly similar to a common spark plug. The cross section shows a hexagonal part consisting of a porcelain or ceramic insulator with a central metallic shaft—the basic components of a spark plug. An X-ray photograph shows what might be the remains of a corroded piece of metal with threads in the upper end of the object. A trained geologist inspected the geode, which was encrusted with fossil shells, and stated that in his opinion the nodule had taken at least five hundred thousand years to attain its present form.

• In 1968, what appeared to be fossilized sandal prints were found in Utah. When the amateur rock hound who discovered it split open a two-inch-thick slab of rock, to his astonishment he saw an almost perfect mold of a shod foot with a trilobite right in the footprint itself. Trilobites, now extinct, were tiny hard-shelled animals that lived in the warm primeval seas. They disappeared about 500 million years ago.

These four examples are among hundreds cited with greater detail in *Mysteries of Time and Space* (Dell/Confucian, 1976) by Brad Steiger. The evidence of prehistoric civilizations that Steiger amasses there is overwhelming, but he is not the only one to have compiled such evidence. There are many others.

If we are not the first, we may not be the last, either. In other words, some pole shift theorists propose a cyclical nature to history that includes the rise and fall of many global civilizations prior to the epoch we consider to be the beginning of history. You will meet these views in the following pages. While none of them are definitely proven, there seems to be more than the mere germ of an idea here. Consider, for instance, a story that appeared in the *New York Herald Tribune* on 16 February 1947.

When the first atomic bomb exploded in New Mexico, the desert turned to fused green glass. This fact, according to the magazine *Free World,* has given certain archaeologists a turn. They have been digging in the ancient Euphrates Valley and have uncovered a layer of agrarian culture 8,000 years old, and a layer of herdsman culture much older, and a still older caveman [sic] culture. Recently they reached another layer . . . of fused green glass. Think it over, brother.

That inveterate collector of unusual information Ivan Sanderson offered this comment in his report on the matter in *Pursuit* (January 1970):

> Bits of green glass, possibly fused in an ancient fireplace, is one thing; areas of fused green glass is something quite else again. And this site is not the only one. There are also the fused forts of the west coast of Scotland and elsewhere, in which one side only has been fused, as if hit from above by intense heat. Lightning occasionally fuses sand, but always in a root-like pattern. . . . So just what produced a whole *stratum* of green glass in various parts of Mesopotamia?

And what are we to think about those persistent tales in myth and legend that hint at a lost technology of high degree? Richard Wingate, an explorer-photographer, notes in his recent book, *Lost Outpost of Atlantis:*

> The very old Indian *Mahabharata* text mentions an iron-tipped thunderbolt which was sent against an enemy city. It exploded, according to the account, with the light of ten thousand suns, and had the destructive force of ten thousand hurricanes. Elephants miles away were knocked off their feet; an umbrella-shaped cloud rose to the sky. The enemy city, as well as its army, was entirely destroyed.
>
> Survivors of the conflagration were instructed to wash themselves in a nearby river and rinse their armor. Hair fell from the victims' heads, flesh whitened, and pottery broke by itself long after the dust had settled.

Again I ask: Fanciful thinking? Mere embroidery on imagination? Recall Mallery's final remark about Egyptian metallurgy and read on. . . .

References and Suggested Readings

DE CAMP, L. SPRAGUE. *Lost Continents*. New York: Ballantine, 1975.

HAPGOOD, CHARLES. *Maps of the Ancient Sea Kings*. Philadelphia: Chilton, 1966.

SANDERSON, IVAN T. "Much About Muck," *Pursuit*, October 1969.

STEIGER, BRAD. *Mysteries of Time and Space*. New York: Dell/Confucian, 1976.

TOMAS, ANDREW. *We Are Not the First*. New York: Bantam, 1973.

WINGATE, RICHARD. *Lost Outpost of Atlantis*. New York: Everest, 1979.

ZINK, DAVID. *The Ancient Stones Speak*. New York: E. P. Dutton, 1979.

PREDICTION

*What Modern Researchers
Have to Say*

Introduction to the Section

The history of science shows that certain ideas have been in vogue and then discarded, only to be resurrected later in a modified form or even wholly as new evidence reopens the question. The theory of continental drift is an example. Originally proposed in 1912 by the German meteorologist Alfred Wegener, it aroused some interest but "fell from grace" and for many decades was credited with being merely a curiosity in the development of the geological theory of the earth. Then, in the late 1950s and early 1960s, it was taken out of the museum of scientific ideas, dusted off and reexamined. The result: tectonic plate theory, which superseded continental drift theory by incorporating it in a more comprehensive view.

The ether theory of physics is another example of death and rebirth in scientific history. Derived from the Greeks and upheld by such giants of science as Newton, Faraday and Maxwell, the concept of a luminiferous ether—the medium that transmitted electromagnetic waves—fell into disrepute when the famed Michelson-Morley experiment of 1887 found no evidence of an ether. Einstein assumed in his theory of relativity that there was no ether, and his developing scientific stature resulted, as with continental drift theory, in the abandonment of the concept by the scientific community for several decades.

In 1957, however, the Nobel physicist P. A. M. Dirac asked (as the title of a paper), "Is there an ether?" He answered affirmatively, and since then other atomic scientists have suggested that the ether may be defined as an energy-rich subquantic medium composed of neutrinos, pervading all space, interpenetrating all matter, and acting as the common denominator in all particle reactions. The question is still being debated, but my point is that the ether concept is another example of scientific thought returning to vogue in a modified form.

Perhaps the same will happen to the concept of polar shifting. In any case, it has a long scientific history behind it, which we examine now as a prelude to the modern proponents of pole shift theory whom this section will present.

According to geophysicists Walter H. Munk and Gordon J. F. MacDonald, in their definitive text *The Rotation of the Earth,*

> The possibility of large-scale wandering and of continental drift has, during the last hundred years, excited the imagination of many geologists. The occurrence of late-Paleozoic glacial deposits near the present equator was largely responsible for the initial interest. Comte de Buffon founded the "Catastrophic School" of polar wandering in the nineteenth century. Lubbock, De la Beche, Evans and others voiced their enthusiasm for the theory (p. 251).

Comte de Buffon, the French naturalist and one of the great minds of the eighteenth century, introduced to science the idea that catastrophic changes in the earth have occurred when its spin axis was altered. The Swiss geologist-meteorologist Jean André Deluc also thought that the poles shifted, as did the French geologist-mineralogist Déodat Dolomieu. Another French naturalist, Georges Dagobert, Baron Cuvier, in his *Theory of the Earth,* published in 1812, interpreted the geological record in catastrophic terms. It is evident, he wrote, that

Every part of the earth, every hemisphere, every continent, exhibits the same phenomenon. . . . the various catastrophes which have disturbed the strata . . . have given rise to numerous shiftings of this [continental] basin. It is of much importance to mark, that these repeated irruptions and retreats of the sea have neither been slow nor gradual; on the contrary, most of the catastrophes which occasioned them have been sudden; and this is especially easy to be proved, with regard to the last of these catastrophes. . . . I agree, therefore, with Mm. Deluc and Dolomieu in thinking that if anything in geology be established, it is, that the surface of our globe has been subjected to a vast and sudden revolution, not further back than from five to six thousand years; that this revolution has buried and caused to disappear the countries formerly inhabited by man, and the species of animals now most known; that contrariwise it has left the bottom of the former sea dry, and has formed on it the countries now inhabited; that since the revolution, those few individuals whom it spared have been spread and propagated over the lands newly left dry, and consequently it is only since this· epoch that our societies have assumed a progressive march. . . .

But the countries now inhabited, and which the last revolution left dry, had been before inhabited, if not by mankind, at least by land animals; consequently, one preceding revolution, at least, had overwhelmed them with water; and if we may judge by the different orders of animals whose remains we find therein, they had, perhaps, undergone two or three irruptions of the sea.

The notion of catastrophic global events in prehistory became the most controversial subject in nineteenth century geology, because it directly contradicted the principle of uniformity, first enunciated by James Hutton, of Scotland, in 1795 and the Chevalier de Lamarck, the French naturalist, in 1800, but set forth in definitive form by the English geologist Charles Lyell in 1838 through his classic text *Elements of Geology*. The principle can be stated thus: the earth's external and internal geological processes have been operating unchanged, and within the same range of rates, throughout the earth's history—and these rates are typified by currently observed processes that are clearly gradual in nature.

The uniformitarianism versus catastrophism debate consumed much ink and paper during the nineteenth century, with many notable figures taking part on both sides, especially as the theory of evolution brought biology versus the Bible into the question of the earth's history. Even Charles Darwin, whose perspective on evolution was essentially uniformitarian-gradualist and who denied the occurrence of continental catastrophes in the past, admitted that the riddle of the mammoth extinction was, for him, an insoluble problem. In his classic journal of his travels, *The Voyage of the H.M.S. Beagle*, Darwin wrote:

> What then has exterminated so many species and whole genera? The mind at first is irresistibly hurried into the belief of some great catastrophe; but thus to destroy animals both large and small, in Southern Patagonia, in Brazil, on the Cordillera of Peru, in North America up to Behring's Straits, *we must shake the entire framework of the globe*. No lesser physical event could have brought about this wholesale destruction not only in the Americas but in the entire world.... Certainly no fact in the long history of the world is so startling as the wide and repeated extermination of its inhabitants.

In 1877 Darwin published an analysis purporting to show that the pole "may wander indefinitely from its primitive position" if the earth were "plastic," but no more than a few degrees at most if it were "sensibly rigid." Darwin favored the latter view, because Sir William Thompson had demonstrated to his satisfaction that the planet was indeed "sensibly rigid."

The permanence of the poles was also questioned by the British physicist Lord Kelvin, who in 1876 asserted it was "highly probable" that in ancient times the axis of rotation, moved slowly by nonalignment with the earth's axis of maximum inertia, may have "gradually shifted through 10, 20, 30, 40 or more degrees without at any time any perceptible sudden disturbance of either land or water."

In 1889 the Italian astronomer G. V. Schiaparelli entered the pole shift debate. In *De la Rotation de la Terre* he pointed out that under certain conditions the earth could enter a condition of strain, causing the axis to wobble. "The permanence of the geographical poles," he wrote, "in the very same regions of the earth cannot yet be considered as incontestably established by astronomical or mechanical arguments. Such permanence may be a fact today, but it remains a matter still to be proven for the preceding ages of the history of the globe."

He went on to hypothesize that a series of geological changes could, by slow but cumulative effect, disrupt the planet's equilibrium. "The possibility of great shifting of the pole is an important element in the discussion of prehistoric climates and the distribution, geographic and chronologic, of ancient organisms. If this possibility is admitted, it will open new horizons for the study of great mechanical revolutions that the crust of the earth underwent in the past. We cannot imagine, for instance, that the terrestrial equator could take the place of a meridian, without great horizontal tension in some regions, that would open great rifts; and in other regions, horizontal compressions would have taken place, such as are imagined today in order to explain the folding of the strata and formation of mountains."

He ended his argument with these words: "Our problem, so important from the astronomical and mathematical standpoint, touches the foundations of geology and paleontology: its solution is tied to the most grandiose events in the history of the earth."

In 1889, also, a little-known work by Marshall Wheeler, *The Earth—Its Third Motion*, was published, which postulated an aperiodic third motion for the planet. Wheeler stated that "the Earth will come to rest at 90° from its starting point, reversing the present position of the poles and the equator." His closing remark is of great interest for our theme: "If they [his arguments for the third motion] prove true, the whole world should be informed of the fact, and means taken to forever perpetuate that knowledge so that, when the dread event transpires, mankind should not lapse again into prehistoric barbarism, but, instead, the rhythm of man's existence be raised to a higher plane of action" (p. 10). This is a remarkable statement, and comes close to meriting credit for Wheeler as the father of scientific pole shift predictions. Technically speaking, however, he does not flatly *predict*—he only says tentatively, pending further research, that a pole shift *might* occur.*

Despite many data in favor of polar shifts, the question was settled—for the nineteenth century, at least—when it was decided that no conceivable force originating within the earth could make it shift on its axis. Both James Clerk Maxwell and Sir George Darwin, son of Charles Darwin, advanced these arguments, and their influence was sufficient for a time to keep others from seriously considering the question.

*For an elaboration on this point, and an alternative view, see Appendix One.

In the twentieth century, however, that giant of science, Alfred Wegener, who formulated the theory of continental drift, came to believe that polar shifts provided the driving force behind the movement of continents. As Walter Sullivan explains it, Wegener adopted the view of the Hungarian physicist Eötvös, who had calculated that, because the earth is spinning and its shape bulges at the equator, there should be a very slight force nudging the continental blocks toward the equator. This force, when combined with a tendency for tidal drag to pull the continents westward, might account for continental movements, Wegener believed, if the spin axis changed from time to time, causing the earth to readjust its equatorial bulge and produce new directions of drift. Wegener called this *Polfluchtkraft,* or "pole-fleeing force." It was soon shown by others, however, that this force was insufficient to do what Wegener proposed.

About the same time, another meteorologist, the Austrian Julius Hann, seeking a cause for climatological mysteries of prehistory, made this statement:

> The simple and obvious explanation of great secular changes in climate, and of former prevalence of higher temperatures in northern circumpolar regions, would be found in the assumption that the earth's axis of rotation has not always been in the same position, but that it may have changed its position as a result of geological processes, such as extended rearrangement of land and water.

Scientists continued to comment on the concept of pole shifts throughout the century. Since the uniformitarian doctrine was, and still is, firmly entrenched—meaning it was, and is, widely assumed by earth scientists that pole shifts were nonexistent and therefore needed no further debate—the bulk of the material published on the subject was favorable to the concept, attempting to reopen the question. Among the publications, we may note the following.

The geologist Damian Kreichgauer, in a 1926 paper, suggested for the first time in scientific literature* that only the lithosphere of the earth moved—a novel concept that we shall see developed to its fullest extent in Chapter 5. Likewise, the major concept to be explored in Chapter 4, the centrifugal effects of the ice caps,

---

*A German writer, Carl Löffelholz von Colberg, first suggested the idea in 1886, according to Isaac Asimov.

was examined in a privately published pamphlet by L. Taylor Hansen entitled *Some Considerations of and Additions to the Taylor-Wegener Hypothesis of Continental Displacement.*

W. B. Wright, in his 1937 *The Quaternary Ice Age,* stated that during geological history there occurred many changes in the position of the planet's climatic zones that cannot be explained except by a shifting of the axis or a displacement of the poles from their present positions.

The English astronomer Sir Arthur Eddington, in his paper "The Borderland of Geology and Astronomy," thought that the ice ages were caused by crustal shifting as a result of tidal friction or the inequality of lunar pull on various layers of the earth. However, the difficulty with Eddington's suggestion was that the lunar influence is applied only in an east-west direction. This could not explain how lands had been moved out of their earlier latitudes to become glaciated.

Although the mechanism Eddington proposed to explain ice ages was inadequate, the concept itself nevertheless appealed to K. A. Pauley, who remarked in an article, "The Cause of the Great Ice Ages," in *Scientific Monthly* (August 1952), "We are fully justified in concluding that the lithosphere was displaced during the great Ice Ages, and that the displacements were the direct cause of the alterations in climates during these periods."

Crustal shifts also appealed to the Dutch geophysicist F. A. Vening Meinesz, who stated in a 1943 *Nederlandsche Akademie . . . Verslagen* article (Vol. 52, No. 5) that they could explain certain geographical features of the planet. If one assumes that the crust has moved clockwise over the core by more than 70°, he said, the expected effect "shows a remarkable correlation to many major topographic features and also to the shearing patterns of large parts of the Earth's features, as, e.g., the North and South Atlantic, the Indian Ocean and the Gulf of Aden, Africa, the Pacific, etc. If the correlation is not fortuitous, and this does not appear probable, we have to suppose that the Earth's crust at some moment of its history had indeed shifted with regard to the Earth's poles and that the crust has undergone a corresponding block-shearing."

In 1955 Dr. Thomas Gold, then at the Royal Greenwich Observatory, reexamined the question closed by Darwin and Maxwell last century. In a *Nature* magazine article (26 March 1955), he postulated that the earth's wobble on its axis could cause a plastic flow of the mantle that would readjust the equatorial bulge. Thus, large polar wandering could be expected to occur over periods of geological time. Gold offered this example to illustrate his thinking:

> If a continent of the size of South America were suddenly raised by 30 meters, an angle of separation of the two axes [spin axis and axis of angular momentum] of the order of one-hundredth of a degree would result. The plastic flow would then amount to a movement of one-thousandth of a degree per annum. The earth would hence topple over at a rate of one degree per thousand years or by a larger angle in about 10^5 years. . . . It is thus tempting to suggest that there have been just a few occasions when the axis has been "free" and has swung around as rapidly as would be given by the stiffness of the earth and the rates of tectonic movement, leading to a timescale of the order of 10^5 or 10^6 years, but scarcely longer.

In other words, Gold concluded that the earth may have rolled over several times during its history—i.e., it may have turned over on its axis so that points formerly near the geographic poles are now near the equator.

In an April 1958 *Sky and Telescope* article on irregularities in the earth's rotation, Gold elaborated his position with this provocative remark: "We can easily have polar wandering without continental drift, but, except under the most artificial assumptions regarding the earth's interior, continental drift cannot occur without polar wander. The material under shifting continents would have to be so plastic that polar wander must certainly be expected as well." He added,

> These investigations [of paleomagnetism] have shown that in geologically recent times the average direction of the magnetic axis of the earth was coincided quite accurately with the rotational axis, even though at any one time there may be a considerable excursion. This provides a strong reason for supposing that the two axes always tend to go together over long periods.

Vening Meinesz adopted the suggestion of J. Tuzo Wilson for a mechanism by which continents might drift—convection currents—to explain crustal shifts. Wilson proposed in the early 1960s that magmatic currents deep in the earth rise to the asthenosphere, where they create circulating cells of magma. These cells create smaller, secondary cells of magma, which are in motion directly beneath the crust. The secondary cells of magma then move the tectonic plates through friction. A paper by the British scientist P. Chadwick, in S. K. Runcorn's 1962 anthology *Continental Drift,* expresses Chadwick's agreement with the idea of lithospheric displacement:

> The balance of the evidence at present appears to favor displacement of the whole crust over the substratum rather than polar wandering of the whole earth. Vening Meinesz has suggested that displacement of this type might be produced by large-scale convection currents in the mantle. It is usually supposed that the orogenic [mountain-building] significance of polar wandering is slight, but the possible effects of the equatorial bulge in movements of the whole crust appear to merit further investigation.

Runcorn himself, while at the University of Cambridge, published an article in *Scientific American* (September 1955) that asserted polar shifts had taken place. Discussing the cause of magnetic field reversals, he declared that "there seems no doubt that the earth's field is tied up in some way with the rotation of the planet. And this leads to a remarkable finding about the earth's rotation itself." The unavoidable conclusion, Runcorn says, is that "the earth's axis of rotation has changed also. In other words, the planet has rolled about, changing the location of its geographical poles."

Munk and MacDonald, in *The Rotation of the Earth,* usefully survey the problem of polar shifting and conclude that "the problem is unsolved" (p. 284). Beginning with an examination of the paleontological and paleoclimatic evidence, they remark, "The classic case for (and against) polar wandering . . . has rested on geologic evidence. This evidence is based principally on the distributions of fossil plants, animals, beds of tillite and physical markings of glacial origin. The literature is vague, confusing and conflicting . . ." (p. 259). They remark that there is "little positive evidence" in the paleoclimatic and paleontological data for polar wandering of the kind suggested by paleomagnetic observations.

Nevertheless, they say, if one assumes that polar wandering involves the earth as a whole and that the equatorial bulge is anelastic—i.e., the bulge cannot stretch and travel around the globe in a wave-like motion—the distribution of the continents and oceans is most puzzling. They are distributed in a manner inconsistent with the present pole position. Theoretically,

> The final position of the pole is one that places the continents as well as possible on top of the equatorial bulge. This puts the pole into the equatorial Pacific, as might be expected. The travel time depends on the anelasticity; it is less than 100,000 years according to the interpretation by Bondi and Gold of the damping of the Chandler wobble. The fact that the pole is not in the Pacific nor traveling toward it at this rate poses a dilemma (pp. 275–76).

According to calculations made by several geophysicists, they say, "The pole [should be] in the vicinity of Hawaii, almost as far from the present pole as it can get" (p. 277).

As for the possibility of a thin upper layer sliding over the interior, Munk and MacDonald conclude that "the continent-ocean system could result in a displacement of the outer shell by a few degrees at most, and that the stress differences thus generated are too small to lead to failure" (p. 282). "A crust gliding over the mantle does not appear to be a reasonable model for polar wandering" (p. 285).

Further examples of scientific comment could be cited here, but this is sufficient to establish the validity of the concept for the scientific community. As M. H. P. Bott remarked after surveying the question in his 1971 textbook *The Interior of the Earth,* "Even if the lower mantle does have a high viscosity, apparently polar wandering by slippage of the lithosphere as a whole remains a possibility." Likewise Walter Sullivan in *Continents in Motion:* "Although today the concept of continental drift—more properly of independent moving plates that carry continents with them—predominates, the idea of gross changes in the spin axis, relative to the earth's crust, is far from dead."

Not only far from dead, but reemerging. Like continental drift, polar shift has been put aside for a long time because science hasn't had sufficient data to evaluate it—and perhaps not sufficient vision, either. As Einstein once said, "The mere formulation of a problem is far more essential than its solution, which may be merely a matter of mathematical or experimental skill. To raise new questions, new possibilities, to regard old problems from a new angle requires creative imagination and marks real advances in science."

A recent book, *We Are the Earthquake Generation,* by geologist-anthropologist Jeffrey Goodman, shows such creative imagination. We will meet Goodman in later chapters. For the moment I simply want to note that in his book he proposes that polar shifting provides a unifying explanation for many geological and archaeological puzzles. These include mountain building and its association with ice ages, the driving force behind continental drift, magnetic field reversals, the erratic occurrence of glaciers, volcanic activity's association with ice ages, frozen mammoths, animal extinctions, glacial deposits in presently tropical areas, and abrupt climatological changes.

We may, therefore, see the concept of pole shift return to vogue. If it does, however, it will certainly become more controversial than ever. That is because, as indicated briefly here and as you will see at length in the following chapters, there is widespread disagreement among its proponents. A host of questions demand further study: What is the mode of pole shift—global capsizing, spin axis migration or crustal slippage? What is the time scale involved? What is the trigger mechanism? When was the last pole shift? And most important of all: When might the next pole shift occur? For notice: except for Wheeler, no one mentioned above even remotely approaches the point of *predicting* one.

The men whom you will meet can be characterized as "radicals" with regard to the mainstream of geological and astronomical orthodoxy because, first of all, with two exceptions they predict a pole shift. And although I have subtitled this section "What Modern Researchers Have to Say," it will become clear that not all the theorists presented here qualify as scientists in terms of formal degrees and professional positions within the field of science. Two are engineers, one is a lawyer, one is a psychiatrist-historian, one is a professor of geography and history, one is a research scientist, and one a self-educated jack-of-all-trades. What they have in common is a claim to have based their theories on scientific research, empirical observations and the inductive method. That is the sense in which I have grouped them together and described them as researchers—to distinguish them from both the psychic and the prophetic groups. It must also be said, however, that they are mavericks who are uniformly at odds with established views about earth's history and planetary mechanics, even with those relatively

unorthodox views described in this commentary. And it will be
clear in some cases that I regard their claims to scientific standing
as largely unwarranted.

The main point of contention among pole shift theorists themselves is the mode of shift and the forces involved. There are, as
noted above, three principal forms of the pole shift concept. Does
the planet capsize while the spin axis migrates oppositely within
the planet, does the planet's crust slide around the interior, or does
the planet flip end over end? And does it occur because of centrifugal force, gravity or electromagnetism? The first three chapters
present the principal advocates of each theory.

Between pole shift proponents and the rest of the scientific
community, however, the main point of contention is the *time
scale* on which these events are portrayed. The time scale proposed
is radically collapsed from notions of geologic time involving
aeons. Rather, with two exceptions, the pole shift theorists presented here maintain that it will happen *soon and fast*. This radical
collapse of the time said to be involved in a shifting of the poles
is the source of their dire predictions of the ultimate disaster in
the near future.

Is there any degree of truth to what they say? That is the all-
important question to be raised in this and succeeding sections.

Now hear the researchers speak.

Chapter Four

Hugh Auchincloss Brown: Cataclysms of the Earth

The Antarctic ice cap glistens blue-white in the sun. It covers nearly six million square miles—twice the area of the contiguous United States, three percent of the earth's surface—with an average thickness of about one mile. Although this incredible continental glacier is only about one thousand feet thick near the seacoast, it is more than two miles thick at the South Pole, and as much as three miles thick in some spots. Larger than all Europe, if the cap were centered in North Dakota it would extend from the Atlantic to the Pacific and from Mexico to the northern extremity of Canada. Its weight is estimated at the astronomical figure of nineteen quadrillion tons—19 followed by fifteen zeros. This ice cap holds 90 percent of all fresh water on the earth. Were it to melt or slide into the ocean, it is estimated that sea level would rise around the world anywhere from two hundred to four hundred feet. Were it to melt or to slide into the ocean *suddenly*, the consequences would be disastrous. Colossal floods would sweep across the continents, drowning and burying civilization. And perhaps the poles would shift.

"Some say the world will end in fire,/Some say in ice," The second of these lines by Robert Frost aptly describes Hugh Auchincloss Brown, the father of scientific pole shift predictions, because he was one of those who foresaw an icy Doomsday. Antarctica, he said, would soon become "mankind's glistening executioner."

It is by no means easy for a balanced person to make such a prediction, thus opening himself to scientific scorn and public ridicule. Yet Brown, persuaded by the evidence he patiently amassed over decades, did just that. When he died, on November 9, 1975, his exceptionally long lifetime's study of the question of polar shifting—though rewarded largely with avoidance and derision by science—had been performed for motives that can only be characterized as reasonable and humanitarian.

Two days after his passing, an obituary appeared in *The New York Times* with the headline, "Hugh Brown, Who Cited Peril From Polar Ice Cap, Dies at 96." Thus ended the life, but not the influence, of the man who formulated the first of the principal modern pole shift theories. The obituary capsulizes his story:

> Hugh Auchincloss Brown, Sr., an electrical engineer who devoted more than 60 years of his life to the promulgation of his theory that a vast polar ice cap would tip the earth over in this century and wipe out civilization, died Sunday night in his home at 115 Prospect Avenue, Douglaston, Queens. He was 96 years old.
>
> Mr. Brown graduated from Columbia University in 1900 [with an engineering degree]. . . .
>
> As early as 1911, Mr. Brown became intrigued by reports that mammoths had been found frozen in the Arctic "with buttercups still clenched between their teeth." This led him to believe that an accumulation of ice at one or both poles periodically, perhaps every 8,000 years, upsets earth's equilibrium and causes it to tumble over "like an overloaded canoe."

A Polar Doomsday

Mr. Brown continued to push his theories of an impending Doomsday, which he believed was overdue, until his death. For years he bombarded members of Congress, editors of newspapers and magazines, government leaders and scientists with written "proof" of his theories.

Concerning his theories, Mr. Brown told a *New York Times* reporter, in 1948, that as an engineer he knew that the bulge of the earth around the equator stabilizes its spin. But, he said, an abnormal amount of Antarctic ice, at that time said to be two or three miles thick, could be enough to topple the spin.

This would cause floods of enormous proportions, earth-quakes, and other phenomena, Mr. Brown said, wiping out civilization. He said such a cataclysm was imminent, and noted that "tales of sudden floods and the mysterious appearance and disappearance of large land masses are found in the folklore and legends of all races and men."

Mr. Brown recommended establishment of a worldwide Global Stabilization Organization, and recommended that it devote $10 million to a study of how to effectively set off atomic blasts in the Antarctic to break up the ice mass there and thus save the world from certain disaster.

Mr. Brown's book, *Cataclysms of the Earth,* expanded on his theories. Published in 1967, it titillated many general readers but failed to raise a great deal of scientific concern, although it contained Mr. Brown's statement that a particularly ominous omen for the earth's future was the wobble in the planet's spin. Such a [wobble] is a scientific fact, and continues to challenge scientists seeking an explanation for it.

Mr. Brown predicted that in a forthcoming cataclysm caused by the earth tipping over, New York would probably wind up 13 miles under water, and so would the rest of the world. Among the few survivors, he theorized, would be the Eskimos, because the polar areas would be the least subject to catastrophic water action. . . .

Except for the misstatement about New York City ending up 13 miles below the ocean (Brown had only said it would be sub-merged), the obituary was correct. Understandably, though, much had of necessity been left out. Because of Brown's dedicated effort, several results are visible today and worth mentioning. The first is a modest publication entitled *Pole Watchers' Newsletter.* Begun in 1969 by Mrs. Dorothy Starr, an advocate of Brown's theory, the informal monthly bulletin is produced single-handedly by her and goes without charge to several hundred people. Al-though Brown lent great support to the newsletter, it is not a forum for his view alone. The editor, known affectionately to her readers as "The Pole Starr," writes in an informal, conversational style about any and all matters related to the theme of impending destruction through a polar shift. Oftentimes readers use the pages to debate various points or simply to offer new information. Dorothy Starr, I am moved to say, has shown high-minded and steadfast devotion to a noble mission.

A second, more visible effect of Brown's work is a popular novel entitled *The HAB Theory,* by author-television writer Allen W. Eckert. Based upon Brown's life and work, *The HAB Theory* is a fictionalized account in which Herbert Allen Boardman carries out a deliberately nonfatal assassination plot against the President of the United States in order to gain publicity for his theory—a theory that predicts the imminent death of civilization due to a tumbling of the earth. Boardman was "well into his ninety-fifth year" at the time he aimed at the President and pulled the trigger of a pistol loaded with wax bullets. He had spent six decades examining the evidence that led him to formulate the HAB Theory.

Although the imaginary assassination plot is unflattering to Brown, he was nevertheless "very excited and very helpful" while the novel was being written, I learned through correspondence with Eckert. Unfortunately, Brown died four months before it was published. The hardcover edition, by Little, Brown and Company, appeared in 1976; a year later, Popular Library brought it out in paperback, with sales to date nearing the three hundred thousand mark. An option to film the story has also been taken by Hollywood.

Brown's masterwork, *Cataclysms of the Earth,* has had only minor sales but is still in print. Moreover, Brown's "bombardment" of editors, as the *Times* obituary put it, resulted in half a dozen popular presentations of his ideas in publications ranging from *The Magazine of Sigma Chi* and *Argosy* to *Mechanix Illustrated* and *The New York Times,* with significant mention in several books as well, including Walter Sullivan's *Continents in Motion.*

Brown's theory of pole shift is stated succinctly in the introductory section of *Cataclysms of the Earth*. There Brown describes "the certainty of a future world cataclysm during which most of the earth's population will be destroyed in the same manner as the mammoths of prehistoric times were destroyed" (p. 4). After noting that scientists have learned the south polar ice cap is growing rather than waning, he declares that this fact confronts us with "an entirely new understanding of the limited time during which our present civilization has been developing, and the precariousness of its continuation. We are faced with the alternative of limiting the growth of the ice cap or accepting a limit to the duration of our present epoch" (p. 10).

> The growing South Pole ice cap has become a stealthy, silent and relentless force of nature—a result of the energy created by its eccentric rotation. The ice cap is the creeping peril, the deadly menace, and the divinely ordained executioner of our civilization. Just as a sword, suspended by a single hair, hung above the head of Damocles at Dionysius's banquet, so today the baneful jeopardy of an impending world flood hangs over us all. . . .
>
> The elemental forces of nature that are involved are now known and they are clearly identified. If we procrastinate and do nothing, the Flood will occur when the present polar areas move away from the earth's axis of spin . . . to latitudes of ten to fifteen degrees, or about 5,500 miles away from the North and South Poles of Spin. The earth will tip over, like an overloaded canoe towed in a circle behind a power boat, in consequence of the wobble of the earth and the resulting eccentric force of rotation of the present South Pole ice cap and its constantly increasing weight. The earth of today may quite readily be compared to a top-heavy, dying out, wobbling, spinning top, getting ready to fall over on its side (pp. 10–11).

This theory had its inception in 1911. Brown—by then a licensed professional engineer, husband and father—had read earlier of the frozen mammoths, especially the Berezovka specimen. The report of well-preserved animal tissue and undigested vegetation stirred his imagination, fired his curiosity. How could this astonishing phenomenon have happened?

"The idea occurred to me," he told *Mechanix Illustrated* in January 1949, shortly after he had first published his theory, "that possibly the perfectly preserved mammoths found in Arctic tundra and ice had been careened from a tropic to a frigid zone." What might have caused the planet to roll over? Since ice was the mystery, ice might also be the answer—large masses of ice—in fact, the polar ice caps. Could they overcome the earth's stabilizing spin and topple it?

If so, Brown thought, then perhaps he could find some evidence to that effect. With that as his starting point, he relates in *Cataclysms of the Earth,* he drew upon his engineering experience and pondered the logical consequences of a capsizing planet caused by an ice cap or ice caps gone astray.

> It seemed common sense to say that the earth would naturally rotate with its heaviest masses stretched out along the equator because the centrifugal force throws the greatest weights to the periphery or equator of any freely suspended rotating mass. And yet, quite to the contrary, the mighty Rocky Mountains and Andes Mountains are not now stretched along the equator.
>
> My theory was that if I could plot a circumference of the earth which might have been the equator when the mammoths were living—just prior to the latest careen of the globe—then I might be able to find some evidence of a polar area 90 degrees of latitude away from that equator, and this would prove the theory to be correct.
>
> I was curious enough to make a trip one Sunday afternoon in 1913 to the Public Library of Erie, Pa., where I then lived. I took along a spool of ⅛-inch-wide red ribbon, and tied the ribbon around a three-foot globe of the earth, which stood in the middle of the main library room. I remember that I felt self-conscious at first, but soon guessed that the other library patrons must think that I was just one of the men who worked there, for they paid no attention to me, and I worked leisurely (pp. 57–58).

What seemed at the time to be merely a routine action for Brown turned out to be quite otherwise, for he comments that "in retrospect [it] looms as a momentous occasion."

> I tied the first ribbon in a great meridian circle, or equatorlike line, along the great ridges of the Rocky Mountains and the Andes Mountains. It divided the globe into two equal halves. On the opposite side of the globe, I was interested to note, the ribbon traversed East Asia and the shallow seas surrounding the Malay Peninsula. I tried to make it represent the meridian and equatorial band traversing the heaviest land areas, and it seemed that the land zones it touched far outweighed any other circumferential belt of the earth's surface that I could have selected.
>
> Then I attached two other ribbons—representing great circles of the earth—at random places, but taking particular pains that they were at exactly a 90-degree angle to the first band. The idea was that I might find some evidence of former polar areas where these two upright ribbons intersected each other, for the intersections would mark the location of polar centers, at the moment that my first ribbon was an actual under-the-sun equator.

One intersection was found to be at Lake Chad, Africa—
which I thought might give me a clue; the other intersection
was located in the Pacific Ocean, and no clue came to mind.
I recall writing "Lake Chad" on a piece of paper, with a resolve
to try to find out something about it. I was looking for evidence
of glacial action, or of a dent left in the earth by an ice cap.
I was astonished to discover just what I was looking for: The
great Sudan Basin of Africa (pp. 58–59).

The Sudan Basin, Brown was to learn, is a huge depression
in the earth covering nearly four million square miles. It is peculiar
for a number of reasons. First, watercourses wander through it in
the most varied directions and have no connection with each other.
Second, Lake Chad, in the southern part of the Sudan Basin, is
a dying lake—a remnant of the past—which loses about one third
of its area every fifty years or so. Interestingly, Lake Chad has
no outlet to the sea, although it is above sea level. Last of all,
while other lakes pick up salts from their tributaries and remain
fresh only by constantly surrendering their waters to the sea, Lake
Chad never contained any salt at all.

For these reasons Brown surmised that the Sudan Basin marked
the location of a previous polar ice cap, which had pressed in the
earth, forming watercourses and Lake Chad by runoff as it melted
long ago under the equatorial sun.

What followed that Sunday afternoon visit to the library were
many years of research, years of patient "digging," before Brown
felt ready to publish his evidence and conclusions. By 1946 he
had "gone public" by talking to reporters. In 1948 he published
(at his own expense) a treatise entitled *Popular Awakening Concerning the Impending Flood*. In the Foreword he stated:

> The centrifugal force of eccentric rotation of the great South
> Pole ice cap tends to careen the globe. That force of rotation
> will roll the earth over and move Antarctica and its ice cap
> almost directly under the sun because in due time it will neu-
> tralize and overcome the centrifugal force of the earth's bulge.
> During the reeling period the earth will experience another of
> its recurrent great deluges when most living things perish.

Since Brown was an engineer, it was natural for him to seek
an engineering solution when presented with a dangerous situation.
His response to "Antarctica's icy menace": reduce the dimensions
of the ice cap through atomic detonations at strategic locations.
The first step, he declared, would be to survey the ice cap to

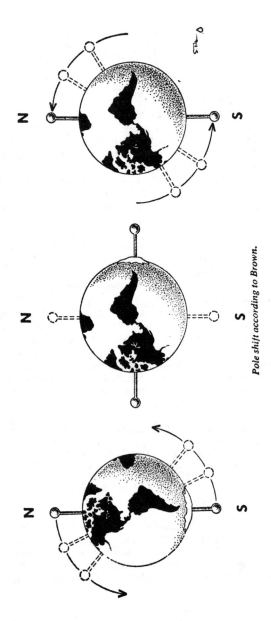

Pole shift according to Brown.

determine exactly how fast it was growing. Several ships with engineers and scientists should circumnavigate the continent, and within several years the data should be sufficient to indicate where explosions would best be placed. Toward that end, Brown proposed the creation of a Society for Global Stabilization, with a $10-million budget.

The reception given *Popular Awakening Concerning the Impending Flood* was largely scornful. Although Brown received some publicity in the public media, the scientific community by and large ignored him. Some scientists at the U. S. Geological Survey scoffed at his warning that the oceans would be sloshed out of their basins by a capsizing planet. It was strongly doubtful, they said, that any existing polar ice masses were likely to top over the globe. Brown was pictured as a modern-day Noah and, as with Noah in his time, very few listened. Walter Sullivan, who listened but did not believe, comments in *Continents in Motion* that the stability of the earth's spin "is so great that it would be difficult to find a scientist who believes the axis is likely to change in any sudden way, even if a lopsided chunk of ice does accumulate in the South Polar area" (p. 18).

Let's look more closely at Brown's theory of Antarctic disaster. I am sorry to say, incidentally, that *Cataclysms of the Earth* is not a well-written book. While generally satisfying as a work of scientific literature, at points it lacks clarity, especially in Part II,* the technical section on the mechanics of careening. This is due in part to the cut-and-splice method of composition that Brown and his editorial assistant (Phillip C. Flayderman, who is acknowledged on the title page) used. Sections of previous publications are occasionally woven together, not always gracefully. The result is sometimes a disjointed narrative and sometimes an incomplete explanation that one must "sniff out" in succeeding passages. The organization and, to a lesser degree, the style keep Brown's masterwork from being a masterly work.

The first part of *Cataclysms of the Earth* is entitled "Evidence of Careenings of the Globe." This long section—more than one hundred pages—includes data listed under headings such as his-

---

Cataclysms of the Earth is actually three short treatises in one. The second half of the book contains Part Two, "The Cause of Gravitation," and Part Three, "Origin of the Earth's Materials." In these, Brown proposes that gravity is an electrical phenomenon and that the earth is solid throughout. We will not concern ourselves here with these theories.

torical writings (about a global deluge), archaeology, mammoths, sea life, fossils, trees and other vegetation, rivers and waterfalls, ice ages, and half a dozen categories of geological data. While I cannot possibly summarize the evidence here, it is interesting to note some of the most dramatic examples cited by Brown as proof that the poles have shifted suddenly by as much as 80° in the past.

• The discovery of mammoths, rhinoceroses and other animals in Arctic regions where they could not possibly survive under present climatic conditions. "Yet just prior to the latest careening of the globe this region was populated with teeming herds of animals. They lived there because an ample food supply existed, and the food supply grew because the climate was warm" (pp. 22–23).

• The story of Noah has its counterpart in the Greek myth of Deucalion. Likewise, the ancient Hindu, Chaldean and Chinese records tell of a great deluge "slightly more than 5,000 years ago." The Chinese tradition "tells of their sudden, flying leap to the Arctic," and Confucius, born about 551 B.C. begins his history of China with a reference to a receding flood that has been "raised to the heavens."

• Ur of the Chaldees, excavated by archaeologists, confirms the historical reality of the deluge myth for its part of the world. Likewise excavations at Cnossus, Crete, show that "the rock substrata below the soil . . . fit into the pattern of the theory of a careening globe."

• Seals found in the Caspian Sea and in Lake Baikal, in Siberia, are the same as the seals that inhabit Alaskan waters. A certain genus of lobster is found only in the icy waters of the Arctic and in the coldest parts of the Mediterranean Sea. These zoological mysteries are explained by global floods transporting some animal survivors across enormous distances.

• Fossil jellyfish have been found entombed in mud. They could not have been preserved except by quick freeze after a global tumble. "Solidification by [flash] freezing of both the sand and the jellyfish, at the moment of a career of the globe, is the simplest scientific answer to this age-old riddle of how a soft jellyfish could become solid rock." The same explanation applies to fossil imprints of even more evanescent events—raindrop splashes and small seabird footprints—which surely would have been washed away by further rain or the next high tide.

• Fossil trees and other vegetation are found in the sea and underground. Obviously they didn't grow there. Others are found in Siberian permafrost—one with fruit and leaves still on it.

• In Yellowstone Park, a gouged-out mountain shows seventeen layers of upright petrified trees. In between each pair of adjoining layers of standing trees is a stratum of volcanic earth. Each layer of standing trees existed in its own geologic epoch of vegetable and animal life.

• The rate of erosion and upstream movement of Niagara Falls indicates an origin about seven thousand years ago due to the most recent careen of the globe. The Falls of St. Anthony, in Minnesota, give similar indications.

Brown's interpretation of the evidence led him to conclude that the planet has a naturally recurring cycle, about seven thousand years in duration, that begins and ends in cataclysm as the weight of the polar ice caps tumbles the globe about 80°. In working out the details of previous historical epochs, he arrived at the following chart showing correlation of the epochs he identified with epochs identified in other geological systems.

Epoch	Duration Years Approximate	U.S. Geological Survey Series of Epochs	Mississippi River Commission Designations for Substrata
Present	7,000		Recent Alluvium
1 B.P.	4,400		Jackson
2 B.P.	7,000	Wisconsin Ice Age	Vicksburg-Jackson
3 B.P.	5,000	Peorian Life Age	Claiborn
4 B.P.		Iowan Ice Age	Wilcox
5 B.P.		Sangamon Life Age	Midway
6 B.P.		Illinoisan Ice Age	Upper Cretaceous
7 B.P.		Yarmouth Life Age	Lower Cretaceous
8 B.P.		Kansan Ice Age	Mesozoic
9 B.P.		Aftonian Life Age	
10 B.P.		Albertan Ice Age	

During Epoch No. 1 B.P. (before the present), which preceded the most recent careen of the globe, Brown says the Sudan Basin was at the North Pole. In Epoch No. 2 B.P., the Hudson Bay

Basin was at the North Pole. Hudson Bay is a shallow dent in the earth, averaging 420 feet below sea level. It marks the ice cap's approximate center, while the heights of land known as the Laurentian Shield, which almost surround Hudson Bay, mark the final edges of lips of the main ice bowl. It is a fact that the last North American glacier was centered in the Hudson Bay area, rather than at the geographical North Pole. Brown resolves the geological puzzle by saying simply: that is because the North Pole was not always at its present location.

Thus Epoch No. 1 B.P. ended at the time of the latest world flood, and the present epoch commenced. During Epoch No. 1 B.P. . . . the continents of North and South America then lay in tandem on one side of the globe, along the equator, and the eastern parts of Siberia and China were on the opposite side. . . .

We are left with 4,400 years as the duration of Epoch No. 1 B.P. . . .

During Epoch No. 2 B.P. the geological dent of the Hudson Bay watershed contained the North Pole and its ice cap, while South America, Africa, Borneo, and India lay along the equator . . . thus the approximate duration of that epoch may be set at 7,000 years.

The identification of this former North Pole area is based on the fact that its distance from Lake Chad . . . is approximately the same as the present distance of Lake Chad from the North Pole. . . . All careening moves cover about 80 degrees of latitude.

The Caspian Sea is located in a large low-lying area—called the Caspian Depression. . . . This sunken Caspian area has geological similarities to the depressed areas in which Hudson Bay in Canada and Lake Chad in Africa are located. . . .

We know that each successive ice cap leaves a depression in the land, and each careen of the globe moves both the North and South Pole ice caps distances of about 80 degrees of latitude—about 5,500 miles. With this information as a guide, we can identify the locations of former poles of many successive epochs (pp. 61–63).

Part II of *Cataclysms of the Earth* is entitled "Mechanics of the Great Deluge." Having shown in Part I that his theory is based on adequate evidence and that it explains many mysteries of the earth, including frozen mammoths, ice ages, mountain building and rifts, Brown goes on to explain in physical-mathematical terms how such cataclysms come about.

Brown's theory of Antarctic disaster rests on two assumptions.

They are 1) the south polar ice cap is growing, and 2) eventually it will develop enough off-center mass to overcome the stability of the earth's spin axis.

Regarding the first point, Brown declares that the evidence is clear: the ice cap is increasing at the rate of 293 cubic miles of ice each year—almost as much as if Lake Ontario were frozen solid annually and added to it. Brown cites the Symposium on Antarctica sponsored by the International Union of Geodesy and Geophysics of Helsinki, Finland, in 1960. There Dr. Pyotr Shoumsky, a Soviet polar explorer, reported the rate just mentioned, and his finding was confirmed by Dr. L. Loewe, of France, and Dr. Malcolm Mellors, of Australia. And even though Antarctica gets so little precipitation—about twelve inches annually, in the form of snow—that it is classified geographically/meteorologically as a desert, that precipitation adds 20 billion tons of weight per year to the ice cap, Brown says.

The mechanics of an earth tumble—Brown's second point—are summarized in a section on the creation of mountains:

> Because of the curvature of the globe, the centrifugal forces of the rotating ice caps which initiate the careens soon reach a maximum and then diminish.
>
> When the ice caps have migrated 45 degrees of latitude, their centrifugal force responds to the combined motions of careening and rotation. Between the sun latitudes of 45 degrees and 0 degrees they change from being upsetting to being stabilizing forces.
>
> Equatorial bulges then start to form, and the centrifugal forces of the ice caps and of the new bulges of the earth are soon working in unison to bring the reeling motions of the globe to a rapid slow-down and stop.
>
> In the meantime, kinetic energy which has developed in the continental land masses because of their weights and velocities, collides with the combined energy of the newly generated bulges of the earth and of the ice caps.
>
> The result of these collisions of forces is that the energy of the moving continents is absorbed by the crushing, elevating, and wrinkling of large land areas whose rock strata are crumpled and bent in ridges at right angles to the forces being dissipated (p. 102).

Elsewhere, Brown explains that a pole shift occurs within a single day. The speed of the ice cap accelerates rapidly during the first 45° of latitude traveled, because then the centrifugal force of its motion would be pulling nearly sideways. However, due to the spherical surface of the earth, this side pull changes to an upward and outward pull. By the time the ice cap has traveled 80°, there is very little side pull, but there is a very great upward stabilizing pull, away from the center of the earth. The ice cap then becomes part of the new bulge of the earth, and begins to melt under the tropical sun.

Essentially, then, the Brown theory of pole shift asserts this: every seven thousand years or so, an eccentric "throw" of the ice cap or ice caps, propelled by centrifugal force, pulls or drags the earth *as a whole* with it as it slides toward the equator. This movement is resisted by the stabilizing tendency of the earth's spin, which asserts itself by rapidly migrating its spin axis and equatorial bulge *through the earth itself* so as to remain in place as viewed from space. That is, the spin axis, as seen from the moon, remains "upright" and the equatorial bulge remains stationary in relation to the spin axis. (This is a crucial point which Brown, regrettably, does not make clear in his book.) Meanwhile, however, the tumbling planet is actually slipping its bulge over itself, like a wave in the ocean, at a rate of speed equal to the speed of tumble—a speed reaching hundreds of miles per hour. An observer standing on the earth between the pole and the equator along the line of careen would see the bulge heading toward him as a ground wave some thirteen miles high! As a consequence of this action, tidal waves are set up and seismic activity commences. The twenty-seven-mile difference between the polar and equatorial diameters of the earth assures that the destruction of civilization ensues. After the careen has ended, the ice caps end up near the equator, having traveled about 80° of latitude, or some 5,500 miles. The polar locations have changed on the surface of the earth, yet the spin axis and equatorial bulge retain their angular positions with regard to the sun. And this shifting depends only on gravitational and centrifugal forces to carry it out.

Toward the end of Part II, Brown comments:

> Following the next careen of the globe the present continent of Antarctica can reasonably be expected to become the center of a land hemisphere—because of the centrifugal force of rotation which will be created by its weight and speed of motion. This transient force will not only pull Antarctica but also its

surrounding ocean floors upward and keep them above sea level, thus creating new land areas.

The area of the globe now occupied by the Arctic Ocean will probably become the center of a water hemisphere—like the Pacific Ocean today. What is now northern Siberia, northern Canada and Alaska will probably become parts of the submerged ocean floor (pp. 133–34).

Thus prepared, we move to Part III of *Cataclysms of the Earth,* "Man: The Past, the Present and the Future." The important data for us are those bearing on the future. During the period of the next great deluge, Brown says, a general chaos involving readjustments of land and water areas will take place. Some existing mountains will be raised, some will be lowered, and new ones will be formed. Earthquakes and volcanic activity will occur as the earth readjusts itself and becomes stable. The oceans will increase in depth as the ice cap melts. Later on, new ice caps will form at the new polar locations.

The four safest areas, in Brown's view, will be Greenland, Antarctica and the two pivot areas on which the planet careens. The eastern pivot point, he feels, will be approximately 45° East— almost on the eastern coast of Africa—and the other in the Pacific. The earth will swivel at those points because the center of mass for the Antarctic ice cap is approximately 350 miles from the South Pole along the 80° East meridian. Thus the ice cap would tend to be thrown in that direction. "According to this hypothesis," Brown writes (p. 151), "Brazil would roll to the South Pole and the Philippine Islands would become the land area nearest the North Pole."

But, he continues, ". . . it would be equally valid speculation to say that some area of the globe within about 2,000 miles of Lake Chad will be at the North Pole during the epoch of time following our own, and that this would occur as a result of the past—namely the Hudson Bay Basin careened to the North Pole Axis of Spin, then Lake Chad moved in, only to be supplanted by the present Arctic Ocean area. This shows a tendency for land areas to roll back to nearly the same position of latitude and longitude that they rolled away from."

The dynamics of the situation are puzzling to Brown, who notes that "something different" may be expected to result when the south polar ice cap, acting alone, initiates a roll-around of the globe. He closes the section with the frank admission that it "has contained speculations and not precise forecasts."

Cataclysms of the Earth ends with an exhortation and a description of how a planetary career might be avoided. The exhortation:

> Awareness of the gigantic power to destroy inherent in the enormous unwieldy weight of the gyrating Antarctic ice cap must be the first step in creating a cooperative group reaction to the deadly peril. People of education and initiative must become awakened to full awareness of the lurking danger represented by this wanton titanic power which is ready, able and destined to end our civilization—if left uncontrolled by man.
>
> It is, therefore, important that the facts be communicated so that the menace will be generally understood and discussed. An awakening to the danger among the people at large is the first requisite!
>
> ... If everybody talked about the careening of the globe, as they do about the weather, then a great many people will try to do something about it. It will indeed become a matter of great personal interest to many people.
>
> Let therefore all nations unite in pooling all their resources of atomic energy, mechanical equipment, and man power. Let us not continue to waste our substance in building giant mechanisms for destroying each other, while the growing ice cap is developing its latent resources to annihilate all of us.
>
> Let there be an international war in which all nations fight as brothers against the common enemy. Let us attack the ice cap (pp. 152–53).

How the attack might be carried out as "an engineering problem" is briefly suggested in the final pages of Brown's book. Essentially, atomic bombs would be used to gouge channels in the peripheral rock masses of the Antarctic coastline. These channels would permit the natural flowoff of the glacial ice, which is presently dammed in by the rocky lip of the south polar depression. The time is urgent, Brown remarks, because in addition to the fact that our present epoch has exceeded the average length of the cataclysm cycle, making a career theoretically imminent, the wobble of the earth appears to be an ominous sign that the career is practically upon us. If the Chandler wobble were to act in conjunction with a nutation of the spin axis, this off-center "throw" might induce the start of the next global tilt. "We will all join in this work," Brown ends, "once we know that our lives are at stake!"

* * *

Everyone is open to criticism, and Brown has had his share. Much of it arises in the best tradition of science's self-correcting tendency through internal criticism. The *Pole Watchers' News-letter* of August 1977 contained a useful analysis of a portion of Brown's mathematics by Kelly Starr, son of the newsletter's editor.

> Mr. Brown did, indeed, make an incredible error when he calculated the weight of his hypothetical ice cap on p. 127 of his book. The correct weight which I just calculated myself is 4,768,948,800 pounds, which is more than ten times greater than Mr. Brown's 424,560,000!
>
> A most tragic mistake by Mr. Brown—and one that, further, is quite mystifying—because (1) one would think that he would double check his work, (2) a discrepancy that large should have been realized by a person of Mr. Brown's expertise without the need for a mechanical rechecking of the figure, and (3) his answer was *not exactly* ten times less than the correct one, meaning that he really butchered a rather simple calculation. . . .
>
> Mr. Brown made a monstrous error when he said that force is measured in pounds, when it actually is measured in pound-als, which are quite different. By definition a pound is a unit of weight or mass, whereas a poundal is a unite of force which equals mass or weight times acceleration ($F = ma$ is a basic formula encountered at the outset of any course in physics).

Starr went on to say that although he found some mistakes in *Cataclysms of the Earth* and doesn't agree with Brown's theory, he still has "the greatest respect for his work." The prodigious effort Brown made to substantiate his concept was most laudable, he says, and it should be gratefully acknowledged by the entire scientific community.

Another crucial point in Brown's position is whether the Antarctic ice cap is growing or shrinking. Brown maintained that it was growing faster than it could shed weight by breaking off icebergs. This was disputed by Walter H. Bucher, professor of geology at Columbia and former president of the Geological Society of America. At a 1955 scientific gathering to examine the concept of pole shift via the Antarctic ice cap, Walter Sullivan tells us in *Continents in Motion,* Bucher "pointed out that measurements of snow accumulation near the coasts could not be taken as an indication of total accretion (it was subsequently found that snowfall along the coast is often ten times what it is inland)" (p. 25).

On the other hand, the *Pole Watchers' Newsletter* of April 1974 points out that some glaciologists say that the total ice mass is increasing in the Antarctic. *The Water Budget in Antarctica,* by F. Loewe, and *The Dynamics of the Antarctica Ice Cover,* by K. K. Markov, take this position, Dorothy Starr writes. This is further support for the Shoumsky-Mellors data that Brown cited in his book, she says.

A more recent newsletter (April 1977) summarizes an article in the 9 January 1977 *Chicago Tribune* by its science editor, Ronald Kotulak, who visited the Antarctic. The article, entitled "Fear Antarctica Ice Collapse Would Flood Coastal Cities," quotes Dr. Richard Cameron, National Science Foundation program manager for glaciology studies of the Ross Ice Shelf and West Antarctic ice sheet. Something dramatic is happening there, Cameron says. An urgent study is needed, because in the past two or three years scientists have realized that the giant ice sheets could break up rapidly, forming catastrophic surges that would raise ocean levels nearly twenty feet.

The mechanism causing this is heat, according to the 1964 theory of J. Tuzo Wilson. Heat radiates from within the earth, melting the bottom of the ice cap. In addition, heat is generated at the bottom of the ice cap because its sheer mass increases molecular action. Finally, as the ice cap slides outward from the center of Antarctica, friction also increases the heat. The result is a slushy layer of water that acts as a lubricant. Studies have shown that such a condition exists, along with numerous lakes, beneath the ice cap. Riding on this layer, the ice cap has in the past uncoupled from its bed and surged outward, spreading itself in a sun-reflecting mantle around the pole, thereby inducing a new ice age around the globe, Wilson theorizes.

There could be even more immediate and disastrous consequences, according to science writer Fred Warshofsky, who reports in *Doomsday:*

> A confluence of catastrophic events—a series of earthquakes, volcanic eruptions and violent storms lashing the earth's surface, all working to tip the globe's spin axis only slightly or merely to increase its wobble—would be enough to send the monster ice sheet plunging into the Pacific and Atlantic oceans.
>
> Even before the great white mass could hurl back enough

sunshine to chill temperatures still further, huge tidal waves would roar down on almost every coastal area on earth. Hundreds of millions of lives would, of course, be lost instantly, and those who remained to pick up the pieces would find a few decades later an ice-covered world with only the lands about the equator warm enough to raise crops.

Impossible? Not at all, and the irony is we may not even give nature a chance. Precisely such a possibility was envisioned in the 1960s by a group of strategists at the Institute for Defense Analyses in Washington, D.C. They, however, did not in their wildest speculations conceive of nature taking so malevolent a course. Their fears were that some nation or terrorist group might use nuclear weapons to shake the Antarctic ice sheet loose from the land and send it plunging into the sea.

"The most immediate effect of this vast quantity of ice surging into the water, if velocities of 100 meters per day are appropriate, would be to create massive tsunamis that would completely wreck coastal regions even in the Northern Hemisphere," wrote Dr. Gordon J. MacDonald, then executive vice-president of the institute (pp. 191–92).

As we have seen, Brown proposed to detonate nuclear devices in the Antarctic. Fortunately for the world, this did not happen. The best-laid plans of mice and men. . . .

The strongest objection to Brown's theory is, surprisingly, one that has never been raised, so far as I have learned. It involves the age of the Antarctic ice sheet. According to the late Richard F. Flint, a geologist at Yale and author of the 1971 *Glacial and Quaternary Geology*, "The K/Ar dates [potassium and argon radiometric measurements] of igneous rocks that overlie, and therefore are younger than, glaciated surfaces or sediments of glacial origin show that glaciers in Antarctica existed at least 10 million years before the present" (p. 62). More recent studies—which I'll quote below—have indicated that it is as much as 20 million years old. Yet, in the course of my research, I've found no one who thought to examine this aspect in critiquing Brown.

It is understandable that Brown himself did little about it, since, quite simply, there were almost no data available when he formulated his theory. "The present ice cap in Antarctica is merely the last of many thousands that have previously existed," he wrote. "Geological records reveal that it is the successor to a long lineage of glistening assassins of former civilizations of this earth" (p. 8). The "geological records" Brown refers to have already been discussed. However, *Cataclysms of the Earth* does not contain data that refer directly to the age of the Antarctic ice sheet except for the following statement:

> A communication from Captain Charles W. Thomas (now a rear admiral retired, U. S. Coast Guard), a noted ice navigator of both the antarctic and arctic regions, states that cores taken from the ocean bottom off the coast of Antarctica and examined by him lead him to conclude that the South Pole Ice Cap is not of great antiquity, but that it is a recent phenomenon, its age being no more than a few thousand years (p. 72).

Brown remarks that the cores showed the ocean bottom to have been formed in layers. The top layer contained cold-water sea life and ice-transported sediments. The next layer did not have the cold-water sea life (diatoms), but they occurred again in a lower layer. Other sea-bottom cores—these from the Arctic—show a similar sequencing of layers that are alternately due to warm and cold water. This, Brown surmises, demonstrates that the Antarctic and Arctic regions were ice-free at nearly regular intervals.

As we shall see in the next chapter, Brown's position was not totally unreasonable. There *is* evidence to support it. However, according to Guy Guthridge, the Polar Information Service's editor of the monthly *Antarctic Journal of the United States*, the accepted view among professionals is that the ice sheet is 11–14 million years old, and perhaps—as indicated by ice cores from the Ross Sea—20 million years old. When I interviewed Guthridge, he directed me to some primary source material—research reports and technical publications—that is usefully summarized by *Science News* (31 March 1973) in an article entitled "Antarctica's Ice Cap: Older Than Believed":

> The deep-sea drilling ship *Glomar Challenger* has penetrated into the sea-floor history of the ocean waters off Antarctica and returned with cores that show the continent has been covered with ice much longer than previously believed.
>
> The results of the first deep drilling ever conducted in the waters near Antarctica also reveal that the Ross Ice Shelf once extended 200 to 300 miles farther than now, that a great increase in glaciation began about 5 million years ago. . . .
>
> The big achievement . . . is the finding that Antarctica has been glaciated for the last 20 million years. This is in contrast to the view held by most scientists that the glaciation began 5

million to 7 million years ago. . . .

The main clue to the early glacial history is the presence in the sediment cores of pebbles, stones and other material that are too heavy to have been carried out to sea by ocean currents. But they are carried out to sea in large quantities by icebergs, and their presence in sediments is a classic indicator of continental glaciation. The glacial pebbles are absent in the portion of cores older (deeper) than 20 million years. . . .

Another indicator of this climatic scenario is the presence in the cores beginning about 20 million years ago of the types of fossil organisms that thrive in cold water.

When faced with data such as these, Brown's theory of polar shifting obviously doesn't deal with all the facts. Small wonder that scientists have generally paid no attention to him.

What can be said about Brown in summation, then? His argument is not without merit, yet in light of the *Glomar Challenger* data, it doesn't have sufficient force of presentation to be fully acceptable. His concept of the earth tumbling in space due to off-center polar ice has an appealing elegance in light of the mystery of the frozen mammoths. Moreover, his identification of "Antarctica's icy menace" has helped direct attention to the south polar ice cap's dangerous condition. He has gathered puzzling data and offered what seems to be the beginning of a solution to a planetary mystery. But as we shall see, major modifications and alternative explanations to Brown exist—and they are more persuasive.

It seems fairest to say aloud on Brown's behalf, as Galileo is supposed to have muttered after being forced by the Inquisition to recant his idea that the earth revolves around the sun, "Nevertheless, it moves."

References and Suggested Readings

"Antarctica's Icy Menace," *The Magazine of Sigma Chi,* Winter 1970.

BROWN, HUGH AUCHINCLOSS. *Cataclysms of the Earth.* New York: Freedeeds Associates, 1968. Reprinted from the original edition by Twayne Publishers, 1967. Distributed by Multimedia Publishing Corp., Blauvelt, NY 10913.

ECKERT, ALLEN W. *The HAB Theory*. New York: Popular Library,
     1977.

Pole Watchers' Newsletter. Available without charge from Mrs.
     Dorothy Starr, ed., 4153 38th Street, San Diego, CA 92105.

SULLIVAN, WALTER. *Continents in Motion*. New York: McGraw-
     Hill, 1974.

Chapter Five

Charles Hapgood:
Earth's Shifting Crust
and the Path of the Pole

In the autumn of 1949, a history teacher named Charles H. Hapgood was astonished when a freshman in his world history class at Springfield College asked him, "Mr. Hapgood, do you believe in the lost continent of Mu?" Henry Warrington was not one of Hapgood's most alert students. In fact, he had a habit of sleeping in class. Moreover, the topic he raised had long since been dismissed by Hapgood as mere legend. But the teacher saw an opportunity to stimulate his student's curiosity into genuine research, so Hapgood replied with a question.

"What, Henry, exactly do you mean?"

"It was a lost continent in the Pacific like the lost continent of Atlantis."

"Well, Henry, there is much to be said on both sides. Why don't you look it up and report to the class in two weeks?"

At the appointed time, Hapgood recently told me as we sat in his New Hampshire home, Warrington came into class with at least ten books on Atlantis—not Mu—in his arms. "My heart sank at the appalling prospect of having to wade through all that junk, but wade through it I did, and while a lot of it was junk, I did see when I got through that there was indeed much to be said on both sides. So I made the study of Atlantis a class project. We neglected the history of Greece and the history of Rome, while members of the class competed in bringing in all sorts of books and articles on Atlantis."

Hapgood soon saw that the real obstacle to the acceptance of the traditional story of Atlantis came not from the field of archaeology but from the field of geology, which upheld the uniformitarian view that the earth's surface has always been subject only to very gradual change and that the poles have always been situated where they are now. So geology became the subject of the class study.

One day, someone brought in an issue of *Argosy* magazine containing an article about Hugh Auchincloss Brown's theory. As Hapgood recalled it, "It was presented sensationally, with scare headlines about civilization being threatened with destruction, but nevertheless I saw that there was much meat in the article. His was an elegant idea, and we decided to investigate it by examining in detail all the lines of evidence he had amassed."

Thus it was that Hapgood, like Brown many years before him, became intrigued with an idea that carried him far afield from his formal training in medieval and modern history. The result was to be a book, *Earth's Shifting Crust* (later revised as *The Path of the Pole*), that stands as one of the most important challenges to conventional notions of earth's history.

Hapgood had earned his master's degree from Harvard in 1932, and by 1934 had completed all requirements for the Ph.D. except his dissertation, which was to focus on the French Revolution and restoration. But the Depression interrupted his academic studies, and the next five years saw him, as he expressed it to me, "explore many windings and turnings." These included teaching for a year in Vermont, directing a community center in Provincetown, and serving as Executive Secretary of President Roosevelt's Crafts Commission. After World War II erupted, Hapgood held a variety of posts ranging from editorships at the COI (later the OSS, now the CIA) and American Red Cross national headquarters to liaison officer between the White House and the Office of the Secretary of War.

When peace was established, Hapgood returned to the groves of academe, and by the time Henry Warrington enrolled in freshman history, was at his second collegiate teaching post, as an assistant professor offering courses in world history, American history, comparative government and intellectual history. Soon after Warrington asked his seminal question, the official history curriculum was left far behind. For the question stimulated Hapgood to challenge the accepted view of our planet's history and mechanics. In the spirit of true inquiry, he invited his students to join the research.

This they did and over a ten-year period contributed many valuable research papers on various aspects of the question of polar shifts. At Hapgood's request, many scientists also assisted by providing special knowledge and searching criticism.

At first the investigation focused on the ideas of Brown, whom Hapgood quickly contacted. Brown made many suggestions for research that proved to be productive, and he generously shared his own research data with Hapgood and his students.

By the third year of inquiry, however, it had become apparent to Hapgood that Brown's theory was untenable for a basic reason: calculations showed the impossibility of the mass of the Antarctic ice cap developing sufficient momentum to capsize the earth. A radical modification was needed if the notion of shifting poles was to be retained.

At that point, the dilemma was solved by a friend of Hapgood, James Hunter Campbell, a mathematician-engineer who helped develop the Sperry gyroscopic compass. Thereafter, Campbell became Hapgood's constant associate in the research, and Hapgood credits him with taking hold of the work when it was still "an amateur inquiry" and transforming it into "a solid scientific project."

Campbell's contribution? The suggestion—and the mathematical-mechanical arguments to demonstrate it—that only the crust of the earth moved, slipping intact over the interior.

In 1958 Hapgood published the results of the decade-long joint inquiry. The Foreword to *Earth's Shifting Crust* was written by none other than Albert Einstein. Hapgood and Campbell had approached Einstein in 1954 and found, just as they had heard, that he welcomed new ideas. During the following months, before he died in 1955, Einstein not only gave his reactions to their presentation but also offered suggestions for their further development. In his Foreword he wrote:

I frequently receive communications from people who wish to consult me concerning their unpublished ideas. It goes without saying that these ideas are very seldom possessed of scientific validity. The very first communication, however, that I received from Mr. Hapgood electrified me. His idea is original, of great simplicity, and—if it continues to prove itself—of great importance to everything that is related to the history of the earth's surface.

Einstein concluded his Foreword by saying:

> The author has not confined himself to a simple presentation of this idea. He has also set forth, cautiously and comprehensively, the extraordinarily rich material that supports his displacement theory. I think that this rather astonishing, even fascinating, idea deserves the serious attention of anyone who concerns himself with the theory of the earth's development.

Despite the endorsement by Einstein and by Harvard professor of geology Kirtley F. Mather, who wrote a Foreword to the British and foreign-language editions of the book, the reception of *Earth's Shifting Crust* was, in Hapgood's phrase to a recent audience, "noteworthy in the negative sense." In other words, he said, "the silence has been deafening. There were very few reviews. . . . I can say that no crude errors have been found in the work, but it is clear that the basic challenge it presents to accepted geological ideas has been too extreme to be taken up by the Establishment."

That is not merely a self-serving appraisal by Hapgood. It is, like the book itself, an objective statement of fact. Supporting this view is Dr. F. N. Earll, of the Department of Geology at Montana College of Mineral Science and Technology, who wrote a Foreword to the revised edition of the book, *The Path of the Pole*, published in 1970. "After carefully reading *Earth's Shifting Crust,*" Earll stated, "I began searching through the technical journals and other likely sources for the discerning criticism that I felt should be forthcoming from experts in the field. . . . A reaction came, of course, and largely it came from men who under ordinary circumstances are both rational and competent, but their reaction could hardly be described as rational; hysterical would be a better description." He continues:

> The fact is that almost without exception Americans commenting on the book couched their discussion in thick and

unwarranted sarcasm, selecting trivia and factors not subject to verification as the bases for condemnation, seeking in this way to avoid the basic issues. Only the European reviewers were gracious enough to be fair, not that they accepted the theory without question, but they were prepared to offer it its day in court. Nowhere, in all that has been written about the book, have I found a single authority who has calmly and rationally offered a clear and documented criticism of the basic theory involved: that uncompensated masses on or in the earth may cause the earth's crust to slip over its core (p. viii).

The extensive revision of *Earth's Shifting Crust* a dozen years later was necessary primarily because of Hapgood's intellectual integrity, which shines throughout the book. In a note to the reader, he says that the most significant change in *The Path of the Pole* is directly related to the question on which both Einstein and Mather expressed their strongest doubts: the ice cap "mechanism" by which he originally proposed to account for displacements of the earth's outer shell. "Their doubts have been vindicated by the progress of earth studies in the past decade," he writes.

Advancing knowledge of conditions of the earth's crust now suggests that the forces responsible for shifts of the crust lie at some depth within the earth rather than on its surface.

Despite this change in the character of the proposed explanation of the movements, the evidence for the shifts themselves has been multiplied many fold in the past decade. The main themes of the book—the occurrence of the crust displacements even very recently in geological history, and their effects in forming the features of the earth's surface—therefore remain unchanged.

With Charles Hapgood, then, we have a view of the nature of polar shifts that is radically different from Brown's. Beginning with Brown's concept, Hapgood felt it necessary on the basis of scientific data to modify it step by step. First, in *Earth's Shifting Crust*, he abandoned the idea that the planet as a whole shifted its position in space. Next, in *The Path of the Pole*, he abandoned the idea that the ice caps provided the trigger mechanism for the shift. This chapter, therefore, will confine itself to Hapgood's theory as presented in its latest form, and quotations will be from *The Path of the Pole*. However, readers should know that Hapgood will soon publish a third edition of his book, and he will restore the original title to it. This revision, he told me, will not alter his basic theory. Rather, it will introduce still newer evidence that further strengthens his position.

Let's look in detail at Hapgood's presentation in *The Path of the Pole*. The four hundred-plus pages, incidentally, are a pleasure to read for their lucid exposition. The book's clear style, logical structure and restrained manner while offering the most detailed and substantial case for polar shifts that has ever been made are, in my judgment, admirable and reflective of a first-rate mind. Nothing in the mountains of material I went through while researching this book, with the exception of what we'll examine in the next chapter, comes close to attaining the stature of *The Path of the Pole*. Even if Hapgood were to be totally and irrevocably refuted someday—which I doubt in the strongest possible terms—

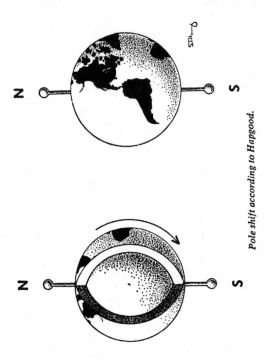

Pole shift according to Hapgood.

his book would be valuable as a model of scientific literature for a lay audience.

The theory of *The Path of the Pole* maintains that the crust of our planet has undergone repeated displacements—at least two hundred since Precambrian time, more than 600 million years ago. These polar shifts have not been near-instantaneous movements, within a day or so, as advocated by Brown and others whom we'll meet. Rather, Hapgood says, the slippage of the crust has, on the average, taken several thousand years each time. Furthermore, it has involved an average distance of no more than 30°—about two thousand miles, or one third of the distance from pole to equator. Thus, the average speed of polar shift would be measured in thousands of feet, not thousands of miles, per year.

Although this position makes Hapgood the most moderate by far of those who propose cataclysmic polar shifts, he is still an "extremist" when judged by reigning geological doctrine. Hapgood himself is well aware of this. After noting many discrepancies in the accepted views of geological history, he says, "I present a theory that can reconcile the facts. It is unfortunate that the theory makes so complete a break with accepted ideas. Nevertheless the truth, if it is the truth, must eventually be faced" (p. 172).

Hapgood would have science face its own ignorance about the fundamental principles and major events of the earth's evolution. It is not insignificant that his dedication of the book to a friend describes him as "a fighter against those blights of the scientific community: smugness, intolerance and materialism."

As noted in the Introduction to this section, Hapgood was not the first to propose crustal shifts. The geologist Damian Kreichgauer did so in 1926. But, Hapgood notes, he "did not have enough influence to impose the idea on his contemporaries."

The Path of the Pole is Hapgood's attempt to impose the idea through rational argument and weighty evidence. It begins with a chapter that defines the basic concepts to be examined: geomagnetism, continental drift and polar wandering. The latter is based on the idea, Hapgood explains, that the outer shell of the earth shifts about from time to time, moving some continents toward the poles and others away from them, changing the climates of the continents. Polar wandering or lithospheric displacement— the terms are synonymous—causes both the movement of the continents and sea-floor spreading. Thus, while Hapgood does not deny the notion of continental drift, he argues that both it and sea-

floor spreading are secondary effects caused by the more fundamental force, crustal shift.

Two kinds of evidence are primarily used by Hapgood to demonstrate the validity of his theory: new knowledge of geomagnetism
and two methods of absolute dating (ionium and radiocarbon dating). The former, he says, has led to the discovery that the poles
have changed their places on the surface of the earth no fewer than
sixteen times in the past million years, and at least two hundred
times since geological history began, more probably twice that
number. The latter evidence allows the climatic history of the earth
to be reconstructed in great detail for the past hundred thousand
years. As a result of his studies, Hapgood asserts in a prefatory
note:

> I have found evidence of three different positions of the
> North Pole in recent time. During the last glaciation in North
> America the pole appears to have stood in Hudson Bay, ap-
> proximately in Latitude 60° North and Longitude 83° West. It
> seems to have shifted to its present site in the middle of the
> Arctic Ocean in a gradual motion that began 18,000 or 17,000
> years ago and was completed by about 12,000 years ago.
>
> The radioactive dating methods further suggest that the pole
> came to Hudson Bay about 50,000 years ago, having been
> located before that time in the Greenland Sea, approximately
> in Latitude 73° North and Longitude 10° East. Thirty thousand
> years earlier the pole may have been in the Yukon District of
> Canada (p. xvii).

To understand how the crust of the planet might slip over the
interior, causing the various geological phenomena that Hapgood
attributes to polar shifts, he gives the reader a brief but very useful
survey of the earth's structure. In general, Hapgood says, the
asthenosphere is far too viscous to allow the necessary sliding
motion for crustal shifts. However, there is a special layer in the
asthenosphere named the "wave-guide layer" by its discoverer, the
Soviet geophysicist V. V. Beloussov, that provides a physical
structure for crustal displacements. Existing at a depth of about
one hundred miles, this layer is a very liquid one. Unlike the rest
of the asthenosphere, which lacks crystalline structure but is nevertheless semisolid due to the extreme pressure on its materials, the
wave-guide layer is characterized by lighter rock. This lighter rock
is created there through a change of phase which allows chemical
reactions not possible on the surface. The creation causes gravi-

tational instability within the layer as the lighter rock tries to rise
to the surface. This can cause imbalances leading to isostatic distortion that might trigger a displacement. "It seems that if the
earth's outer shell does slide as a unit over the interior," Hapgood
writes, "this is the most likely level at which the movement can
occur" (p. 23). In a later passage he elaborates:

> The wave-guide . . . is of great advantage for the concept of
> displacement. It suggests an easy zone of shear for the move-
> ment, wherein all frictional effects will be minimized. Actually
> the displacement would take place, according to this thinking,
> at a level where the viscosity of the asthenosphere would be
> reduced to its lowest point by the fluid wave-guide layer, and
> so the lithosphere would in effect be borne along on a stream
> flowing in a liquid, much as the Gulf Stream flows over the
> deeper waters of the ocean. The movement might be the equiv-
> alent of a flow of liquid over liquid. Friction would be mini-
> mized, while viscosity would present no bar to a comparatively
> rapid displacement.
>   The last point is one of great importance, for the field studies
> I am presenting . . . indicate that the shifts of the lithosphere
> have at times attained extraordinary speeds as compared with
> the speeds of subcrustal currents now estimated by geophysi-
> cists. The combination of the geometrical progression of cen-
> trifugal effects with the zone of easy shear in the wave-guide
> layer opens up the possibility of extremely rapid movements
> of the earth's outer shell (pp. 42–43).

Since *The Path of the Pole* was published, Hapgood told me,
a new structure has been discovered by satellite probe that may
also account for displacements of the lithosphere. It is a truly liquid
layer immediately beneath the lithosphere. Hapgood regards this
as equally probable to be the interface of crustal shifts, and feels
that both should be investigated. This discovery, incidentally, removes the semantic problem created when Hapgood says that
crustal movement takes place as "a unit" at the wave-guide layer,
which is some seventy miles below the crust proper. If the lithosphere alone shifts over the newly found liquid layer, then Hapgood's use of the term "crust" becomes precise. With the wave-
guide layer conception, the term must be understood to include
a large part of the asthenosphere, in addition to the lithospheric
crust.

As noted earlier, due to insufficient evidence Hapgood had to

abandon the idea that shifts of the crust might have been caused
by the centrifugal effects of polar ice caps. "It is necessary to
admit," he writes, ". . . that at the present time there is no satisfactory explanation of the *modus operandi* of displacements of the
lithosphere. The purpose of this book is simply to present the case
for the assumption that such shifts have occurred and to show how
the assumption explains numerous unsolved problems in geology
and in the evolution of life" (pp. 40–41).

Although a specific cause for displacements is not yet in sight,
Hapgood says, there are indications of the general direction from
which it may ultimately come: gravitational imbalances—within
the lithosphere or immediately below it—that give rise to the
centrifugal or centripetal effects originally postulated by Campbell
in his elaboration of the ice-cap trigger mechanism.

That such imbalances exist is unquestioned; whether they are
sufficient in magnitude to initiate a pole shift is not. Hapgood
refers to changes in surface mass loading (due to erosion, land
submergence, etc.) that exceed the limits of isostasy, the tendency
of the surface to seek equilibrium without lateral movement. The
earth's surface occasionally gets out of balance, and if what Hapgood calls "a large surplus or deficiency of mass" were to be
brought into existence near the surface, the result would be an
equatorward movement in the case of surplus masses and a poleward movement in the case of deficiencies. No candidate for this
role is presently visible, and it is hard to imagine one greater than
the abandoned one, the Antarctic ice cap.

Despite Hapgood's inability to identify a specific trigger mechanism for polar shifts, the evidence he amasses in favor of the
concept is truly staggering. In his second chapter, "The Failure
to Explain the Ice Ages," he reviews the evidence that ice sheets
have existed in many parts of the world, not always in the present
polar zones. Hapgood declares that these ice sheets are "misplaced," according to geological orthodoxy.

These ice caps, he says, have refused to have anything to do
with the polar areas of the present day, except in quite incidental
fashion. An example is the ice sheet that covered India 280 million
years ago. This extraordinary glacier moved northward from a
center in southern India for a distance of 1,100 miles. Since an
ice sheet on level ground will extend in all directions, did it push
as far to the south of Latitude 17° N. as to the north? If so, it
would have extended to the equator. While the latter case has not

been demonstrated, it is nevertheless a queer fact that an ice sheet should appear to *originate* near the equator and move northward. Continental drift cannot account for it, because, according to current theory, India was in the temperate zone at the time. Nor can the notion of lowered worldwide temperatures sufficient to allow glaciation in the temperate zone, for if the earth's temperature was cold enough to form glaciers there, how could tropical plants and animals have survived?

It is also a fact, Hapgood points out, that the ice caps of all geological periods in the southern hemisphere were eccentric to the South Pole, just as the Pleistocene ice caps were eccentric to the North Pole. "Is it not extraordinary," he asks, "that the Antarctic ice cap, which we can actually see because it now exists, is the only one of all these ice caps that is found in the polar zones?"

Hapgood also examines the alleged causes of the ice ages, and shows that none of the theories offered up to now are sufficient to explain all the facts. One group of theories postulates the onset of ice ages because of a worldwide lowering of temperatures due to various causes such as volcanic dust in the atmosphere, variations in the sun's rate of particle emission, etc. Another group attempts to explain the ice ages as a result of changes in the relative positions of the earth and sun.

In the latter category would fall the theory of the team headed by James D. Hays, mentioned in Chapter 1. I asked Hapgood to comment on this, and he remarked that his position was unchanged by the announcement. In *The Path of the Pole* he had written, "We have seen that ice ages existed in the tropics and that great ice caps covered vast areas on and near the equator. This happened not once, but several times. The question is, if the temperature of the whole earth fell enough to permit ice sheets a mile thick to develop on the equator, just where did the fauna and flora go for refuge? . . . It is small wonder that W. B. Wright insisted, over a quarter of a century ago, that the Permo-Carboniferous ice sheets in Africa and India were proof of a shift of the poles" (pp. 53-54).

As for the most recent theory, Hapgood told me, ice ages cannot be sufficiently explained by it. The Hays team is using selected data to fit a model they prefer, but they are overlooking important data about ice ages, such as the fact that the last North American ice cap has been shown to have disappeared from New York and New England in less than three thousand years. Evidence shows that it reached its maximum extent and thickness—more than a mile—about seventeen thousand years ago and then melted so fast that by fourteen thousand years ago temperate forests had been established there. How does that datum fit into a theory that maintains the ice ages were slow, regular events? Not at all, Hapgood remarked.*

Concerning the last North American ice cap, which disappeared completely by 9,000 B.C., consider some of the anomalous data that Hapgood cites in his book. First, the ice cap did not cover the northern hemisphere uniformly. It has been shown that some northern countries were *never* covered by ice during the last glacial period, and that in reality there were several more or less distinct ice sheets starting from local centers and expanding in all directions, north as well as south, east and west. Siberia was not covered, nor were most of Alaska, the Yukon and northern Greenland, while northern Europe, despite its relatively mild climate now, was buried under ice as far south as London and Berlin. Furthermore, most of Canada and the United States were covered with ice, which reached as far south as Cincinnati. This ice cap, curiously, was thicker and extended farther south on the lower central plains than on the mountainous highlands, while in Europe the ice was thinner and did not extend as far south.

What can account for these problems of geology? A different north polar position, Hapgood suggests—one situated in Hudson Bay. He points to evidence that the middle of the Arctic Ocean was "a temperate-climate refuge...for the fauna and flora that could not exist in Canada or the United States" at precisely the time of the great advance of the Wisconsin continental glacier three thousand miles *to the south*. But the evidence is not limited to the Arctic. From Japan, Mexico, Africa, South America, Tasmania and Antarctica comes evidence—too voluminous to summarize here—that Hapgood cites to demonstrate that various climatic zones were established around the globe and that they center on a spin axis whose north pole is in Hudson Bay and whose south pole is off the MacRobertson Coast of Queen Maud Land, Antarctica.

---

* An even more dramatic exception is given in a *National Geographic* article (November 1976) entitled "What's Happening to Our Climate?": "One drop from fairly warm times into full ice-age cold took place in Greenland 89,500 years ago—apparently in less than a hundred years" (p. 590).

One of the most dramatic pieces of evidence for Hapgood's theory came not from geology but from archaeology—the ancient maps of an ice-free Antarctica mentioned in Chapter 3. The full story is contained in Hapgood's *Maps of the Ancient Sea Kings* (Chilton, 1966), which will soon appear in an updated version. I will summarize this magnificent piece of archaeological research as Hapgood reported it in his writings and as he told it to me.

Hapgood's initial contact with Mallery's work came in 1956, when someone told him of the Georgetown Forum broadcast. It was of extreme interest to Hapgood because the Piri Re'is map evidenced a warm, deglacial period in Antarctica in fairly recent times—something called for in his theory of earth's shifting crust. So he contacted Mallery with an offer to work cooperatively on the matter. However, Mallery proved to be a "loner," Hapgood told me, and so there was no collaboration between them. Nevertheless, because Mallery provided the original inspiration, *Maps of the Ancient Sea Kings*, with its subtitle, *Evidence of Advanced Civilization in the Ice Age*, is dedicated to him.

As with the geological research, Hapgood turned the question of ancient maps into a class project. Research started in 1956 and continued for nine years. In the course of it, Hapgood and his students turned up nearly two dozen ancient maps that indicated the existence of an advanced worldwide culture long before the rise of Greece, Rome, Babylonia and Egypt—a culture with a higher level of technology than was attained at any time in history until the modern era. The period of that civilization? At least six thousand years before the present, and probably much longer ago. Hapgood put it this way in an article he wrote for *Argosy* magazine (October 1973):

> The fact that Antarctica was shown ice-free strongly suggested that the original of the map must have been drawn not much later than 17,000 years ago, when the whole crust began to move, because if the climate warmed up as fast as it seems to have in North America, it must have cooled off at a similar rate in Antarctica. The continent could have become uninhabitable as early as 14,000 years ago.

The map referred to here is not the Piri Re'is, but the Oronteus Finaeus map, of 1531. Published in Europe that year by the French geographer Oronce Fine, it was Hapgood who first recognized the sensational import of that ancient chart. The discovery happened this way.

In the course of the Piri Re'is investigation, Hapgood and his students searched for other charts of the Middle Ages and the Renaissance that might show Antarctica. During the Christmas recess of 1959–60, Hapgood traveled to the Library of Congress for just that purpose. Several hundred maps were made ready for his inspection, and by working long hours he was able to make a dent in the enormous mass of material. Then, one day, he writes in *Maps of the Ancient Sea Kings,*

> I turned a page and sat transfixed. As my eyes fell upon the southern hemisphere of a world map drawn by Oronteus Finaeus in 1531, I had the instant conviction that I had found here a truly authentic map of the real Antarctica.
>
> The general shape of the continent was startlingly like the outline of the continent on our modern map. . . . The position of the South Pole, nearly in the center of the continent, seemed about right. The mountain ranges that skirted the coasts suggested the numerous ranges that had been discovered in Antarctica in recent years. It was obvious, too, that this was no slapdash creation of somebody's imagination. The mountain ranges were individualized, some definitely coastal and some not. From most of them rivers were shown flowing into the sea, following in every case what looked like very natural and very convincing drainage patterns. This suggested, of course, that the coasts may have been ice-free when the original map was drawn. The deep interior, however, was free entirely of rivers and mountains, suggesting that the ice might have been present there (p. 79).

As Hapgood described it to me, upon seeing the map he felt as if he'd been stunned with a straight right to the jaw. "The point was that when the map was first published, nobody knew what Antarctica looked like, and by the time they had discovered what it looked like, about 1920, everybody had forgotten about the old map. It was when I put the two together that I saw they were the same."

Study of the map showed, in Hapgood's judgment, that it was based on an authentic ancient source map of Antarctica compiled from local maps of the coast drawn before the Antarctic ice cap had reached them. Several years of research enabled Hapgood and his students to identify more than fifty geographical features of the continent within an average two degrees of accuracy, including mountain ranges only four degrees from the South Pole.

They then sought to have their conclusions about several of the maps tested by an independent source. The Cartographic Section of the Strategic Air Command, at Westover Air Force Base, in Massachusetts, consented to review the work, and in August 1961 the chief of the section wrote to Hapgood:

> . . . The agreement of the Piri Re'is Map with the seismic profile of this area made by the Norwegian-British-Swedish Expedition of 1949, supported by your solution of the grid, places beyond a reasonable doubt the conclusion that the original source maps must have been made before the present Antarctic ice cap covered the Queen Maud Land coasts.
>
> c. It is our opinion that the accuracy of the cartographic features shown in the Oronteus Finaeus Map (1531) suggests, beyond a doubt, that it also was compiled from accurate source maps of Antarctica, but in this case of the entire continent. Close examination has proved the original source maps of Antarctica, but in this case of the entire continent. Close examination has proved the original source maps must have been compiled at a time when the land mass and inland waterways of the continent were relatively free of ice. The conclusion is further supported by International Geophysical Year teams in their measurements of the subglacial topography. The comparison also suggests that the original source maps (compiled in antiquity) were prepared when Antarctica was presumably free of ice. The Cordiform Projection used by Oronteus Finaeus suggests the use of advanced mathematics. Further, the shape given to the Antarctic continent suggests the possibility, if not the probability, that the original source maps were compiled on a stereographic or gnomonic type of projection (involving the use of spherical trigonometry).
>
> d. We are convinced that the findings made by you and your associates are valid, and that they raise extremely important questions affecting geology and ancient history, questions that certainly require further investigation (pp. 244–45).

As is often the case with radically new ideas, however, there was almost no notice taken of the findings by the scientific community, despite a Foreword by one of the most eminent geographers of the day, Dr. John K. Wright, who was for many years director of the American Geographical Society. Wright stated his opinion in the Foreword: "Whether or not one accepts [Hapgood's] 'identifications' and 'solutions,' he has posed hypotheses that cry aloud for further testing. Besides this, his suggestions as to what might explain the disappearance of civilizations sufficiently ad-

vanced in science and navigation to have produced the hypothetical lost prototypes of the maps that he has studied raise
interesting . . . questions."

Although official science has ignored Hapgood's findings, they
have not gone unnoticed by some who labor in what has been
called parascience or, less generously, pseudoscience. Notable
among them are writers who claim that *Maps of the Ancient Sea
Kings* shows Antarctica was mapped from the air. Some claim this
was done by a terrestrial civilization that had achieved high-altitude
air travel; others attribute it to "ancient astronauts" from extraterrestrial civilizations.

Hapgood wholly rejects this conclusion, claiming it ignores
important evidence. In an article in *Info Journal* (November 1974),
he points out that Antarctica as drawn on the Oronteus Finaeus
map is distorted in comparison with aerial survey photos. The
reason, he says, is that the ancient mapmaker used a cartographic
projection unsuitable for the polar regions. Aerial photography
would have yielded a more accurate map. This is even more evident
in the case of the Piri Re'is map, which is composed of about
twenty separate maps that were apparently drawn at different times
and finally compiled together in a general map. There are errors
in the surviving fragment of the Piri Re'is map that would not
appear in an aerial survey. Among the errors Hapgood notes:

> In our research we established the fact that about 900 miles
> of the eastern coast of South America were simply omitted from
> the map, that the Amazon was represented on it *twice,* that the
> Caribbean area was drawn with a north almost at right angles
> to the north of the main part of the map, and that the Andes
> Mountains (which had not been discovered in 1513, when the
> Piri Re'is map was drawn) were drawn on a different scale, to
> a different north, and in the wrong longitude. These are the
> errors of earth men.

To return to *The Path of the Pole,* Hapgood's final comment
on the Oronteus Finaeus map is that no map of this accuracy could
have been drawn in modern times until the invention of the chron-
ometer, about 1780, since it was this instrument that first made
possible the accurate determination of longitude.

The remainder of the book is too much to summarize here.
Suffice it to say that Hapgood deals with a host of fascinating
problems in geology and evolution, and he does so in a most

thoughtful way. He shows that ice ages in low latitudes and warm ages near the pole, *correlated in time,* are different sides of a single coin—a coin that can also explain how the earth's geological features are shaped and how animal extinctions can happen, allowing new forms to evolve rapidly in the "life niches" thus opened. That coin, of course, is crustal displacement.

Two important questions remain for us in this examination of Hapgood's theory. First, how does he deal with the riddle of the frozen mammoths? Second, what does he have to say about a future pole shift?

In a thirty-page chapter on the extinction of the mammoths and mastodons, Hapgood offers both consummate scholarship and provocative insights—some of which I presented without attribution in Chapter 2. He pays particular attention to the Berezovka mammoth and gives this interpretation of the data:

> The evidence shows that the animal suffered a very severe fall, severe enough to break his pelvis and leg. We learn also that the food in his stomach and mouth did not match the vegetation around him at the spot he was found. He did not fall into water because, as was ascertained by another investigator, large masses of blood were found under him. The blood would, of course, have been washed away had he tumbled into a river. The fact that his penis was found to be erect indicates that he was not instantly killed by his fall, but that he froze to death. He was certainly plunged suddenly into extreme cold.
>
> I think we can see how this might have happened. With a high level of earthquake activity, large fissures could be opening in the crust in considerable numbers, as they commonly do in many earthquakes. Let us assume that the mammoth fell into one of these.
>
> We must remember that according to our theory a long period of intense cold had gripped the Siberian coast until only a short while before. This would have been the time of the pole position in the Greenland Sea. The situation of the pole just north of Norway would have logically involved an ice age in the region of the Beresovka River. The frozen ground, or permafrost, of this ice age might have extended down thousands of feet, as it does today in some places in the Arctic. When the pole moved to Hudson Bay, the climate in the Beresovka region would have become about like that of Minnesota today, where the winters are severe enough to prevent, or greatly delay, the deep melting of a permafrost extending down thousands of feet.
>
> We may suggest, then, that the Beresovka mammoth fell

into a deep crevasse of fracture in the earth's crust, perhaps several hundreds of feet deep. He might have tumbled down a sloping wall of the crevasse a long way without actually killing himself, but of course at the bottom loose earth dislodged by his fall could have cascaded down upon him and buried him alive. According to biologists I have consulted, the erection of the penis could have resulted from the poor animal's emotions of terror and from his pain.

The mammoth might have frozen to death and afterwards been gradually frozen through in the manner I have suggested in the preceding pages. The fissure would very likely have been largely filled in as the result of continuing earth shocks, landslides and the like, and then gradually the temperatures of the body would have been reduced to the low temperatures prevailing deep in the permafrost.

And what of that great fissure? What is the existing evidence of it? Why, the valley of the Beresovka River itself! The valley, or channel, of the present river may have been created by the filling in of the fissure.

But how then, you may ask, did the mammoth come to the surface? The answer may be that erosion in the valley was rapid during the ensuing warm period because the river must have been much better fed by its tributaries than it is now. Moreover, it is generally thought that the coast stood higher then, than now, so that the New Siberian and Liakov Islands were connected with the mainland. The result of these factors would probably have been that the river was much larger and flowed much faster than now, and consequently in the 30,000 years or so of the warm period could have eroded the valley to a very considerable depth.

The Beresovka mammoth, and the other bodies we have of about the same age, might thus have been brought nearer the surface but not actually uncovered until after the climate again grew cold with another poleward shift of Siberia (pp. 270–71).

As plausible as this seems, there is a major flaw to it, in the judgment of Dwardu Cardona, who mentions in his *Kronos* article that there seems to be no evidence that the Berezovka Valley owes its origin to an earthquake. If so, the riddle of the frozen mammoths remains unsolved by Hapgood. When I mentioned this to Hapgood, he responded, "In any great quake, a fissure can open and close in a few moments. The probability is that the origin of the valley of the Berezovka was there for ages before the displacement that caused the fissures. They are unrelated."

Interestingly, Hapgood feels that the riddle was not solved by

Sanderson, either. Sanderson's presentation is incorrect, he told me—a sensationalized portrait that vastly overstates the time scale involved in crustal shifts. Although Sanderson was a good friend of his, Hapgood remarked, "He simply didn't do the necessary painstaking work he should have on this problem. The evidence of a northward movement of Siberia is overwhelming, but it certainly did not take place overnight."

Our second question concerns Hapgood's views of a future pole shift. Two lines of evidence lead him to conclude that the lithosphere may be in motion at the present time—and accelerating! First, the rising curve of great earthquakes since 1900 indicates that they are getting both more frequent and more severe. (This point will be given dramatic statement in Part III, where psychics predict a decade of cataclysmic earthquakes at the end of the century, culminating in a pole shift.)

The other line of evidence that suggests to Hapgood that a pole shift may be under way now comes from astronomical observations. At the end of the first chapter, he writes:

> We have two observations of a movement of the North Pole with reference to the earth's surface. The first of these is cited by Deutsch on the authority of Munk and MacDonald. It suggests that the North Pole moved 10 feet in the direction of Greenland during the period from 1900 to 1960. This (according to Deutsch) would be at a rate of 6 centimeters (about 2.5 inches) a year. The other findings, cited by Markowitz, based on later data, suggest that the pole moved about 20 feet between 1900 and 1968 . . . and that it is now moving at the rate of about 10 centimeters (4 inches) a year. . . .
>
> A second point, possibly even more interesting, is that if both these observations were accurate when made, as we have every right to expect in view of the eminence of the scientists involved, then we have evidence here of a geometrical acceleration of the rate of motion. If the pole moved 10 feet between 1900 and 1960, but 20 feet between 1900 and 1968, then it moved 10 feet between 1960 and 1968, which would suggest an acceleration by a factor of about 8 (p. 44).

Hapgood commented on this in his *Argosy* magazine article, saying that it would be inevitable that a movement so vast as that of the earth's whole outer shell would start very slowly, but if the movement has already started, and if it is accelerating at the rate suggested, then it is quite possible that within a century a state of chaos may embrace civilization. For if the movement has started, he declares, it is now too late for man to stop it or slow it down.

> There is only one hope of salvaging civilization from a total wreck, a wreck that has perhaps occurred several times in the past, and this is through some form of centralized world government that could organize the shifting of populations from one part of the globe to another, until the earth's surface was again stabilized. Let us hope man wakes up before it's too late!

Is there motion to the pole right now—motion other than the presently recognized effects of Chandler wobble, nutation and precession—and which is accelerating? In seeking to resolve this question, I found an article in *Sky and Telescope* (August 1976) that contains part of the answer. "Polar Motion: History and Recent Results," by William Markowitz, reports that the International Polar Motion Service (IPMS), which has observed polar motion through its five participating observatories since 1900, recognizes a secular motion or progressive drift of the North Pole. From 1965 to 1973, the IPMS data show, the North Pole drifted slightly more than one meter, and from 1903 to 1973 it wandered erratically toward Greenland slightly more than eight meters over seventy years.

I was still unsatisfied with these data, because they did not confirm or deny Hapgood's contention that polar motion might be accelerating, so in August 1978 I wrote to Markowitz, at Nova University, in Florida, asking, "Could you state for publication, in layman's terms, what these data mean with regard to Hapgood's thesis that a radical polar shift may now be starting?"

Markowitz's reply arrived a few weeks later, and directly answered my question:

> The position of the pole cannot be obtained directly, say, by going to the pole and measuring its location with respect to some fixed point. It is found, instead, from observations of latitude made some thousands of kilometers away. These observations include effects other than the motion of the pole itself. In consequence, different investigators have derived different rates of drift for the mean pole, even for essentially the same time interval. A comparison of rates of secular motion by two investigators reflects the differences in data used and the assumptions made by each investigator. The astronomical observations provide no basis for believing that the secular motion of the pole is accelerating.

* * *

Thus, pole watchers may breathe a little easier to know that a shift has not begun—yet. Nevertheless, we should bear in mind Hapgood's statement to readers at the beginning of *The Path of the Pole* in which he withdraws advocacy of the ice cap as trigger mechanism. "Despite this change in the character of the proposed explanation of the movements," he writes, "the evidence for the shifts themselves has been multiplied many fold in the past decade. The main themes of the book—the occurrence of the crust displacements even very recently in geological history, and their effects in forming the features of the earth's surface—therefore remain unchanged."

We turn now to a man who is the most famous pole shift theorist of all. His perspective is limited to previous pole shifts; he makes no predictions of a coming one. But what he has to say about pole shifts happening "very recently in geological history" is, in comparison even with Hapgood and Brown, absolutely sensational.

References and Suggested Readings

HAPGOOD, CHARLES. *Earth's Shifting Crust*. Philadelphia: Chilton, 1958 and forthcoming revised edition.

——. *Maps of the Ancient Sea Kings*. Philadelphia: Chilton, 1966 and forthcoming revised edition.

——. *The Path of the Pole*. Philadelphia: Chilton, 1970.

——. "Shifting Continents," *Argosy*, October 1973.

——. "The Maps and the Galaxies," *Info Journal*, November 1974. Published by the International Fortean Organization, 7317 Baltimore Avenue, College Park, MD 20704.

Chapter Six

Immanuel Velikovsky: Worlds in Collision, Earth in Upheaval

Dr. Velikovsky's claim that there have been changes in the structure of the solar system during historical times has implications which apparently he has not thought through; or perhaps was unable to convey to me in our brief conversation. If in historical times there have been these changes in the structure of the solar system, in spite of the fact that our celestial mechanics has been for scores of years able to specify without question the positions and motions of the members of the planetary system for many millennia fore and aft, then the laws of mechanics which have worked to keep airplanes afloat, to operate the tides, to handle the myriads of problems of everyday life, are fallacious. But they have been tested competently and thoroughly. In other words, if Dr. Velikovsky is right, the rest of us are crazy. . . .

The author of that statement, Dr. Harlow Shapley, a respected astronomer, was director of the Harvard College Observatory when he wrote it in a letter to a colleague in 1946. Immanuel Velikovsky, a psychiatrist-historian, had sought out Shapley and asked him to review a manuscript in which he proposed a radical theory of changes in the constitution of the solar system. Shapley demurred, passing the matter to a mutual acquaintance of theirs, Dr. Horace Kallen, requesting his opinion before going ahead with some simple tests of the theory.

Kallen replied favorably about Velikovsky's work, but Shapley nevertheless withdrew his offer to cooperate. In doing so, he made the sarcastic remark above, and thereby began his shameful role in what has come to be known as "the Velikovsky Affair." For many of his colleagues, Shapley's opinion amounted to divine edict. So "the word" went out about Velikovsky. And although the passing years have shown that Velikovsky was right in many respects, they have also shown with sad irony that Shapley, too, was right. He had spoken truer than he understood: the rest of them *were* crazy.

"Them" refers to a large segment of the scientific establishment at that time, and "crazy" is meant in the sense that some of the most irrational, unprincipled and venomous acts in intellectual history were committed in a conspiratorial fashion by supposedly reasonable, highly educated scientists and scholars—all for the purpose of suppressing a revolutionary idea.

The idea? Venus, Velikovsky argued, had in recent time been born as a comet ejected from Jupiter, and had wandered the solar system, nearly colliding with the earth in the fifteenth century B.C., causing the earth to rock on its axis twice, and slip its crust as well. Eight centuries later, Venus again disrupted the solar system, causing Mars to wander erratically into near-collision with the earth, again with several displacements of the terrestrial axis and accompanying lithospheric slippage. The forces involved in accomplishing this, Velikovsky asserted, were primarily electromagnetic, not gravitational. And to support his assertions, Velikovsky offered as primary proof the ancient world's myths, legends and histories of cataclysm.

In one breath, it seemed, an unknown outsider had profaned the most sacred doctrines of celestial cause and effect—doctrines jealously guarded by the high priests of technological society: scientists. Because of that cosmological heresy, Velikovsky would not be allowed a hearing. In the name of science, *scientia*, knowing, his data would be denounced and—even worse—maliciously distorted, his views would be peremptorily blocked from presentation on the podium and in professional journals, his books would be boycotted and scurrilously reviewed, his supporters summarily fired from positions they had held responsibly for many years, his integrity impugned, his reputation blackened, his very name made anathema. Savage verbal attacks would be made against both the man and the work by people who bragged in the same breath that they had not read his books. And all because he had challenged a dogma.

Brown and Hapgood had also challenged dogma, but the response to them was principally silence and sequestration. Velikovsky was to experience something different: unbridled outrage. Whereas Brown and Hapgood were to be largely ignored, Velikovsky would feel the full wrath of some virtual (but not virtuous) pillars of the scientific community. In the process, much would be shown about the fragile assumptions and cliquish associations upon which presumed truth is often built. Much would also be shown about the state of mind prevailing in the scientific community—a state characterized by intellectual arrogance. Consciousness is the key to understanding new knowledge such as Velikovsky brought to light. The state of consciousness prevailing in the scientific community, as displayed in the Velikovsky Affair, was simply incapable of handling it, and was decidedly unbecoming.

Immanuel Velikovsky, a Russian Jew born in 1895, studied medicine at the University of Moscow and graduated from it in 1921 with a medical degree. Earlier he had studied in Scotland. After becoming a doctor, Velikovsky moved to Berlin, where he helped found a scholarly publication, *Scripta Universitatis*. Conceived as a cornerstone for what would become Hebrew University, at Jerusalem, it contained articles by Jewish scholars. Albert Einstein edited the mathematical-physical volume of *Scripta Universitatis*. Years later, Einstein would enter the Velikovsky Affair as one who disagreed with Velikovsky's views but who defended his right to be heard without prejudice and abuse being heaped on him.

In 1923 Velikovsky moved to Palestine and began medical practice in Tel Aviv. Soon, however, he became interested in psychoanalysis and went to Vienna to study under a pupil of

Freud's. Thereafter, his practice in Jerusalem was limited to psychoanalysis. Freud and Velikovsky met in Vienna and later exchanged letters on scientific matters. Velikovsky also contributed articles to Freud's psychology journal, *Imago*. In 1930 he published a paper that proposed for the first time that the recently discovered phenomenon of electrical brain-wave patterns could be applied in the diagnosis of epilepsy because epileptics would have a particular brain-wave "signature."

When, in 1939, Freud published a series of essays that were to be his last book, *Moses and Monotheism*—an attempt to discern the psychological and historical origins of Judaism—Velikovsky began a complementary study of three subjects in Freud's work: Oedipus, Akhnaton and Moses. The research required Velikovsky to come to the United States, so in 1939, a few weeks before the Second World War began, he brought his family to New York on a year's sabbatical leave. War soon imposed a longer stay. For eight months he worked in libraries researching what would become his fourth book, *Oedipus and Akhnaton*. It was nearly completed when other research led to discoveries that drastically altered the course of his life.

According to a short biography in *Pensée* magazine (May 1972), in April of 1940 Velikovsky "was first struck by the idea that a great natural catastrophe had taken place at the time of the Israelites' Exodus from Egypt—a time when plagues occurred, the Sea of Passage parted, Mt. Sinai erupted, and the pillar of cloud and fire moved in the sky. Velikovsky wondered: Does any Egyptian record of a similar catastrophe exist? He found the answer in an obscure papyrus stored in Leiden, Holland—the lamentations of an Egyptian sage, Ipuwer" (p. 5).

In other words, Velikovsky realized that the Old Testament story of Moses was not allegory or embroidered legend, but was the documentary account by primitive people of actual historical events. The Ipuwer papyrus parallels the Book of Exodus, describing the same catastrophe, the same plagues. As a result of this discovery, Velikovsky began to reconstruct Middle Eastern history, *Pensée* tells us, taking this catastrophe—which brought the downfall of the Egyptian Middle Kingdom—as a starting point from which to synchronize the histories of Egypt and Israel. The result was to be his second book, *Ages in Chaos,* published in 1952. But there was a furor well before that, arising from the 1950 publication of his first book, *Worlds in Collision.* This was the one that so infuriated the scientific establishment.

Worlds in Collision began one afternoon in October 1940 when Velikovsky noticed an important fact: the Book of Joshua describes a destructive shower of meteorites occurring before the sun "stood still" in the sky. "Could this be a coincidence," the *Pensée* biography remarks, "or were the ancients recording a cosmic disturbance that must have shaken the entire Earth and might have been related to the upheavals approximately 50 years earlier during the Exodus? A survey of other sources around the world convinced Velikovsky that a global cataclysm had indeed overtaken the Earth, and that Venus played a decisive role in that cataclysm" (p. 5).

With this discovery, Velikovsky asked himself how a disaster of such magnitude could be blotted from human memory. His psychoanalytic training suggested the answer. If individual memories could submerge painful experiences from normal recall, so also might the human race blot out recollection of a devastating catastrophe that virtually destroyed society. "He called such a process collective amnesia," Fred Warshofsky reports in *Doomsday* (p. 40), "and began a monumental life's work, a reconstruction of ancient history according to his catastrophe theory."

Velikovsky continued for ten years to research the subject. In 1950 he published the first of half a dozen books to date that offered a radical revision of planetary mechanics and early human history. Velikovsky's theory was as extraordinary for its method as for its results. For he was no experimental scientist in a laboratory, adjusting sophisticated machines to get decimal-point readouts. He was a scholar-psychologist reconstructing history from long-ignored data found in myth and literature, and was supporting his reconstruction with an incredible array of evidence from the physical, biological and social sciences. The scope of his interdisciplinary research and the boldness of his conclusions were breathtaking. Moreover, he presented it all with a prose style that is dramatic and often spellbinding. I recall my own first exposure to Velikovsky, through the pages of *Reader's Digest*, which condensed his 1955 book *Earth in Upheaval*. The images of frozen mammoths in Siberia and gigantic mounds of bones in Alaska impressed themselves indelibly on me.

Impressionable teenager though I was, the same thing happened to many others more mature and more educated than I. Less than two months after publication, *Worlds in Collision* climbed to the top of the best-seller list. This, however, only added insult to injury, in the eyes of the scientific establishment. What seemed to be a campaign led by Shapley had begun among college professors and scientists to pressure Velikovsky's publisher, Macmillan, not to bring out the book at all. When it nevertheless appeared and quickly rose to sales prominence, the pressure tactic changed to threats of boycotting all Macmillan books. Fearing for its textbook sales, Macmillan took a step that was unprecedented in publishing history. While *Worlds in Collision* was holding the number one position on *The New York Times* list, Macmillan gave it to Doubleday, which had no textbook division and thus was immune to the boycott threats. Macmillan also fired the editor who had purchased *Worlds in Collision*. He was one of several people who would be sacrificed by various institutions to appease the wrath of the high priests of science.

The further history of the Velikovsky Affair has been widely reported, and we therefore need not go over it here. Interested readers should peruse *Pensée* (May 1972), Alfred DeGrazia's *The Velikovsky Affair*, and other writings listed at the end of this chapter. Suffice it to say that the treatment given Velikovsky constitutes an ugly record of intellectual dishonesty, moral cowardice and scientific hubris. Except for a handful of courageous men who stepped forth to defend Velikovsky's right to be heard in a fair and professional manner, without emotional attacks, Velikovsky was shunned by the scientific and intellectual communities for a decade.

Velikovsky's work was primarily a reconstruction of early history based on the testimony of early civilizations. From that reconstruction he inferred certain astronomical events, which he claimed would be proven by scientific experimentation. In the 1960s, when space research began to give startling new data about the nature of the solar system, many of Velikovsky's predictions were shown to be correct.

For example, in 1950 the cloud surface above Venus was known to be $-25°$ C. on the day and night sides alike; yet Velikovsky declared in *Worlds in Collision* that the surface temperature of Venus would be much higher—in the range of incandescence. This was entirely discordant with prevailing theory. By 1961, however, theory had to be revised because Mariner II flew past Venus and recorded a surface temperature of $800°$ F., two hundred degrees above the melting point of lead. More exact measurements since then have raised the figure to $900°$ F.

Likewise, in 1953 Velikovsky predicted that Jupiter would be found to send out "radio noises as do the sun and stars." This, too, was contrary to theory at the time. So certain was he of the logic of his deductions and the coherence of his theory that Velikovsky asked his old scientific colleague Einstein, also living in Princeton, to use his influence to have Jupiter surveyed for radio emissions. Warshofsky recounts for us what happened:

"Space," Velikovsky had declared, "is not a vacuum; and electromagnetism plays a fundamental role in our solar system and the entire universe." Although some stars were known to give off radio waves, the idea of a noisy space, crackling with radio waves, pressed by magnetic fields and riven by electrical charges and radioactivity, was not a widely accepted part of the astronomy of 1950. Thus, few astronomers gave any credence to Velikovsky's claim in a 1953 lecture at Princeton University that Jupiter was emitting radio noise.

Our picture of Jupiter has been vastly expanded since then.

Albert Einstein was sympathetic to some of Velikovsky's fundamental concepts but vigorously opposed his theory that space was permeated by magnetic fields, that the sun and planets are charged bodies and that electromagnetism plays a role in celestial mechanics.

In June 1954 Velikovsky offered in writing to stake the outcome of his debate with Einstein on the question of whether Jupiter emits radio noises, as he had claimed. Einstein replied, as was his custom, by making marginal notes, one of which discounted the idea.

Ten months later, early in 1955, astronomers at the Carnegie Institution were shocked to hear strong radio signals pouring in from Jupiter. When Einstein heard the news, he emphatically declared that he would use his influence to have Velikovsky's theory put to experimental test. Nine days later he died—a copy of *Worlds in Collision* open on his desk (p. 45).

Before looking in detail at Velikovsky's presentation, I must make quite clear that his presence in this section is not based upon any prediction of his concerning a future pole shift. He has made no such prediction and thus should not be thought of as one who prognosticates Doomsday through a recurrence of the ancient events he describes. Moreover, when I visited him at his home in Princeton, New Jersey, where he resided from 1952 until his death in January, 1980, he made clear that he places no credibility in psychics who forecast a pole shift (although, he remarked, he accepts in principle the possibility of ESP and once wrote a paper on it).

Why, then, is Velikovsky considered here? The answer is simply that he has done more than anyone else to place the question of pole shift before the public and the scientific community. His assertion that our planet has been suddenly, massively and repeatedly rocked on its axis by extraterrestrial forces, like a fishing bob on the water, and his delineation of the electromagnetic dynamics by which that occurred, are deeply thoughtful and provocative. Velikovsky has given us the third principal variant of the pole shift concept. Even though he does not forecast a pole shift, any examination of the subject must consider his historical reconstruction of previous ones.

The cosmic scenario projected by Velikovsky is succinctly stated in a condensation of Warshofsky's *Doomsday* chapter on it appearing in *Reader's Digest* (December 1975) as "When the Sky Rained Fire."

> At some time more than 4000 years ago, according to Velikovsky's interpretation of ancient texts, the giant planet Jupiter—about 320 times more massive than Earth, underwent a shattering convulsion and hurled a planet-size chunk of itself into space. The blazing new member of the solar system—the protoplanet Venus—hurtled down a long orbit toward the sun, on a course that would eventually menace the Earth. . . .
>
> To Velikovsky, it was clear that this fiery birth of Venus had been recorded by peoples the world over. "In Greece," he wrote in *Worlds in Collision,* "the goddess who suddenly appeared in the sky was Pallas Athene. She sprang from the head of Zeus-Jupiter." To the Chinese, Venus spanned the heavens, rivaling the sun in brightness. "The brilliant light of Venus," noted one ancient rabbinical record, "blazes from one end of the cosmos to the other."
>
> In the middle of the 15th century B.C., Velikovsky theorized, Earth in its orbit around the sun entered the outer edges of the protoplanet's trailing dust and gases. A fine red dust filled the air, staining the continents and seas with a bloody hue. Frantically, men clawed at the earth seeking underground springs uncontaminated by the red dust.
>
> "All the waters that were in the river were turned to blood. . . . And all the Egyptians digged round about the river for water to drink," says Exodus 7:20–24. "The river is blood. . . . Men shrink from tasting—human beings thirst after water," confirms the Egyptian sage Ipuwer.

As Earth continued to move through the cometary tail, Velikovsky claims, the particles grew coarser and larger, until our planet was bombarded by showers of meteorites that were recorded all around the world. Exodus: "There was hail and fire mingled with hail . . . there was none like it in all the land of Egypt since it became a nation . . . and the hail smote every herb of the field, and broke every tree of the field." Ipuwer concludes: "Trees were destroyed. No fruits or herbs are found. That has perished which yesterday was seen." These things happened, say the Mexican Annals of Cuauhtitlán, when the sky "rained not water but fire and red-hot stones."

Then an even more terrifying event took place. *Popul-Vuh,* the sacred book of the Mayas, tells the story: "It was ruin and destruction . . . a great inundation . . . people were drowned in a sticky substance raining from the sky." What happened, says Velikovsky, is that gases in the protoplanet's tail combined to form petroleum. Some of this rained down unignited, but some mixed with oxygen in Earth's atmosphere and caught fire. The sky seemed to burst into flames, and a terrible rain of fire fell from Siberia to South America.

Earth now penetrated deeper into the comet's tail, on a near-collision course with its massive head. Great hurricanes pummeled Egypt and other lands. A violent convulsion ripped the Earth, tilting it on its axis. In the grip of the protoplanet's gravitational pull, the terrestrial crust folded and shifted. Cities were leveled, islands shattered, mountains swelled with lava, oceans crashed over continents. Most of the Earth's animal and human populations were destroyed.

"Then the heavens burst, and fragments fell down and killed everything and everybody. Heaven and earth changed places," states the tradition of the Sashinaua of Western Brazil. Plagues of vermin descended on China, and the land burned. Then the waters of the oceans fell on the continent and, according to an ancient text, "overtopped the great heights, threatening the heavens with their floods."

Earth turned part-way over. Part was now in extended darkness, part in protracted day. The Persians watched in awe as a single day became three before turning into a night that lasted three times longer than usual. The Chinese wrote of an incredible time when the sun did not set for several days while the entire land burned.

The catastrophe was also responsible, according to Velikovsky, for the most memorable drama in the Old Testament—the Exodus of the Israelites from Egypt. The awful devastation toppled the Egyptian Middle Kingdom, and Moses led the people of Israel, erstwhile slaves, out of the ruined land. As they fled across the border, before them moved the huge pillar of fire and smoke.

For the fleeing Israelites it marked the way to Pi-ha-Khiroth, near the Sea of Passage. Behind them raced the angry and vengeful pharaoh and his army. Ahead lay the seabed, uncovered, its waters piled high on either side by the shifting movement of the Earth's crust and the gravitational and electromagnetic effects of the protoplanet. The Israelites hesitated, then rushed across the seabed, which, according to rabbinical sources, was hot. As the pharaoh's armies followed, an incredibly powerful electrical bolt passed between Earth and the protoplanet. The walls of water collapsed.

Through the world, populations were all but annihilated. The survivors were threatened with starvation. And then yet another phenomenon, recorded from Iceland to India, as well as in the Old Testament, took place. The hydrocarbons in the comet's tail that had drenched the Earth in petroleum were now being slowly changed within the Earth's atmosphere, possibly by bacterial action, possibly by incessant electrical discharges, into an edible substance—the manna of the Israelites, the ambrosia of the Greeks, the honey-like madhu of the Hindus.

The close approach of the protoplanet Venus produced gravitational dislocations that reversed the direction of the Earth's axis. To the shocked and dazed people of Earth, the sun was rising in the western sky and setting in the eastern sky. Seasons were exchanged. "The winter is come as summer, the months are reversed, and the hours are disordered," states an Egyptian papyrus. In China, the emperor sent scholars to the four corners of the darkened land to relocate north, east, west and south, and to draw up a new calendar. For a generation Earth was enshrouded in an envelope of clouds—the Shadow of Death of the Scriptures, the *Götterdämmerung* of the Nordic races. It endured for 25 years, according to Mayan sources.

Slowly Earth and its people began to recover. But only 50 years later, around 1400 B.C., according to Velikovsky's interpretation of ancient sources, Venus made a second pass at Earth. The terrestrial axis again tilted, and Earth heaved and buckled. The few rebuilt towns flamed up and collapsed in heaps of rubble. The book of Joshua records that "The Lord cast down great stones from heaven upon them" (The Canaanites). On the other side of the world, Mexican records speak of a lengthened night. Once again Earth was wracked by earthquakes, global hurricanes, continental shifts, and by universal destruction.

The peoples of the world who survived the second holocaust bowed down before the dreaded Venus, goddess of fire and destruction, and each in a manner dictated by cultural heritage placated her, with human sacrifices and bloody rituals, with prayers and incantations. Cuneiform tablets found in the ruins of the library palace in Nineveh, the Assyrian capital, record the erratic behavior of Venus. The fearful Babylonians pleaded with the errant queen of the heavens to leave Earth in peace: "How long wilt thou tarry, O lady of heaven and earth?" . . .

Martian Threat

Several centuries after Venus had twice menaced Earth, according to Velikovsky's reconstruction of events, it nearly collided with Mars. "Mars, being only about one-eighth the mass of Venus—was no match for her." In the eighth century B.C. or earlier, the smaller planet was pulled from its orbit and flung into a new path about the sun, one that threatened Earth. The annals and sacred books of antiquity record a violent turmoil in the sky as Mars drew near. Earth staggered in its own orbit. Again, cities collapsed, earthquakes split the surface, and men died amid geophysical upheavals. The prophets Isaiah, Hosea, Joel and Amos recorded these catastrophes, and they are also described in Homer's *Iliad*. The effects of the close passage of Mars did not equal those of Venus, but they were great enough to again shift Earth's axis and orbit. The old calendar of 12 months of 30 days each, adding up to a 360-day year, was no longer accurate. All over the world, throughout the eighth and seventh centuries B.C., calendars were reformed.

Mars returned every 15 years. During one near encounter, according to Velikovsky, at the moment of closest passage when the gravitational attraction between Mars and Earth was at its greatest, the sun—as viewed by the Israelites, and as recorded in Midrashic sources*—seemed to hurry to a premature setting. It dropped below the horizon several hours before it normally set. The Greeks and other nations and races observed the same phenomenon and described it.

Although Mars did far less damage than Venus had some seven centuries earlier, it now became a dominant, fierce god in the pantheon of man's heavenly forces. Velikovsky believes that its last cataclysmic approach took place in the spring of 687 B.C. In that year, the Assyrian king, Sennacherib, marched against Hezekiah, king of Judah, planning to capture Jerusalem. On the evening of March 23, the first night of the Hebrew Passover, Mars unleashed "a blast from heaven" that, according to the Books of Kings and Chronicles, left 185,000 men of the invading army dead.

* And also in the Bible, Velikovsky notes in *Worlds in Collision*.

That same night, the Chinese recorded a great disturbance in the sky. "In the night," the *Bamboo Books* report—and they give the date—"stars fell like rain. The Earth shook." French scholars calculated that the event took place on March 23, 687 B.C. To the Romans, March 23 became the festival of *Tubilustrium*, a major celebration in honor of the god of war, Mars.

In some longitudes, as Mars made its last terrible pass at the Earth, the rising sun dipped back below the horizon. This retreat of the sun was caused by a tilt in the Earth's axis, a tilt that nearly corrected the one that occurred a generation earlier. "So the sun returned ten degrees, by which degrees it was gone down on the sundial of Ahaz," recorded Isaiah.

At long last, the heavens became more peaceful. Mars was cast out beyond the range of danger to Earth. And Venus, which had assumed a dominant role in the heavens, soaring up to the zenith, dropped back to become a morning and evening star that never rises to zenith.

With the "big picture" before us, let's look more closely at Velikovsky's theory of pole shift. We can begin with his own statement in the Epilogue to *Worlds in Collision* (Pocket Books edition): "In the middle of the second millennium before the present era, the terrestrial globe experienced two displacements; and in the eighth or seventh century before the present era, it experienced three or four more" (p. 388). A few pages earlier he states:

> We claim that the earth's orbit changed more than once and with it the length of the year; that the geographical position of the terrestrial axis and its astronomical direction changed repeatedly, and that at a recent date the polar star was in the constellation of the Great Bear. The length of the day altered; the polar regions shifted, the polar ice became displaced into moderate latitudes, and other regions moved into the polar circles.
>
> We arrived at the conclusion that electrical charges took place between Venus, Mars, and the earth when, in very close contacts, their atmospheres touched each other; that the magnetic poles of the earth became reversed only a few thousand years ago . . .(pp. 379–80).

Velikovsky, then, claims that the earth experienced *both* shiftings of its axis and displacements of its crust. In addition, "the earth was removed to an orbit farther from the sun" (*WC*, p. 132), and its diurnal motion was temporarily slowed. Altogether, the

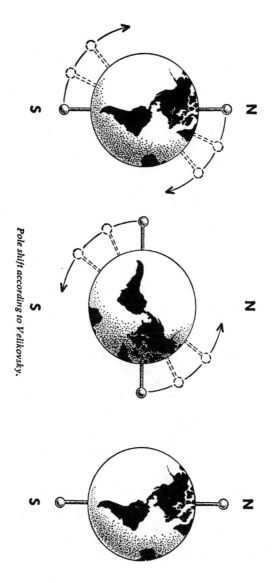

Pole shift according to Velikovsky.

cardinal points of a compass changed, the seasons changed, the climates changed and the length of the day and year changed. "It is probable," he says, "that twenty-seven centuries ago, or perhaps thirty-five, the present North Pole was at Baffin Land or close to the Boothia Felix Peninsula of the American mainland" (*WC*, p. 329). "In like manner, the old South Pole would have been roughly the same 20° from the present pole (in the direction of Queen Mary Land of the Antarctic continent)" (*WC*, p. 329).

What evidence does Velikovsky offer in support of his contention? It is enormous and varied. In the first part of *Worlds in Collision*, which deals with the Venusian encounter, he shows what the effects of a global tilting would be and then demonstrates through documentary evidence from ancient sources that this is precisely what happened. One effect would be a period of extended darkness in part of the world, extended daylight in others. Records show just that. Exodus 10:22 relates, "And there was a thick darkness in all the land of Egypt three days." Nations and tribes in many places of the globe corroborate this, Velikovsky says. In some traditions, the sun didn't shine for several days; in others, it didn't set for an equal length of time.

The Greek historian Herodotus relates his conversations with Egyptian priests of the fifth century B.C., remarking that their records assert an amazing fact: within historical ages and since Egypt became a kingdom, "four times in this period (so they told me) the sun rose contrary to his wont; twice he rose where he now rises." Several Egyptian papyri tell of "the earth turned upside down," "land upside down; happens that which never (yet) has happened," and "the south becomes north, and the earth turns over." Velikovsky interprets these to mean two radical displacements of the axis, that "the turning over of the earth is accompanied by the interchange of the south and north poles" (p. 120).

Not only do papyri tell of this, a panel on the tomb of Senmut, the architect of Queen Hatshepsut, shows it. The celestial sphere with the signs of the zodiac and other constellations of the southern sky are reversed in orientation.

Velikovsky quotes Plato's dialogue "The Statesman," in his *Politicus:* "I mean the change in the rising and the setting of the sun and other heavenly bodies, how in those times they used to set in the quarter where they now rise, and used to rise where they now set. . . . At certain periods the universe has its present circular motion, and at other periods it revolves in the reverse direction. . . . Of all the changes which take place in the heavens this reversal is the greatest and most complete." Of this reversal of the movement of the sun, Plato adds, "There is at that time great destruction of animals in general, and only a small part of the human race survives" (p. 124). That this was not a local event, Velikovsky says, is shown by the fact that virtually all peoples record this in their myths and legends. Even the Eskimos of Greenland told missionaries that in an ancient time the earth turned over.

Another effect of the Venusian encounter was interference with the earth's rotation. The rotation may have been completely reversed, since it is "referred to in the written and oral sources of many peoples" (p. 131). Later in *Worlds in Collision* Velikovsky notes that cessation of the diurnal rotation could be caused most efficiently by the earth's passing through a strong magnetic field; "eddy currents would be generated in the surface of the earth, which in turn would give rise to magnetic fields, and these, interacting with the external field, would slow down the earth or bring it to a rotational stasis" (p. 386).

How might the earth start rotating again? "If the magma inside the globe continued to rotate at a different angular velocity than the shell, it would tend to set the earth rotating slowly" (p. 386).

Thus both axis tilt and crustal slippage occurred. "This [rotational slowing] would cause friction between the various liquid or semifluid layers, creating heat; on the outermost periphery the solid layers would be torn apart, causing mountains and even continents to fall or rise" (p. 60). "...did the earth change the direction if its rotation at that time?" Velikovsky asks. "If we cannot assert this much, we can at least maintain that the earth did not remain on the same orbit, nor did its poles stay in their places, nor was the direction of the axis the same as before" (p. 128). He also admits frankly, "Whether there was a complete reversal of the cardinal points as a result of the cosmic catastrophe of the days of the Exodus, or only a substantial shift, is a problem not solved here" (p. 130).

Part II of *Worlds in Collision* deals with the Martian encounter from about 747 to 687 B.C. Here, too, Velikovsky assembles a massive amount of evidence to demonstrate that Mars moved the earth from its pivot. Among the facts indicating this are reforms of the lunar calendar, changes in the orientation of temples and obelisks (which were placed in such a way as to be celestial timekeepers), changes in the sundials and water clocks of the day, and explicit written records that indicate that the earth changed its orbital circumference (and thus the length of year), its angle of inclination to the ecliptic, the geographical position of the poles, and the velocity of its axial rotation (and thus the length of the day).

"According to Seneca," Velikovsky writes, "the Great Bear had been the polar constellation. After a cosmic upheaval shifted the sky, a star of the Little Bear became the polar star" (p. 317). Hindu sources confirm this sudden change, noting that "the earth receded from its wonted place by 100 yojanas, a yojana being five to nine miles. Thus the displacement was estimated at from 500 to 900 miles" (p. 318). Babylonian records agree: the latitude of Babylon shifted to the south by 2½° (or about 170 miles—the latitude shift not being equal everywhere).

Worlds in Collision presents evidence mainly from historical documents, celestial charts, calendars, sundials and water clocks, classical and sacred literature of the East and West, mythology and epics of the northern races, and the oral traditions of primitive peoples from Lapland to the South Seas. However, *Earth in Upheaval,* Velikovsky's 1955 sequel to *Worlds in Collision,* excludes all references to ancient literature, traditions and folklore. In answer to the outcries of his critics, who said scientific evidence was lacking, Velikovsky states at the end of his Preface, "Stones and bones are the only witnesses. Mute as they are, they will testify clearly and unequivocally. Yet dull ears and dimmed eyes will deny this evidence, and the dimmer the vision, the louder and more insistent will be the voices of protestation. This book was not written for those who swear by the *verba magistri*—the holiness of their school wisdom; and they may debate it without reading it, as well." This was to be the case.

Earth in Upheaval begins with the riddle of the frozen mammoths—and mastodons, super-bison, horses and many other extinct fauna. Citing fact after fact about the enormous numbers of animals and plants found frozen in the Arctic muck, Velikovsky asks:

> What could have caused a sudden change in the temperature of the region? Today the country does not provide food for large quadrupeds, the soil is barren and produces only moss and fungi a few months in the year; at that time the animals fed on plants. And not only mammoths pastured in northern Siberia and on the islands of the Arctic Ocean. On Kotelnoi

Island "neither trees, nor shrubs, nor bushes, exist . . . and yet the bones of elephants, rhinoceroses, buffaloes, and horses are found in this icy wilderness in numbers which defy all calculation." (Pocket Books edition, p. 6).

He notes a passage from a nineteenth-century book, *Travels in Siberia*—the passage which so impressed my memory as a young boy:

> In New Siberia [Island], on the declivities facing the south, lie hills 250 or 300 feet high, formed of driftwood, the ancient origin of which, as well as of the fossil wood in the tundras, anterior to the history of the Earth in its present state, strikes at once even the most uneducated hunters. . . . Other hills on the same island, and on Kotelnoi, which lies further to the west, are heaped up to an equal height with skeletons of pachyderms [elephants, rhinoceroses], bisons, etc., which are cemented together by frozen sand as well as by strata and veins of ice . . . (p. 7).

What, indeed, could have caused this? And what, Velikovsky asks, could have placed erratic boulders three hundred feet in circumference, and one fully three miles long and one thousand feet wide by two hundred feet thick, in isolated places tens of miles from their native bedrock? What could have crushed and ground the bones of thousands of sea animals in caves hundreds of feet above sea level? What brought glaciers to Brazil and Madagascar? corals to Spitzbergen? whales to Michigan? coal to Alaska? seashells and fish skeletons to the top of the Himalayas?

In chapter after chapter Velikovsky gathers geological and archaeological evidence to support his theory. In chapter after chapter he shows that his concept unifies many data and solves many long-standing problems of science.

Let us assume, as a working hypothesis, that under the impact of a force or the influence of an agent—and the earth does not travel in an empty universe—the axis of the earth shifted or tilted. At that moment an earthquake would make the globe shudder. Air and water would continue to move through inertia; hurricanes would sweep the earth and the seas would rush over continents, carrying gravel and sand and marine animals, and casting them on the land. Heat would be developed, rocks would melt, volcanoes would erupt, lava would flow from fissures in the ruptured ground and cover vast areas. Mountains would spring up from the plains and would travel and climb on the shoulders of other mountains, causing faults and rifts. Lakes would be tilted and emptied, rivers would change their beds; large land areas with all their inhabitants would slip under the sea. Forests would burn, and the hurricanes and wild seas would wrest them from the ground on which they grew and pile them, branch and root, in huge heaps. Seas would turn into deserts, their waters rolling away.

And if a change in the velocity of the diurnal rotation— slowing it down—should accompany the shifting of the axis, the water confined to the equatorial oceans by centrifugal force would retreat to the poles, and high tides and hurricanes would rush from pole to pole, carrying reindeer and seals to the tropics and desert lions into the Arctic, moving from the equator up to the mountain ridges of the Himalayas and down the African jungles; and crumbled rocks torn from splintering mountains would be scattered over large distances; and herds of animals would be washed from the plains of Siberia. The shifting of the axis would change the climate of every place, leaving corals in Newfoundland and elephants in Alaska, fig trees in northern Greenland and luxuriant forests in Antarctica. In the event of a rapid shift of the axis, many species and genera of animals on land and in the sea would be destroyed, and civilizations, if any, would be reduced to ruins (pp. 124–25).

"This catastrophic shifting of the axis, once or a number of times," Velikovsky declares to his critics, "is presented here only as a working hypothesis but, without exception, all its potential effects have actually taken place" (p. 126). For Velikovsky, pole shifts provide the key to solving archaeological and geological mysteries.

The evidence is . . . overwhelming that the great global catastrophes were either accompanied or caused by shifting of the terrestrial axis or by a disturbance in the diurnal and annual motions of the earth. The shifting of the axis could not have been brought about by internal causes, as the proponents of the Ice Age theory in the nineteenth century assumed it was; it must have occurred, and repeatedly, under the impact of external forces. The state of lavas with reversed magnetization, hundreds of times more intensive than the inverted terrestrial magnetic field could impart, reveals the nature of the forces that were in action. . . .

Many world-wide phenomena, for each of which the cause is vainly sought, are explained by a single cause: The sudden changes of climate, transgression of the sea, vast volcanic and seismic activities, formation of ice cover, pluvial crises, emergence of mountains and their dislocation, rising and subsidence of coasts, tilting of lakes, sedimentation, fossilization, the provenience of tropical animals and plants in polar regions, conglomerates of fossil animals of various latitudes and habitats, the extinction of species and genera, the appearance of new species, the reversal of the earth's magnetic field, and a score of other world-wide phenomena (pp. 239–40).

The mechanism that accomplished this was not simply inert celestial bodies, nor gravitation alone. Rather, it was electrically charged bodies—a point that contradicted scientific thinking of the 1940s, when Velikovsky wrote *Worlds in Collision.* In *Earth in Upheaval,* Velikovsky quotes the British cosmologist Harold Jeffreys, who says in his book *The Earth* that the earth's axis has definitely changed its angle of inclination to the plane of its orbit, and that "this can change in direction only through couples acting on the earth from outside." What could have played the role of couples, or a vise, acting from outside, Velikovsky asks rhetorically, and was it a gradual change or a sudden displacement?

He proceeds to answer his question like this: After examining various theories of pole shift, he discards the theory of the sliding lithosphere as inadequate to account for his data. He also says—in an apparent reference to Brown—that "the theory that would explain the displacement of the crust by an asymmetric growth of the polar icecaps is quantitatively indefensible; this theory uses the same phenomenon—the growing icecaps—as the cause *and* the effect of ice ages" (*EU,* p. 116). Not only must the causative agent have been more powerful than these: "Sudden the agent must have been, and violent; recurrent it must have been, but at highly erratic intervals; and it must have been of titanic power" (p. 117).

We cannot imagine any cause or agent for this, unless it be an exogenous agent, an extraterrestrial cause. For the removal of the poles from their places, or the shifting of the axis, also, only an external agent could have been responsible (p. 121).

That "extraterrestrial cause" we already know. Velikovsky explains:

The earth is itself a large magnet. A charged cloud of dust or gases [i.e., a comet], moving in relation to the earth, would be an electromagnet. An extraneous electromagnetic field that would produce a thermal effect on the earth would also shift the terrestrial axis and change the rotational velocity of the earth. This, in turn, would have a thermal effect, since the energy of motion would be converted into heat, and possibly into other forms of energy—electrical, magnetic, and chemical, as well as nuclear—with ensuing radioactivity, again with thermal effect.

An extraneous mechanical or electromagnetic force would produce both phenomena, which are prerequisites of a glacial period: the astronomical or geographical shifting of the axis and the heating of the globe (p. 122).

What about the riddle of the frozen mammoths? Velikovsky answers it thus:

The sudden extermination of mammoths was caused by a catastrophe and probably resulted from asphyxiation or electrocution. The immediately subsequent movement of the Siberian continent into the polar region is probably responsible for the preservation of the corpses.

It appears that the mammoths, along with other animals, were killed by a tempest of gases accompanied by a spontaneous lack of oxygen caused by fires raging high in the atmosphere. A few instants later their dying or dead bodies were moving into the polar circle. In a few hours northeastern America moved from the frigid zone of the polar circle into a moderate zone; northeastern Siberia moved in the opposite direction from the moderate zone to the polar circle. The present cold climate of northern Siberia started when the glacial age in Europe and America came to a sudden end (*WC*, p. 330).

Velikovsky has been criticized on many grounds, most of them irrational but some of which appear to offer cogent counterarguments to aspects of his theory.† His contention about the mammoths is one such instance. Radiocarbon dating techniques were developed in the late 1940s and were applied to geology and archaeology too late for them to be of much use to Velikovsky while writing *Earth in Upheaval*. Radiometric measurements are hardly mentioned in the book; the most significant instance is

†See, for example, Walter Sullivan's comments in *Continents in Motion* and Donald Goldsmith's *Scientists Confront Velikovsky*. (The latter is rebutted by Lewis Greenberg's *Velikovsky and Establishment Science*.)

Velikovsky's citation of plants associated with mastodons in Mexico being dated by C^{14} as only thirty-five hundred years old.

Although that dating fits remarkably well into Velikovsky's thesis, as does the Mexican mammoth bone dated 690 B.C. that I mentioned in Chapter 2, we must remember, first, that Mexico has not produced *frozen* mammoths and, second, as also noted in Chapter 2, that the many mammoth datings since *Earth in Upheaval* was written have fallen, by a great majority, far outside the time frame he proposes. The Berezovka mammoth, one of Velikovsky's prime evidences, has been dated tens of thousands of years older than the Exodus events. Walter Sullivan, a friendly critic of Velikovsky, notes in *Continents in Motion* that the dates of the Alaskan mammoths are scattered rather uniformly through the past twenty thousand years of the Wisconsin Ice Age, which is thought to have terminated ten millennia ago. "There is no hint of one or two cataclysmic events," he writes, "and all ages are far greater than proposed by Velikovsky" (p. 37).

This is a serious blow to Velikovsky—though by no means fatal, for he himself declares in *Worlds in Collision* a degree of uncertainty about the matter. "A problem the archaeologists will have to solve is that of clarifying whether the extermination of life in these regions of northwest America and northeast Asia, resulting in the deaths of mammoths, took place in the eighth and seventh or fifteenth century before the present era (or earlier)—in other words, whether the herds of mammoths were annihilated in the days of Isaiah or in the days of the Exodus" (p. 332).

This problem was taken up by Dwardu Cardona in his *Kronos* article cited in Chapter 2. There he points out that on the basis of Velikovsky's own evidence, we are forced to assume that the mammoths could only have been killed and frozen during the last cataclysm, that of the eighth and seventh centuries B.C. Yet radiocarbon dating of the Siberian mammoths does not confirm extermination at that time, or even around the 1500 B.C. cataclysm. ". . . one is forced to assume," he concludes, "that the mammoths in question could not have been the victims of the same cataclysm" (p. 83).

A possible solution was suggested to me by C. L. Ellenberger, a chemical engineer who has spent much time studying the Velikovskyan thesis. Ellenberger noted in a letter that the reliability of radiometric measurements in general are open to question because in recent years evidence centering around a geological phenomenon called radiohalos has been found that suggests the decay rate of radioactive elements may not have been invariable through the earth's history. Moreover, the evidence suggests that some radioactive materials, usually thought to have been formed when the planet originated, may instead have been formed later, even up to very recent times. And the factor that may account for all this is large electrical discharges affecting the entire earth.

Radiohalos are microscopic, ringlike discolorations observed in thin sections of certain minerals when they are examined by polarized light. They result from radioactivity emitted from tiny impurities in the rock called radionuclides and are centered in them and surround them like halos. Although it was originally assumed that radiohalos were produced by steady radioactive decay of the radionuclides, findings now indicate that the rates of decay may have been far from steady. If so, the entire field of radioactive dating would be upset.

Velikovsky, of course, proposed that the close passage of electrically charged bodies—Venus and Mars—created severe electrical disturbances in the earth. Gigantic interplanetary sparks arced between planets, altering the earth's state of electrification. In such a case, writes Ralph E. Juergens in a *Kronos* article on the subject (Fall 1977), "it would seem to follow that decay rates for radionuclides might well differ radically from today's norms" (p. 12). They would accelerate, appearing older than they really are.

Elaborating upon the possibility that radioactivity might be altered in such conditions, Ellenberger suggests, as an example, that an interplanetary thunderbolt touching down in the far north may have altered earth's potential in Siberia. "Perhaps the mammoths were exposed to the discharge in such a way that their C-14 decay rates were altered differentially according to their position with respect to the point of spark touchdown. This would explain the range of dates if the carcass distribution dated by C-14 exhibited a concentric pattern of isochrons [lines connecting events with the same date]."

This is an intriguing speculation, worthy of study. One would hope to see it, and other hypotheses based on accelerated radioactive decay rates, speedily examined. If only Siberian mammoths display early dates, while European and Mexican specimens have much later dates, an ancient local electromagnetic or electrical anomaly in Siberia becomes more plausible. Until such time as research is done, however, it must be said that on the basis of present radiometric datings, the riddle of the frozen mammoths has not been fully resolved by Velikovsky.

Despite that, it must also be acknowledged that he has made an enormous contribution to scientific thinking through his challenge to uniformitarian dogma. The Velikovsky controversy continues in the pages of scientific publications, and although the vitriolic attacks have largely diminished, the level of skepticism remains high. Velikovsky proclaimed a reigning ideology false; what followed has done much to reformulate and refine scientific perspectives about the nature of the cosmos and the record of human history. Our understanding of astronomy, geology, archaeology and mythology has been advanced through the give-and-take of the often bitter struggle. Yet Velikovsky does not declare himself to be the final authority. He acknowledges in his writing that many problems remain to be solved. And at one public meeting in 1972 he remarked to his audience, "What I have written and said is given to examination, to criticism, to variance; and I accept the verdict of facts. . . . Do not accept my work as ultimate truth."

This nondogmatic stance is characteristic of the true scientist. Even if "the verdict of facts" should not finally be given to Velikovsky, we must admit with Walter Sullivan that "Velikovsky, by any mode of measurement, is an extraordinary man."

References and Suggested Readings

Catastrophism and Ancient History, 3431 Club Drive, Los Angeles, CA 90064. A magazine devoted to discussion, pro and con, of Velikovsky's ideas.

Catastrophist Geology, Caixa Postal 41.003, Rio de Janeiro, Brazil. A magazine dedicated to study of discontinuities in the earth's history.

DeGrazia, Alfred; Juergens, Ralph; and Stecchini, Livio; eds. *The Velikovsky Affair: Scientism vs. Science*. New Hyde Park, NY: University Books, 1966.

Goldsmith, Donald, ed. *Scientists Confront Velikovsky*. Ithaca, NY: Cornell University Press, 1977.

GREENBERG, LEWIS M., ED. *Velikovsky and Establishment Science*, A special issue of the Journal *Kronos* (Vol. 3, No. 2, 1977).

Kronos, Editorial Office, Glassboro State College, Glassboro, NJ 08028. A journal of interdisciplinary synthesis devoted to Velikovskyan themes.

Pensée, P. O. Box 414, Portland, OR 97207. A series of ten issues of the journal of the Student Academic Freedom Forum of Lewis and Clark College that are devoted to the theme "Immanuel Velikovsky Reconsidered." The magazine is now defunct but available in some libraries.

RANSOM, C. J. *The Age of Velikovsky*. New York: Dell, 1978.

ROSE, LYNN, ED. *Velikovsky Reconsidered*. New York: Doubleday, 1976.

Society for Interdisciplinary Studies Review, Flat 6, Jersey House, Cotton Lane, Manchester, 20, England. A journal of interdisciplinary synthesis devoted to Velikovskyan themes.

SULLIVAN, WALTER. *Continents in Motion*. New York: McGraw-Hill, 1974.

VELIKOVSKY, IMMANUEL. *Earth in Upheaval*. New York: Pocket Books, 1977.

——. *Worlds in Collision*. New York: Pocket Books, 1977.

WARSHOFSKY, FRED. *Doomsday*. New York: Readers Digest Press, 1977.

Chapter Seven

Peter Warlow:
Geographic Reversals Explain
Geomagnetic Reversals

A note in the British publication *New Scientist* (9 November 1978) with the title "Does pole-flipping account for Earth magnetism?" first alerted me to a very important—and the most recent—report on pole shifts. The note began:

> A highly speculative theory which suggests that at many times in the past the Earth has flipped over so that the "north" and "south" poles changed place, has recently been suggested by P. Warlow, an amateur theoretician (*Journal of Physics A,* vol. 11, p. 2107)* (p. 436).

After condensing the twenty-three-page article into five paragraphs, the note ended with this editorial remark: "Catastrophists of the Velikovsky school are bound to take heart from the ap-

* The specific issue is 7 October 1978.

pearance of such ideas in a reputable scientific journal."

Velikovskyan catastrophists should indeed be glad to see Warlow's statement because he has derived, through independent research, dates in near-agreement with Velikovsky's findings. But the *New Scientist*'s editorializing shows that an anti-Velikovsky mentality is still entrenched in high places. For such people, opposing a school of thought becomes more important than an unbiased investigation. Would that they understood Dr. Linus Pauling's observation: "Science is the search for truth—it is not a game in which one tries to beat his opponent, to do harm to others."

Peter Warlow, I think it fair to say, is searching for truth about the earth's history in a way that Pauling would approve. His article, "Geomagnetic reversals?" which I will summarize below, is sober, rational yet open-minded. And although his work is independent of Velikovsky, he has come to conclusions quite similar to the gray-haired Princeton resident's. Best of all, he gives us even greater understanding of how this planet may previously have tumbled in space. Also like Velikovsky, he makes no predictions of another pole shift and is highly skeptical of those who do.

Although the *New Scientist* describes him as an amateur theoretician, Warlow is actually employed full time in a technical research position. In a 1974 letter printed in *Nature* he described himself as "a qualified and practising scientist." Moreover, he achieved honors in mathematics and physics at Keele University, in England. He is also a member of the Society for Interdisciplinary Studies, a British organization whose aim is, according to an S.I.S. brochure, "to bring a rational and objective approach to the study of Velikovsky's theories and encourage the detailed evaluation which is their due in the light of the evidence accumulating in their favour." S.I.S. has organized seminars and conferences at various universities in the United Kingdom and is building an archive of reviews, reports and other materials dealing with Velikovsky specifically and, more generally, with topics related to his investigations. It also publishes a quarterly journal, the *S.I.S. Review,** which contains scholarly and scientific articles intended to offer "reasoned speculation and informed discussion of Velikovsky's work at a general level."

Although "Geomagnetic reversals?" doesn't mention Velikov-

---

* Available from the Secretary of the Society, R. M. Amelan, 6 Jersey House, Cotton Lane, Manchester, 20, England.

sky or his books, there is an indirect reference to him. There are
also references from *Kronos*. Thus, in accepting the paper for
publication, the Institute of Physics must be acknowledged as
sharing Pauling's view that science is the search for truth. Warlow
does not emphasize the association with Velikovsky's work, presumably as a strategic ploy to get the ideas more readily accepted
by his readers, but in any case he brings together in short space
an astonishing array of geological, astronomical, mathematical and
historical data, nearly all of which are more recent than Velikovsky's. With it Warlow firmly challenges the dogma of stability
and uniformity, and explains—as his abstract puts it—that "A
reversal of the Earth itself in a particular manner is sufficient to
account for the behavior of the [geomagnetic] field in detail during
a reversal, and for explaining the links with the various other
phenomena" (p. 2107).

Warlow begins by noting that present theories of the origin of
the earth's magnetic field do not adequately explain field reversals.
Furthermore, there are baffling phenomena strongly associated
with such reversals, such as massive faunal extinctions, climatic
changes and periods of extensive volcanic activity. There are also
phenomena that, although not yet clearly linked with magnetic
field reversals, nevertheless appear to Warlow to be highly relevant. These include evidence of rapid changes in the stratification
of African lake waters, similar changes that the Black Sea has
undergone on the same time scale, major fractures in the earth's
crust and, of course, the ice ages.

A feature of many of these recent investigations is the ra-
pidity of change. Many workers are forced to conclude that
sedimentation, glaciation, climate, water temperature, polar
wander, and outbursts of volcanism occurred or changed suddenly, and often such events occurred on a worldwide scale.

The theory proposed here could account for all of the above
data and observations. Ironically, perhaps, and seemingly contradictory, one of its essential features is *not* to have a magnetic
reversal at all. Instead I propose a *geographic reversal*. Not
only can this explain the above data, but it also brings into
perspective many otherwise enigmatic archaeological, astronomical, historical and other data (p. 2108).

Warlow spends three pages explaining the concept of geographic reversals, using illustrations drawn from the motion of a

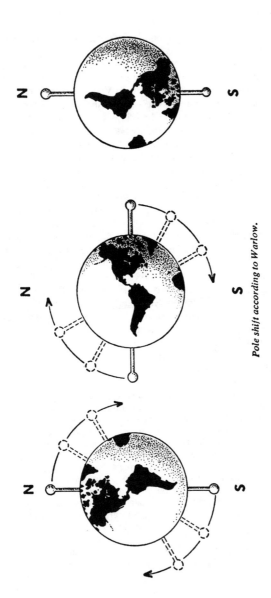

Pole shift according to Warlow.

spinning "tippe top," a child's toy. He shows that the angular momentum, or primary spin, of the rotating planet would be maintained throughout an inversion of the poles. The energy requirement for inversion, he notes, is small relative to the primary spin energy. Furthermore, primary spin can provide, under proper conditions, the initial energy needed for inverting the planet. The gyroscopic stability of the earth, he points out, is less inherently stable than that of a tippe top, although the required torque, or turning force, necessary to invert the earth is considerable. "To produce the torque required . . . we would need a large cosmic body on a near-miss course," he writes, ". . . and there is ample evidence for such near misses in the recent past" (p. 2111).

If conventional models of the planetary magnetic field are not adequate to explain reversals, what, then, might? Warlow notes that a rotating sphere with a magnetic field, if inverted by fast precession, offers a means of effectively reversing the field, "not by a direct action on the field itself, but indirectly through a geographic reversal."

> For such a reversal the primary spin continues with little disturbance and the magnetic field therefore remains more or less fixed with respect to space whilst the Earth is turned upside down so that, for an observer on the surface, there is an effective reversal of the magnetic field with respect to his frame of reference (p. 2112).

Further, Warlow observes,

> During the period of secondary rotation, as the Earth is a viscous body, the motions within the core and other layers may not all follow the general inversion at the same rate. The field will almost certainly diminish in strength but it is very unlikely to reach a zero value. We have, in fact, the exact conditions to cause the field to topple as a number of investigators have suggested, although it is actually the Earth rather than the field that moves (p. 2112).

Because the earth is not a rigid body, however, and not perfectly symmetrical, it seems to Warlow that its delicate balance need not be inverted a full 180° each time. What is likely, he writes, is that a series of events occur, each varying in the degree of completeness of the reversal, and including minor tilts, so that the planet is left rotating about different poles between events. These intermediate polar positions would then be recorded in rock magnetism, accounting for the apparent magnetic polar wandering.

At what speed might a polar reversal occur? The most recent evidence, Warlow says, indicates a time scale much shorter than 10^4 years (ten thousand years), and the results of a scientist named Mercanton (cited in one of Warlow's references) indicate that the period must be reduced to 10^2 years (one hundred years) or less. "I suggest," Warlow writes, "that the actual reversal takes place in a matter of days—even as little as one day" (p. 2114).

If a one-day reversal did take place, what might result? Warlow calculates that the maximum acceleration at the earth's surface is only about 1 cm/sec^2—about one thousandth of the force of gravity. ". . . the effects on oceans and other large bodies of water, and on the air masses would be considerable," he says. "The land masses too, resting on their viscous support, would show marked stress, but the total effect need not be as devastating as one might first imagine. Indeed, a minor tilt of, say $10-20°$ would have surprisingly limited immediate effects" (p. 2114).

The next section of the paper links geographical reversals with other phenomena. Such a motion, Warlow remarks,

> produces massive tidal waves, carrying vast quantities of debris and sediment from the ocean floors and depositing them over the land. The inevitable severe storms as the atmosphere tries to adjust to the new positioning of surface features will be augmented by ash and gas ejected by the equally inevitable volcanic activity. The continental plates themselves would be set in motion, thus giving rise to the volcanism at their boundaries, and with enough impetus to collide eventually and form mountain chains. Some fauna, large or small, could be rendered extinct, others could survive; flora might have a somewhat better chance of survival. Indeed, of all the events that take place on such an occasion, the reversal of the magnetic field is probably the least significant. Only long after the event does it become important in that it has acted as a recorder of history (p. 2115).

All these phenomena, including the magnetic reversal itself, are most economically explained by moving the whole earth. Likewise, ice ages. The center of the north polar ice cap in the recent ice ages does not coincide with the present North Pole, but, rather, is centered on Baffin Bay at about 75° N., 60° W. By treating ice ages as a simple displacement of the pole, Warlow points out, a vexing question is resolved: how is water transported from temperate and equatorial regions to polar regions in order to increase the ice caps when evidence shows that the earth was not cooled overall during ice ages? As I pointed out in Chapter 1, it takes a lot of heat to make an ice cap. Water must be evaporated in great quantities to fall as snow in the polar regions. "That heat," Warlow says, "would then prevent the necessary cooling to form the additional ice. If it be argued that the Earth for some reason enters a simple cooling phase, then water will freeze *in situ*. . . . Displacement of the polar cap requires neither heating nor cooling . . ." (p. 2116).

Although it does not require a significant change in the total quantity of ice present on earth, sea levels would nevertheless be changed—and very drastically if the tilt were only partial. This can account for the many stranded beaches known around the world, some of them hundreds of feet above present sea level. It may also explain undersea guyots and plains as much as one or two kilometers below present sea level.

Still other enigmatic phenomena are explained by the earth capsizing, according to Warlow. Among them are frozen mammoths and other animal extinctions, variation of ocean climate and reversal of ocean currents, and atmospheric flow patterns affecting ancient climates. Perhaps the most important, though it seems to be the least explored, is the relationship between magnetic reversals and cultural changes. Warlow points out:

> Kopper has demonstrated that in the evolution of man, each of the key subspecies introductions or cultural boundary horizons seems to correspond with an important magnetic reversal or excursion. The Gothenburg event appears to mark the end of the Neanderthals (p. 2118).

He then notes that data on the alignments and astronomical significance of megalithic circles such as Stonehenge indicate a shift in interest over the centuries from midsummer rising of the sun to midwinter setting of the sun—an effect that may be linked to a change of spin axis position.

Historical and legendary data cited by Warlow also support the concept of polar shifting and inversion. According to the ancient historian Herodotus, for example, the Egyptians claimed that the sun had reversed its direction four times within 341 generations, or about 11,340 years. "These four Sun-reversals within the period from about 12,000 B.C. to 450 B.C.," Warlow notes, "are in both numerical and overall temporal agreement with Dury's climatic reversals for the British Isles, and within the period covering the Gothenburg, Laschamp and Folgheraiter magnetic events" (p. 2120).

Likewise, Polynesian and Micronesian legends tell of a reversal of summer and winter. There are many such records of aberrant events, Warlow says, and these records fit a pattern—a pattern that is being revealed again in present scientific studies of the earth.

The big question in all this is: What might the trigger mechanism be for initiating a pole shift? As already noted, Warlow proposes a large cosmic body. Data show that they do in fact enter the solar system from time to time. Comets and asteroids are the primary candidates. However, he says, it may be that planets also have entered the solar system from time to time, enlarging it to the present number. Venus, for example, may be a recent addition to the solar system because ancient Hindu and Babylonian astronomical records dated to around 3000 B.C. or earlier refer to Mercury, Mars, Jupiter and Saturn, but not Venus.

Moreover, whatever the actual cosmic body or bodies may be, "it seems that electrical effects have a greater contribution than is normally recognized..." (p. 2119). If cosmic bodies carry charges, then in a near-miss encounter between two bodies of differing net charge, some discharge between them would be likely. The discharge, if concentrated on the surface, would leave a crater with debris scattered radially. The debris would perhaps be in the form of tektites, which have been artificially produced by subjecting earth materials to a beam of electrons and also by immersing cold glass spheres in the plasma jet from an electric arc. Some ancient records say explicitly that tektites, or "fire-pearls," fell from the sky.

Warlow cites R. K. Hartmann, who in 1977 stated in *Scientific American* that there is good reason to believe that the largest bodies ever to have struck a planet in the solar system were one thousand kilometers or more in diameter. Warlow does not consider actual impact situations as the trigger mechanism, but a near miss with a cosmic body one thousand kilometers in diameter would have considerable gravitational and electrical effect. In a six-page-long discussion of these effects, using technical and mathematical arguments that I will omit here, Warlow shows that the maximum value for the torque needed to completely invert the earth is on the order of 4×10^4 times greater than that for the equinoctial precession—a value that is possible if the earth were to meet a cosmic body under certain circumstances, which he describes as follows:

> We may reasonably use the Moon, then, at its present distance from the Earth and at a declination of 12°, as a datum for the comparison of torques. The postulated body passing near the Earth is very likely to pass through a declination angle of 45°, and from that position the torque necessary to bring about an inversion of the Earth would be obtained from, for example, a body of ten times the mass of the Moon at a (centre to centre) distance of about 3.3×10^4 km, or for a body of Earth mass at a distance of about 6.6×10^4 km (p. 2125).

In summarizing his paper, Warlow declares that his theory describes a type of motion for the earth that "is not only possible but probable in the event of a meeting with another cosmic body of comparable size and in close proximity for a brief period" (p. 2127). The data show, first, that such meetings can occur at relatively frequent intervals and, second, that such events have in fact occurred. Many fields of knowledge indicate this.

> Data from studies of geomagnetism, volcanism, ice ages, palaeontology, oceanography, archaeology, history, astronomy and even mythology could be explained by such events, not only individually but particularly in the otherwise puzzling interlinking and coincidences of phenomena across these diverse studies (p. 2127).

Warlow's theory provides a unifying answer for many questions existing in science. It explains the link between magnetic reversals and faunal extinctions. It also accounts for "the displaced poles at various epochs, the probable existence of preferred polar wander paths and the fact that the magnetic field does not pass through a zero state" (p. 2128), as well as periods of massive volcanism and the ice ages.

Warlow ends his paper by remarking that the idea presented in it came to him when a little toy—the tippe top—fell out of a harmlessly exploding Christmas "cracker."

> In a sense, that Christmas "present" may prove to be the key to the past. I . . . did not expect such a thing to lead to a ques-

tioning of established concepts, not only in geology but in astronomy and other fields.

Perhaps it is not just a coincidence that the world we live on, like Christmas crackers, may not be destined to end with a whimper—but with a bang (p. 2129).

Chapter Eight

Chan Thomas:
The Adam and Eve Story

Cataclysmology is the term coined by a Los Angeles pole watcher named Chan Thomas to describe his approach to the investigation of polar displacements. Cataclysmology, Thomas says, is an interdisciplinary approach that draws from a multitude of sciences — from anthropology to cosmology, with the earth sciences in between. His correlation of data between the sciences, he claims, proved the reality of pole shifts. "Not only did it verify that the events have happened," Thomas writes, "but disclosed *when* the last five cataclysms were, and what positions the shell of the Earth had been in for the last 35,000 years."

The preceding quotation comes from *The Adam and Eve Story*, a booklet (fifty-five pages long) that Thomas published in 1963. A Postlude to it, published in 1971 and paginated in continuation with the main text, brought the number of pages in his work to ninety-two. In them, Thomas gives an interpretation of the Book of Genesis that is based on the concept of polar shifting. In his view, for example, the last pole shift occurred sixty-five hundred years ago and produced a global inundation that is the basis for the biblical story of Noah's flood.

The Adam and Eve Story is like the little girl in the nursery rhyme who had a curl in the middle of her forehead. When it is good, it is very, very good, but when it is bad, well . . . We'll look at the strengths and weaknesses of Thomas's work later in this chapter. Before we do that, however, listen to the imaginative and gripping account of the next pole shift with which *The Adam and Eve Story* begins:

With a rumble so low as to be inaudible, growing, throbbing, then fuming into a thundering roar, the earthquake starts . . . only it's not like any earthquake in recorded history.

In California the mountains shake like ferns in a breeze, the mighty Pacific rears back and piles up into a mountain of water more than two miles high, then starts its race eastward.

With the force of a thousand armies the wind attacks, ripping, shredding everything in its supersonic bombardment. The unbelievable mountain of Pacific seawater follows the wind eastward, burying Los Angeles and San Francisco as if they were but grains of sand.

Nothing—but nothing—stops the relentless, overwhelming onslaught of wind and ocean.

Across the continent the thousand-mile-per-hour wind wreaks its unholy vengeance, everywhere, mercilessly, unceasingly. Every living thing is ripped into shreds while being blown across the countryside; and the earthquake leaves no place untouched. In many places the Earth's molten sub-layer breaks through and spreads a sea of white-hot liquid fire to add to the holocaust.

Within three hours the fantastic wall of water moves across the continent, burying the wind-ravaged land under two miles of seething water coast-to-coast. In a fraction of a day all vestiges of civilization are gone, and the great cities—Los Angeles, San Francisco, Dallas, New York—are nothing but legends. Barely a stone is left where millions walked just a few hours before.

A few lucky ones who manage to find shelter from the screaming wind on the lee side of Pike's Peak watch the sea of molten fire break through the quaking valleys below. The raging waters follow, piling higher and higher, steaming over the molten earth-fire, and rising almost to their feet. Only great mountains such as this one can withstand the cataclysmic on-slaught.

North America is not alone in her death throes. Central America suffers the same cannonade—wind, earth-fire and in-undation.

South America finds the Andes not high enough to stop the cataclysmic violence pounded out by nature in her berserk rage. In less than a day, Ecuador, Peru and western Brazil are shaken madly by the devastating earthquake, burned by molten earth-fire, buried under cubic miles of torrential Pacific seas, and then turned into a frozen hell. Everything freezes. Man, beast, plant and mud are all rock-hard in less than four hours.

Europe cannot escape the onslaught. The raging Atlantic piles higher and higher upon itself, following the screeching wind eastward. The Alps, Pyrenees, Ural and Scandinavian mountains are shaken and heaved even higher before the wall of water strikes.

Western Africa and the sands of the Sahara vanish in nature's wrath, under savage attack by wind and ocean. The area bounded by the Congo, South Africa and Kenya suffers only severe earthquakes and winds—no inundation. Survivors there marvel at the sun, standing still in the sky for nearly half a day.

Eastern Siberia and the Orient suffer a strange fate indeed—as though a giant subterranean scythe sweeps away the earth's foundations, accompanied by the wind in its screaming sym-phony of supersonic death and destruction. As the Arctic basin leaves its polar home, eastern Siberia, Manchuria, China and Burma are subjected to the same annihilation as South America: wind, earth-fire, inundation and freezing. Jungle animals are shredded to ribbons by the wind, piled into mountains of flesh and bone, and buried under avalanches of seawater and mud. Then comes the terrible, paralyzing cold. Not man, nor beast, nor plant, nor earth is left unfrozen in the entire eastern Asian continent, most of which remains below sea level.

East of the Urals, in Western Siberia, a few lucky people survive the fantastic winds and quakes.

Antarctica and Greenland, with their ice caps, now rotate around the earth in the Torrid Zone; and the fury of wind and inundation marches on for six days and nights. During the sixth day the oceans start to settle in their new homes, running off the high grounds.

On the seventh day the horrendous rampage is over. The Arctic ice age is ended—and a new stone age begins. The oceans—the great homogenizers—have laid down another deep layer of mud over the existing strata in the great plains, as exposed in the Grand Canyon, Painted Desert and Badlands.

The Bay of Bengal basin, just east of India, is now at the North Pole. The Pacific Ocean, just west of Peru, is at the South Pole. Greenland and Antarctica, now rotating equatorially in the Torrid Zone, find their ice caps dissolving madly in the tropical heat. Massive walls of water and ice surge toward the oceans, taking everything—from mountains to plains—in gushing, heaving paths, creating immense seasonal moraines. In less than twenty-five years the ice caps are gone, and the oceans around the world rise over two hundred feet with the new-found water. The Torrid Zone will be shrouded in a fog for generations from the enormous amounts of moisture poured into the atmosphere by the melting ice caps.

New ice caps begin to form in the new polar areas. Greenland and Antarctica emerge with verdant, tropical foliage. Australia is the new, unexplored continent in the North Temperate Zone, with only a few handfuls of survivors populating its vastness. New York lies at the bottom of the Atlantic, shattered, melted by earth-fire, and covered by unbelievable amounts of mud. Of San Francisco and Los Angeles, not a trace is left.

Egypt emerges from its Mediterranean inundation new and higher—still the land of the ages. The commonplace of our time becomes the mysterious Baalbek of the new era.

A new era! Yes, the cataclysm has done its work well. The greatest population regulator of all does once more for man what he refuses to do for himself, and drives the pitiful few who survive into a new stone age.

Once more the earth has shifted its 60-mile thick shell, with the poles moving almost to the equator in a fraction of a day. Again the atmosphere and oceans, refusing to change direction with the Earth's shell, have wiped out almost all life (pp. 3–6).

Who is Chan Thomas and where does he get his information? According to his (auto)biography in *The Adam and Eve Story*, Thomas is a geologist-engineer who attended Dartmouth College and Columbia University, receiving his degree in electrical engineering from the latter in 1943. He also claims to have become "the world's leading authority in the field of cataclysmic geology and its relationship to uniformitarian geology." A brief profile in the book states:

In 1959 he applied his findings to the possibility of earthquake prediction, and at a seminar in November 1959 issued the results of his studies. He then accurately forecast the months, years and locations of the major African and Chilean earthquakes of 1960, the Iranian earthquake of 1962, the Jugoslavian earthquake of 1963, and further predicted that California would have no major earthquakes for the following five years (p. 55).

Such accomplishments, if true, would certainly merit the title "world's leading authority," etc. However, my research has not turned up any corroborating evidence.* Moreover, Thomas gives no references that can be examined to verify his claim of having predicted earthquakes. We never learn where the 1959 seminar was held, who attended, where the proceedings are available for examination, and where the earthquake forecasts were registered to document his predictions.

This lack of documentation and proper references is—like the little girl in the nursery rhyme—one of the "horrid" aspects of *The Adam and Eve Story*. Thomas claims to be a trained scientist, yet his book fails to show the faintest awareness of proper procedures regarding publication of scientific and scholarly data. The ideas he presents are so scantily developed, so insubstantial in terms of supporting arguments and data, that the actual body of his original presentation takes only twenty-one pages!

Equally disappointing, *The Adam and Eve Story* has no bibliography. It has only one footnote, and even that is incomplete, because the page number of the work noted is not given. And although the 1971 Postlude does include a single page entitled "Recommended Reading," it consists of only sixteen book titles, including at least one of highly dubious value for pole shift researchers, *Sex and Family in the Bible*. Moreover, the sixteen book titles and their authors are the only information given. Thomas doesn't list publisher, date of publication or anything else that convention requires and that a trained scientist wishing to facilitate research for his colleagues could possibly disregard.

If that were all that is unprofessional about *The Adam and Eve Story*, we could grit our teeth and overlook it, being more anxious to consider the substance than the form of the work. But that is not the case. Thomas's interpretation of the Book of Genesis is essentially based on data drawn from the work of Hugh Auchincloss Brown, meaning that the ice caps careen the earth, except that he sides with Charles Hapgood on the question of whether

---

*This included many attempts at personal communication, including a special visit to Thomas's home in California, but he steadfastly refused any contact.

the crust slips over the molten interior or the planet as a whole tumbles in space. However, Thomas acknowledges this in just the slightest way. The only mention of their names occurs in a short introductory paragraph, along with Velikovsky's. Thomas does not even include Brown, Hapgood and Velikovsky in his recommended reading list.

An article in the now-defunct tabloid newspaper *National Tattler* (10 March 1974) entitled "Expert Gives Cataclysm Cycle Theory: We'll All Perish on One Day in Year 2000" reveals that Brown was the source of much of Thomas's data. According to the *Tattler*, "Thomas unearthed the theory, which dates to the 18th century, in 1949 by reading a text by geologist Hugh Brown." Thomas then admitted to the reporter, "It explained every anomaly in geology I'd ever heard of. It so intrigued me, it wouldn't leave me alone. I decided this field was so challenging that it needed verification or refutation once and for all, let the chips fall where they may."

Thomas's failure to acknowledge the major sources of his knowledge—Hugh Auchincloss Brown and Charles Hapgood— is a significant violation of scientific protocol. At the very least, it is most ungrateful to the men who pioneered this investigation, and shows why the author profile describing Thomas as "the world's leading authority in the field of cataclysmic geology and its relationship to uniformitarian geology" is unwarranted self-glorification.

At this point you may well be asking yourself why I have included Thomas in this book at all, if he is merely a secondary source in pole shift literature. My answer is this: the work is greater than the man. Despite major flaws in *The Adam and Eve Story*, Thomas must be acknowledged for making an original contribution to pole shift theory through his identification of a possible trigger mechanism. It bears deepest consideration.

In *The Adam and Eve Story* prior to the Postlude, Thomas wrote:

> Now what about the *trigger?* This turned out to be the most elusive piece of the whole puzzle. We couldn't rely on some supernatural explanation—like sometime happenings in the heavens of a vague character which actually violated the laws of nature; no, it had to be something natural, a part of nature's ordinary structure, which disrupts the Earth's inner electrical and magnetic structures whenever it happens. . . .

We found out that... apparently once every few thousand years neutral matter escapes from the 860-mile-radius inner core into the 1300-mile thick molten outer core, and there is a literal atomic explosion inside the Earth. The explosion in the high energy layer of the outer core disrupts completely the electrical and magnetic structure in both the molten outer core and the outer 60-mile thick molten layer. Finally the ice caps are allowed to pull the shell of the Earth around the interior, with the shallow molten layer lubricating the shift all the way (pp. 13–14).

Atomic explosions in the core of the planet did not satisfy Thomas for long as the trigger mechanism. It is to his credit that he revised his views in the Postlude. On page 82 he writes:

As the years went by and we remained dissatisfied with our concepts concerning the trigger, we concentrated on that part of the puzzle. It has taken almost twenty years to find a satisfactory solution—one which answers all of the facts.

That solution, Thomas says, comes from the work of the Swedish physicist Hannes Alfvén, who discovered the effects of a combination of magnetic, electrical and physical forces that he termed "magnetohydrodynamic" energy—MHD, for short.

The earth, we have seen, is not a solid body. Beneath the solid crust is a layer that is plastic, semisolid. That is, it has qualities of both a liquid and a solid. Like a liquid, it can flow under steady pressure, and like a solid, it can shear under sudden pressure.

Thomas identifies the magnetohydrodynamic energy of the earth, when affected by "galactic-scale null zones" of "zero magnetic energy," as the means whereby the magma layer changes its properties in the direction of greater liquidity and freer flow. These null zones exist between concentric spheres of magnetic energy that fill the galaxy, presumably from the center of the galaxy. Their effect on the earth's magnetohydrodynamic energy in turn makes the magma layer a lubricant for the solid crust, allowing it to slip around as it is pulled by the mass of the ice cap.

We'll examine the notion of "galactic-scale magnetic null zones" shortly, but first consider magnetohydrodynamic energy. Here is the example Thomas gives to explain the effects of magnetohydrodynamic energy.

Take a glass cylinder of mercury at room temperature—so that it is molten as it is in your thermometer—and float a mirror on top of it. (First put some scratches in the mirror, so that when you shine a light on it, the reflection on the ceiling will show images of the scratches.)

Next put an agitator—a miniature version of a washing machine—in the bottom of the glass cylinder and extend the shaft or axle of the agitator through the bottom, so that you can put a handle on it.

Now you are ready to agitate the mercury to see what happens. This is what Alfvén did, in a more refined way, in his laboratory. The results? The mirror continued to float placidly without motion, even though the agitator was moving. This happens because the mercury has a molecular structure that allows the motion caused by the agitator to be absorbed before it reaches the surface.

However, if you wind a wire around the glass cylinder and connect it to a battery—thus making the apparatus an electromagnet—things change dramatically. Now the mercury, affected by the magnetic field generated from the wire and battery, undergoes a change. It loses its internal "slipperiness," becoming nearly solid. The mirror, in turn, moves as the agitator does. Thomas writes:

> Hannes Alfven found that he had discovered the existence of a kind of energy, traveling from the agitator to the mirror, which was previously undetected by any scientist. His rigorous mathematical work in expanding James Clerk Maxwell's three ingenious equations for expressing electromagnetic radiation . . . showed that there were electrical, magnetic and physical force fields acting as one between the agitator and mirror.
>
> Alfven expanded his mathematical research to show that space is literally a sea of MHD energy, and that, as weak as the magnetic field of any blue-white star is, it is strong enough to support an internal MHD energy structure within the star.
>
> Alfven's work also applies to any planet with an organized magnetic-field—that is, with one North and one South pole. Its field is strong enough to support an MHD energy structure in the planet. Moreover, I have built several earth current measuring stations, and know from personal observation that the corresponding electric currents in the earth are strong enough to support our planet's inner MHD energy structure (pp. 84–85).

* * *

The dynamic balance of the components of the magnetohydro-dynamic energy structure is what locks the earth's crust with the underlying magma, preventing slippage, Thomas says. That is because the magma is kept magnetohydrodynamically in a near-solid state. However, the potential for becoming like a liquid is inherent in the planet's balance of forces, and when it does, Thomas told the *Tattler,* "it acts as a lubricant for the shell of the earth to find a new dynamic balance around the interior. The ice caps, which have been forming for thousands of years, pull the shell around to where they are at the equator or near it. It happens in less than a day. It's a real hellraiser!"

Thomas postulates the following in the Postlude:

> . . . at the time of a cataclysm the entire solar system passes through a magnetic null in the Milky Way galaxy. These nulls are sometimes popularly called "reversals." Some physicists are beginning to suspect we are heading into another null zone at an accelerating rate. In any case, when going through a null, our planet's inner MHD energy structure is diminished to the extent that the outer, shallow molten layer is allowed to act as a free liquid. No longer does it bind the shell to the Earth's interior. . . .
>
> The trigger, then, is our planet's passage through a galactic-scale null zone, diminishing the earth's inner MHD energy to so low a level that the shallow molten layer is allowed to act as a liquid lubricating layer between the Earth's shell and interior.
>
> During each cataclysm the shell finds its new dynamic balance, which is resolved when the shell has shifted to a position with the ice caps rotating equatorially and melting in the heat of the Torrid Zone. As they melt relatively fast, and they usually total around eight million miles of ice (as they do today), the oceans the world over rise about 200 feet with the new-found water.
>
> New polar ice caps form on the areas moved into the polar regions; they will not be centered with the axis of rotation, so a new, growing imbalance is created, to be resolved when the Earth, with the entire solar system, passes through another null zone (pp. 85–86).

The existence of "galactic-scale magnetic null zones" is, so far as I know, entirely without scientific basis. There is not the slight-est evidence for them. It is true that space probes apparently have discovered a region near Mars in which electromagnetic communications are disrupted. This loss of communications probably originates from meteorite hits on the spacecraft, rather than from some inherent properties of space itself. Since the matter is not well understood, it has given rise among space scientists to the fantastic notion—only jokingly accepted—of a "great galactic ghoul" that gobbles up spacecraft.

This fact, however, cannot be invoked in support of Thomas's concept of magnetic null zones, because, first, the region was found *after* Thomas asserted the null zones as fact and, second, the zone in which communications are disrupted is apparently a stable feature of the solar system, traveling through space along with the planets, rather than remaining in a fixed location as the solar system passes through it.

It would be quite different if Thomas had simply *postulated* the existence of magnetic null zones, but he doesn't—he asserts them as fact. To compound matters, he casually mentions that some physicists, unnamed and unreferenced, suspect that we are heading into another null zone *at an accelerating rate*. That statement is pseudo-scientific because it means that the earth's speed through the galaxy is increasing—another "fact" that science knows nothing about. These points are still more examples of the "horrid" aspects of Thomas's pole shift theory, and therefore we have to recognize a large nonsense factor operating in *The Adam and Eve Story*.

Although I am describing an aspect of Thomas's theory as ridiculous, my purpose is not to ridicule. Rather, it is to examine critically all the evidence pointing to the possibility of the ultimate disaster. And therefore I must say, in fairness to Thomas, that despite great gaps in his argument, he seems to have identified a factor—magnetohydrodynamic energy—that could provide one of the missing pieces in our attempt to see whether a plausible theory of pole shift can be—or has been—constructed. We can reject the idea of magnetic null zones through which the earth passes but retain the insight that a change in the earth's magnetohydrodynamic energy structure could provide the conditions whereby crustal slippage might occur. As I will show later, another factor may more reasonably be invoked to allow magnetohydrodynamic alteration, giving the same results as Thomas describes.

An article about Thomas entitled "The End of the World" appeared in *Saga* magazine (May 1970) in which the author, Peter Gutilla, asked Thomas when the next cataclysm will happen.

Thomas replied, "We must first face the fact that a cataclysm is a normal part of the earth's life cycle.... It has happened about 300 times in the past and will happen about 300 times more before our solar system enters the deep sleep of being reborn." Then he answered Gutilla specifically:

> Sometime between 30 and 500 years from now.... There are signs indicating the approach of a magnetic null zone at a rapidly accelerated rate.... With adequate funding we could mathematically tie down the *exact* time and make suitable preparations for it....
>
> When it does happen, we will have an Adam and Eve story similar to that of 11,500 years ago, and a Noah story similar to that of 6,500 years ago also.
>
> The survivors will be driven into another Stone Age like the Old Stone Age of 11,500 years ago, and the New Stone Age of 6,500 years ago which followed the last two cataclysms....

From Thomas's interview with the *Tattler* we learn additionally:

> Less than one percent of all life survives. What do these people have? Neither a pencil to write with nor a shovel to dig with. Their clothing lasts only three months, so they're forced to use stone tools, skins for clothes, caves for living.
>
> They are intelligent enough to do what it takes to survive by going back to a Stone Age to survive. It's their offspring that they don't have time to educate who get stupid. And their grandchildren and their grandchildren's grandchildren.

When the next cataclysm occurs, Thomas remarked to the *Tattler* reporter, the Indian Ocean will become the North Pole. "The South Pole will be situated off the coast of Peru. A few lucky people will be high up on Pike's Peak on the eastern side when it happens. They'll survive the holocaust. The highest survival probability area is a triangle in Southeast Asia. That area will rotate north and east. It won't have its environment changed much."

In another one of those "horrids" that compromise so much of *The Adam and Eve Story,* Thomas asserts that he has discovered the "half-life" of the universe. Such a stupendous assertion, coming from someone who knows how to observe the canons of science, would be presented fully and documented with sufficient references to establish credibility in the scientific community. Not so

Thomas's. Rather, he offhandedly mentions (p. 87) that "Hale's mathematics shows the null zone vs. time structure to be helicoid; and, as the universe approaches its half-life point, cataclysms occur at an increasing frequency, with time periods between them increasing in a mirror-image pattern of the first half-life of the universe."

Thomas then presents the following chart, adding, "Of course, there were many eras preceding the Wisconsin era, and there will be an equal number following the Unknown-area era." In other

North Polar Eras (Areas at N. Pole)	Start	End	Duration (Years)
	(Years to and from now)		
Unknown	10,530	25,280	14,750
Bay of Bengal	30	10,530	10,500
Arctic Ocean	6,970	30	7,000
Sudan Basin	11,520	6,970	4,550
Hudson Bay	18,520	11,520	7,000
Caspian Sea	29,020	18,520	10,500
Wisconsin	43,770	29,020	14,750

words, our era—the Arctic Ocean era—will be followed by the Bay of Bengal era, lasting 10,500 years, which in turn will be followed by another (Unknown) era, lasting 14,750 years.

Thomas ends his book on the note that many civilizations before ours—whose existence he supports throughout the book with a meager handful of evidence—have achieved even higher levels of technological development than the present. And yet they perished.

If we look at our technical accomplishments—which have taken us slightly over 6,900 years to achieve—think what we could do if we had 10,500 years. We would be in space as commonly as we walk around the block. Fossil fuels would belong to the dead past; controlled gravity and natural magnetism would be the means for propulsion and power generation, as we would have learned the processes of nature sufficiently to duplicate them in controlled fashion for our uses.

It appears from the legends passed on to us of the Caspian era that man did just that with the 10,500 years he had. Valmiki writes of vimanas and space chariots, of the Brahma Weapon and Indra's Dart, of "celestial chariots" and more. Legends of Mu and Atlantis, of great technical achievements . . . spring from this era. Some of the legends carry over into the 7,000-year Hudson Bay era, showing some retention of knowledge through the cataclysms of 18,500 years ago (pp. 88–89).

Thomas's final thought, as expressed to Peter Gutilla in the *Saga* magazine article, is this: "We are on the brink of realizing a Golden Era, and may be stopped cold by the next cataclysm before our era can mature. . . . It need not be so. What are we willing to do to prepare for the next one?"

Unfortunately for the human race, Thomas has nothing to say about what specifically might be done to prepare, and we are left with only the image of those previous supertechnological civilizations that—for all their high development—could not prevent or avoid the ultimate disaster.

Of all the pole watchers, Thomas is surely the most pessimistic. Let us simply take from him what is useful—some of the data indicating eras of high culture before ours, and the identification of magnetohydrodynamic energy as crucial to the trigger mechanism—and seek a more hopeful view of our situation.

Our next two pole shift theorists, however, are hardly any more optimistic than Thomas. First we will meet Adam Barber, then Emil Sepic. Both are charter members in the latter-day-prophets-of-gloom-and-doom club.

References and Suggested Readings

GUNTHER, MARTY. "Expert Gives Cataclysm Cycle Theory," *National Tattler*, 10 March 1974.

GUTILLA, PETER. "The End of the World," *Saga*, May 1970.

ROSENBERGER, JOSEPH. "Earth's Coming Cataclysm," *Fate*, September 1973.

THOMAS, CHAN. *The Adam and Eve Story*. P. O. Box 45154, Los Angeles, CA 90045: Emerson House, 1965.

Chapter Nine

Adam Barber:
The Coming Disaster
Worse Than the H-Bomb

> ... the next shift causing a great flood will occur on December 21st or June 21st of any year within the next fifty years—maybe next year—any year within fifty years. There is some danger, however, that it might occur on any other day within the fifty years. It must shift on June 21st or December 21st within that time, but the gyroscopic pressure could possibly be sufficient on any intermittent [sic] day to cause the shift, and the risk is so great that the public should be on the alert at all times ... (p. 8).

This prediction, made in 1955, came from a Washington, D.C., attorney named Adam D. Barber, who published that year a very short work entitled *The Coming Disaster Worse Than the H-Bomb*. Only thirty-seven pages long in its first edition, Barber printed it privately and distributed it himself in a crude form that was typewritten, not typeset, and had newspaper articles produced in their original form as he had clipped them from newspaper pages. The book's physical appearance did not inspire credibility.

Barber claimed a print run of five thousand copies for the first edition of what he called the "Disaster Book." Shortly thereafter he revised the text and printed another five thousand copies. In 1956 he added two dozen pages more, making a second edition of his book (although he called it the third, counting the second printing as an edition) and printed another five thousand copies. (The 1956 edition is the source of most of the information given here.) By 1957 the book had expanded, in its "fourth" edition, to ninety-five pages and a print run of fifteen thousand, and Barber had given up his practice of law to devote full time to "this flood matter."

In the opening pages, Barber states that he sent copies of the first edition to "every congressman, senator, member of the cabinet, president and other government agencies," as well as to "every observatory in the United States and many laboratories and colleges." With later editions he included the state boards of education throughout the country, libraries and high schools, as well as the National Academy of Science and the United Nations. He also advertised it in various magazines and newspapers such as *Fate, Psychic Observer,* and the *New York Enquirer.*

Adam Barber's book has long been out of print and nearly unobtainable. Barber himself died about 1963. His prediction of a pole shift is now half through the fifty-year time period he calculated. And while it may come true during the second half of that period, his most dramatic statement, given in a one-page bulletin entitled "Flash!!! Warning!!!" has proved wrong. Shortly after the book was published, an article in the *Los Angeles Examiner* about the earth's geographic equator being as much as three degrees out of alignment with the geomagnetic equator led Barber to state:

> This brings a precise confirmation of my theory that the gyroscopic forces that will cause the shift are increasing in intensity daily, and as soon as these become as great as the power of the north and south magnetic fields that hold the axis in line, then the great shift and flood will occur. Whether this will be next month, or one year, 2, 5, 10 or 20 years from now I cannot ascertain nor calculate, but from the astronomical data at hand *I am certain it will be very soon* [emphasis added].

A year later, Barber declared in News Bulletin No. 3, issued by the Barber Scientific Foundation (a one-man organization—Barber himself—with only a post office box), in Washington, D.C.: "Recent events should convince everyone that our time is running short—possibly only a year left, perhaps 2, 3, 5 or 10, but it would not surprise me if the shift and flood came within four or five years. In fact, I would be surprised if it holds off for more than two years." In May 1958 the Barber Scientific Foundation issued News Bulletin No. 5 (apparently the last published word from Barber), "Live Through or Die in the Great Flood," which declared, "I will be surprised if we do not have the shift and flood within five years—possibly within a year or two."

Barber's urgent warnings were heeded by very few. Some readers wrote in to thank him and to say they were either building boats (which he recommended) or heading for high ground. Others, such as J. H. Whale, secretary for the Royal Greenwich Observatory, in England, commented tersely to a supporter of Barber, "... you should have more confidence in your professional astronomers and less in Mr. Barber." Likewise, an astronomer at the Mount Wilson and Palomar Observatories, writing on official letterhead (which Barber reproduced in the addenda to his book) said even more bluntly, "... it sounds like one of the numerous prophecies of doom circulated by people without scientific training. There is no danger of a catastrophic 'twist' of the axis of the earth ... for many generations to come, so that you do not have to worry about it. You will find, in the Public Library, the *Encyclopaedia Britannica;* it contains many good articles on astronomical subjects; read them; do not read booklets by Barber." And in a follow-up letter later that year (1956), again on Mount Wilson and Palomar Observatories letterhead (and again reproduced by Barber), an astronomer whose name was deleted—apparently the same one as before, however—gave what was intended to be the ultimate refutation of Barber:

> All the observatories receive, all the time, dozens of books or pamphlets per year dealing with "great discoveries," coming disasters, flying saucers, etc. These books and pamphlets are written by people who are not familiar with science, who have read too much science fiction, who are suffering from slight or severe mental disturbances. The pamphlet enclosed with your letter was written not by a gyroscope expert but by a man without scientific background who is trying to scare people with

his prophecies of doom; his arithmetical juggling with "key numbers" is absolutely meaningless (p. 6P).

The amateurish appearance of *The Coming Disaster Worse Than the H-Bomb*—violating nearly every principle of scientific publication—makes Chan Thomas's work look positively professional by comparison. The "Disaster Book" is disjointed and totally lacking in references and proper citations—a mélange of scientific, quasi-scientific, pseudoscientific and theological data. (In Chapter 16 we will examine the biblical data that Barber presents.) For example, the cover of *The Coming Disaster Worse Than the H-Bomb* proclaims that it is "astronomically, geologically and scientifically proven." News Bulletin No. 5 also declares that the book contains "the astronomic, gyroscopic, mathematical and geological proofs, written plainly." Yet when one reads this statement (p. 29), "Geological proof that the earth has suddenly shifted many times in ages past is so abundant I do not consider it necessary to further elaborate on it," the credibility of Adam D. Barber sinks very low.

It may therefore seem pointless to examine *The Coming Disaster Worse Than the H-Bomb*. But let us not take the unreasonable position of the Mount Wilson astronomer who in his letter said, while passing judgment, "I have not seen the booklet by Adam D. Barber. . . ." This is *prima facie* evidence of prejudice. My purpose here is to examine in a comprehensive fashion predictions and prophecies of a pole shift. It may be that Barber is—to use an apropos expression—all wet, but if there is the slightest truth in what he has to say, an out-of-hand dismissal is both foolish and dangerous.

There are two bases for Barber's position. First, he acknowledges Hugh Auchincloss Brown and Immanuel Velikovsky as the sources of his introduction to the pole shift concept. Second, Barber claims (p. 7) to have spent twenty years studying the nature and operation of gyroscopes in order "to reproduce the action of the earth around the sun." Thus, a "gyroscope technician by hobby" made and remade an apparatus—about seventy-five times, Barber says—until, "somewhat by accident," he finally achieved the desired result. "From there on it was merely a matter of six months of intensive mathematical and astronomical calculations. . . ."

When the work was done, Barber believed he had discovered three basic new facts of astronomy. First, as the earth orbits the sun, it makes a second, small orbit of 9,095,621.106 miles along the larger orbit, taking about the same time to complete both. Second, when the axis of the earth comes at right angles to the planes of the large and small orbits simultaneously, "the earth makes a sudden shift of about 135°." This happens about every nine thousand years. Third, the axis shift is caused primarily by "gyroscopic pressure at right angles to the orbits of the earth."

> I do not claim it to be a new discovery that the axis of the earth and earth make periodical sudden shifts, as geologists have written on that subject years ago, touching glacial periods, the ice caps, etc. I *do* claim as new, however, my discovery that these shifts are caused by the gyroscopic action of the earth, and that the periods of their occurrence may be calculated and thus foretold with reasonable accuracy. Likewise, I claim, and have been informed by astronomers* that my discovery of the "small orbit" of the earth, and the calculations relating to it and other matters, are entirely new (p. 6A).

Barber says that his theory depends entirely on whether these claimed discoveries are so. The demonstrations he offers in support of his claims involve complicated mathematical computations—which Barber often carried out to fifteen decimal places!—that simply cannot be reproduced here or even summarized. I will, however, try to quote pertinent sections.

Regarding the presence of a small, secondary orbit imposed on the primary annual orbit of our planet (and, Barber adds, all other planets), this explains why the earth is nearer the sun in December than in June.

> In other words, the large orbit is practically a true (and possibly exactly true) circle, with the sun off center by about 3,000,000 miles. The distance to the sun at 91,000,000 miles

*The astronomers are never named, but Barber quotes a letter from Charles B. Smallwood, who was for twelve years an assistant to Dr. Vannevar Bush, president of the Carnegie Institution of Washington and a member of the team that developed the atomic bomb. Smallwood writes, "Being familiar with gyroscopic action and some astronomy, the treatise of Mr. A. D. Barber on the shift of the earth and coming flood interested me greatly. Anyone going over his calculations and having a gyroscopic demonstration made should be convinced that his theory of the shift, flood and small orbit of the earth is correct."

in round numbers is so great that when the sun is viewed at a mere difference of 3,000,000 miles, there is no perceptible variation to the naked eye in the size of the sun's disc.

Astronomers have never discovered this small orbit, but adhere to the centuries-old theory that the three million miles difference in distance to the sun at different periods of the year is caused by gravitational pull of the moon, sun, planets and other heavenly bodies, and that the orbit of the earth is an ellipse . . . with the sun at the large end. My view is that it is a true circle, eccentric . . . as to the sun, and which is proved by the figures hereafter shown (p. 12).

As for Barber's second claim, he states that astronomers say that about every twenty-five thousand years the axis of the earth describes a cone of about 45°—this is the precession of the equinoxes—but "my calculations show this period to be every 36,119.51 years, and the cone to be 45° 44′ 48″ or twice the approximate 23.5° slant of the axis. Astronomy books do not, however, accredit this 45° cone to a small orbit or to the loss or lag of one circuit of the small orbit by the earth every 36,000 years" (p. 14). Of this lag Barber says:

> The earth lacks 251 miles each year of completing its small orbit of 9,000,000 [miles], thus requiring 36,000 years to complete the loss of one circuit. Ninety degrees is one-fourth of a circle, so as the shift takes the axis away from dead center by 90°, then the shift occurs at one-fourth of 36,000 years, which is 9,000 years (p. 12).

The next shift will occur on December 21 or June 21 of any of the next fifty years, which Barber calculates as "a sure range," because the axis of the earth can assume right angles to both the large and small orbits only on those days, provided the earth is nearest or farthest away from the sun on such a day. This occurs once at the end of every 9,029.87 years, according to Barber's figures. And after the next pole shift, the sun will rise in the west and set in the east for another 9,000 years.

Several times Barber emphasizes that examination of his discoveries must involve use of a gyroscope. "It is absolutely impossible to explain this complicated process in words alone so anyone can understand it," he states (p. 20). "A model is necessary so that one may see the gyroscopic action and direction of forces." Since I cannot include a gyroscope in this book, and since I cannot summarize the mathematics Barber presents, I must reluctantly pass over the main body of proofs in Barber's argument. However, let us note his statement (p. 18) that the key to understanding pole shifts is the mechanism behind the precession of the equinoxes:

"Books do not explain why the axis changes. This is the crux of all my discoveries. It takes the earth 672 seconds longer to make the small orbit than the large orbit, which 672 seconds is the time it takes the earth to travel the 251.82024 [annually uncompleted] miles of the small orbit. This 251 miles causes the 57.823422 fog cone* to be described by the axis at the pole about each year, and to describe the cone of 45° 44′ 48″ each 36,119.51 years as to the large orbit." Barber adds, "A mechanical model is almost absolutely necessary in order to fully comprehend this strange physic [sic] result, but when thus demonstrated becomes perfectly clear."

In a section entitled "The Shift, the Wave and the Flood," Barber describes how the shift will take place. First he gives a theoretical description, saying that the axis will pivot, at the center of the earth, from a plane at right angles to the two orbits to one that will be parallel with the large orbit. The new position of the North Pole will "most likely be on about the 45th parallel of latitude as now known." But new lines of latitude will be established after the shift, because, as the axis is shifting, the planet is also shifting away from the established poles. ". . . the axis changes with reference to its position as to the orbits and the earth changes position with reference to the poles" (p. 8).

After the theoretical description, Barber dramatizes the situation by giving an imaginative account of what might happen during the pole shift. Likening the oceans to water in a pan, which remains still when the pan moves suddenly, Barber gives this statement of "the coming disaster worse than the H-bomb."

> The shift will commence very gently and a person sitting quietly will merely feel a surge such as when a train starts. The sensation of the surge will continue for perhaps five minutes, during all of which time the earth is accelerating in the velocity of the shift. . . .
>
> A person residing at the seashore will have about three or four minutes after the warning bell within which to get in his

*The diameter, at the earth's surface, of the cone the North Pole describes.

boat. At 1,000 miles inland, perhaps a half-hour or hour. While the person at the seashore will be safe in any ordinary boat, the one inland must have a very sturdy and especially-shaped boat, saucer-shaped, to withstand the thrust of the oncoming wave, racing at about 2,000 mph.

The wave will continue over the land for about 3,000 miles, so if one were in the center of a continent 6,000 miles across, he would find himself undisturbed.

Depending on which side of the earth one is on at the commencement of the shift, the water will appear to be flowing north or south and in 15 or 20 minutes will appear to flow east and west. This in on account of the peculiar curve the axis makes during the shift and the corresponding change of direction of the gyroscopic force.

Let us visualize the flood as from an airplane a mile high. The shift is starting; the pilot of the plane notices no change as the air is not noticeably disturbed. He is on the alert, however, and notices a commotion in Times Square, New York. Warning bells have been ringing for a quarter of a minute; vehicular traffic has come to a standstill; everyone is hurrying to the flood boats which for a few years had been resting at street corners and on tops of buildings.

The water is beginning to creep over the streets and in three or four minutes automobiles are completely submerged. In another few minutes, the shift gaining in velocity, the water is forty feet deep, having the appearance of rushing like mad through the city. In fact, however, the earth is moving and the water is practically still.

A large portion of the people are safely in boats, now floating a few miles from the mooring places. Many people refused to heed the warning bells, would not abandon their autos, and perished.

Buildings now topple and the boats with watertight covering which had been on the roofs are fully loaded with people and land in the water with a great splash, jarring some almost into insensibility. The water is soon 400 feet deep, and the Empire State building topples over with a splash that can be seen from a plane 100 miles away. What was a bustling city is now a mass of destroyed buildings and debris under the water.

Inland 100 miles, a great wall of water, almost vertical, 75 feet high, rushes toward a city, striking it with a terrific force and in one blow scatters buildings and skyscrapers like so many match boxes.

At this inland city people had ample warning and are snugly in round boats or "floating saucers" with sturdy watertight tops withstanding a pressure of 100 feet under water. These are all moored behind concrete abutments so as not to receive a sudden jar from the onrushing wave. They quickly come to the surface, bumping turned over buildings on the way, and all is quite well—thanks to a rubber bumper ring around the boat, three feet thick, made of old automobile tires.

The water on the surface is fairly quiet, although the earth below is sliding under it at more than 2,000 miles per hour. No one dares open a door or window as the mountain in the distance may in a few moments cause a swell that might capsize the boat.

Oxygen tanks keep the air in the boats from becoming unbearable.

About one hour after it started, the shift has stopped. The peaks no longer come rushing toward us; instead the water begins to flow away from them. It is now safe to open the windows and the mental tension is somewhat eased.

The earth and the axis have shifted. In half an hour a broadcast from a mother boat not far away tells that astronomers announce the new location of what was New York is now 3,000 miles closer to the equator. We have drifted with the water 500 miles, so that places us only 500 miles closer to the equator and 2,500 miles from the new latitude line that went through our town. This, of course, means the new equator.

Ten days later the water has found its gravitational level. During this ten days every person by walkie-talkie radio has sent his serial and boat number to the mother ship, with the serial number of any near relative who was unable to get on the family boat. These were all tabulated and published, the small boats going to the mother ship for copies. Boats came to rest on land, high and dry, in long rows, in numerical order strictly according to instructions previously radioed. With the printed lists, families quite readily found missing members and within a few weeks more are united.

Boats on the other side of the world which had been moored on land are now 2,500 miles out at sea, and it is several weeks before they reach land.

It will be nearly two years before any substantial crops can be raised in our new location, on account of the salt water injuring the grounds, and during that time food will consist of Agriculture Department produce stored on the mother ships—wheat, rice, dried eggs, beans, etc. and fish.

In Michigan a great ravine, rivaling the Grand Canyon of the Colorado, is washed out.

New York is now under 100 feet of mud and debris and is

uninhabitable. Some immense spaces that had been ocean bottom are now wastes of desert. This is on account of the fact that the earth is flat at the poles and eighteen miles greater in diameter at the equator than from pole to pole.

Ocean bottoms near the equator are now 3,000 miles away from it and are still deeper. Land to the other side of the equator becomes ocean over the new line of equator.

Slowly, over many years, from the centrifugal force caused by the rotation of the earth on its axis, the poles will again be flattened and some of the land under water at the new equator will reappear.

The dense jungles of South America and Africa are covered over with hundreds of feet of gravel, sand and soil, thousands of years later to be mined as coal by the people then living, just as we today mine coal in Pennsylvania, which was once covered with tropical forests and became the present coal beds as a result of a shift of the earth and flood about 27,000 years ago.

The ice caps at what had been the poles begin breaking up and melting, causing great glaciers similar to those that swept the northern United States as a result of previous shifts of the earth.

On each of the two opposite sides of the earth, midway between the old poles and the new equator, there is a spot about 500 miles in diameter which is not materially affected by the shift. One is at the interior of China and the other in the Atlantic Ocean.

A great but peaceful chaos follows. A one-world government which had previously been organized by the United Nations for this catastrophe, supervises the allocation of lands and resources to all alike, joining those of the same tongue as nearly as possible.

To return people to the lands they previously occupied is impossible as in many instances that is no longer habitable.

As there was only a few minutes' warning of the flood, nearly all prisoners perished in it.

Now the greatest struggle for survival is on. There is no time for quarrels. The full manpower and resources are necessary for a rehabilitation. Those engaging in war or serious quarrels are promptly put to death.

The mother ships have been stored amply with modern machinery and scientific apparatus, and it is estimated that in forty years little effect of the flood will be felt.

As the flood is over, only two classes of people remain— dead ones who depended on the biblical story of the rainbow and live ones who heeded the warning of science (pp. 9–12).

Barber shows both great imagination and great naïveté in this scenario. The latter can be seen in the planning and execution of pole shift preparations that would horrify civil defense officials. Imagine how many boats would be found operable after sitting for several years, completely outfitted with food and equipment, on a street corner in New York City. Imagine how many would be found at all!

Barber's credibility is reduced even further by errors of fact such as the statement that Pennsylvania's coal fields were formed twenty-seven thousand years ago.

Last of all, when Barber suggests that the flood can be prevented altogether by "deflecting the axis of the earth away from 'dead centers'" by constructing two "great atomic jets"—one on each side of the earth—the situation begins to sound about as believable as a fairy tale.

Unrealistic as these matters may be, they are secondary to Barber's major statement of having discovered a small orbit to the earth and having determined that this orbit is the cause of pole shifts, the precession of the equinoxes and—a point he mentions only briefly (p. 26)—the tides.

Regarding this point, he says, there is no logical basis for attributing the tides to gravitational pull by the moon.

> If the tide were high on one side of the earth and low on the directly opposite side, it would be very logical, but how could it be, for instance when the sun and moon are both on one side of the earth, that the pull from them could raise tides on opposite sides of the earth in exactly opposite directions? Books state that the gravitational pull takes the earth away from the oceans on the further side, which would mean that it practically leaves them "hanging in the air."
>
> With the tides, the gyroscopic action forces the water parallel to the surface of the earth, and it piles up on the shores. If there were no land on earth, the tides would be very slight— only such as would be caused by gravitational pull.
>
> With the great flood and shift of the earth, the action is just reversed. The earth shifts and the oceans stand practically still. When the tides occur, the earth does not shift because the axis has not yet reached "dead centers" as to both of the orbits (pp. 26–27).

The view of official science on these matters was stated succinctly by an unnamed astronomer at the U.S. Naval Observatory in Washington, D.C., whose letter Barber reproduced. The astronomer said to an advocate of Barber's theory that he had no time to read Barber's book but noted that the motions of the earth are well understood and that none of them is called a "small orbit" by astronomers. "Either there is no 'small orbit' or else it is a new term used for a motion that is already well known," he wrote. "Nothing happens suddenly to the earth or its axis every 9,000 years; the motions are regular. The cause of the tides is gravitational and not gyroscopic" (p. 6N).

If Barber were the sole voice announcing the ultimate disaster, such arbitrary rejection of his request *merely to examine his claims* might be allowable. But Barber obviously is not alone, and it is the central premise of this book that enough data exist to warrant a full-scale examination of the question in its various guises, even though the data sometimes come from unorthodox sources in forms that admittedly are quasi-scientific or nonscientific, Barber's being one.

The closest anyone with scientific training has come to doing that for Barber's claims appears to be a statement from Dr. Malcolm H. Tallman, whom Barber describes as a noted astronomer. In a letter dated 1957, written from a Brooklyn, New York, address (which Barber quotes on page 3 of his News Bulletin No. 4), Tallman says of Barber's book:

> ...I find in it much valuable data as well as interesting reasoning which substantiate the conclusions. Those who would condemn the book should first qualify themselves to do so by using first-year algebra and explain why a top apparently defies the law of gravitation.
>
> We know that the geographic axis of the Earth has shifted in the past. This has been verified by the sudden deep-freezing of live mammoths in northern Siberia and Alaska "in the flesh."...Precedents have been established. What has happened in the past can happen again.

Another critical point about Barber's theory was raised by an unnamed chemist who entered into correspondence with Barber. The chemist made this point (p. 30): If the "astronomical fraternity" (his words) has overlooked the small orbit, it would not be the first time truth and new knowledge have been opposed. "Look at the transition from the Aristotle-Ptolemy system to that of Galileo and Copernicus's discoveries. They really caught hell." But, he asks, if your discoveries are correct, why has not the phenomenon been observed and measured on other planets in this solar system, "since it is axiomatic that they are gyroscopic in action?" This action, he feels, would be readily observed and detected in the other planets.

Barber replied, "True, all planets operate on the gyroscopic system, but they are so far away that an astronomer with a large telescope could not detect a [pole] shift on them, even if he happened to have his eye on the planet at exactly the right moment, which he most likely would not have, even if he tried millions of times."

As for "Antarctica's icy menace," Barber declared, "I do not attach any danger to the north or south ice caps causing the earth to shift suddenly, or even slowly. As they increase in weight, they merely press the earth in at the poles and cause it to bulge more at the equator."

Barber modified this position ever so slightly after receiving a letter from Hugh Auchincloss Brown along with an article by Brown on how to prepare for world flood control. Barber reprinted the one-page article in later editions of his book, and commented tersely, "While I admit the heavy ice caps might assist the shift a trifle, I do not believe that will initially cause it. However, I wish to cooperate with anyone in a program to save civilization" (p. 37).

All public pronouncements from Barber appear to have ceased after News Bulletin No. 5, in May 1958. At that point he declared, "If this program does not sustain itself [financially], I will drop it, go to a high mountain with those friends who wish to follow me, build a boat or ark, and let the rest of the world drown." Whether Barber got to that mountain is unknown to me, but the program did not sustain itself as he wished and public awareness of his theory and warning quickly faded—except for one man who took on the task of warning about the coming disaster worse than the H-bomb. His name is Emil Sepic, and we will look at his work next.

References and Suggested Readings

BARBER, ADAM D. *The Coming Disaster Worse Than the H-Bomb.* Washington, D.C.: privately printed, 1955.

ROSENBERGER, JOSEPH. "The Coming Deluge," *Search*, July 1972.

Chapter Ten

Emil Sepic: The Imminent Shift of the Earth's Axis

I first learned of Emil Sepic in 1977, when a friend sent me a copy of his short treatise *The Imminent Shift of the Earth's Axis*. Published privately in 1960, its nineteen oversize pages are filled with long lines of closely spaced small type that pack huge amounts of data into an otherwise slender publication.

Because of certain parallels in their work, I thought at first that Sepic was a student of Barber's, or rather, his successor. Like Barber, Sepic claims to have discovered that the earth makes an extra, minor orbit yearly. This orbit has a diameter of about 3 million miles. He also claims that as the earth makes this small orbit, there is a lag and drag resulting in continual change in the axis' angle of inclination to the ecliptic and that gyroscopic forces, not gravitational or electromagnetic, are primarily responsible for this. His final major claim is that as a result of all this, the earth makes sudden but periodic shifts of its axis—not crustal displacements but global tilts of up to 90°. These occur every 7,500 years or so. Another shift is due soon, he says. It could happen "in approximately 20 years, more or less" (p. 8) and "at the height of the shift the movement could probably attain a speed in excess of 1,500 miles per hour" (p. 11).

Also, like Barber, Sepic gives no references or bibliography in what the cover proclaims is "a scientific, mathematical and geological thesis concerning the earth's present and future status." Scientists will gnash their teeth at this, but Sepic couldn't care less. The book, he states (p. 2), is "the result of 46 years' interest in the earth's motion and 18 years of intensive study and research." He also states that he wrote it intending to present "a minimum amount of unnecessary reading" (p. 1). And anticipating criticism on these grounds, he leaves the reader with these words:

> I have included in this thesis everything that I believed to be important. The principal factors are given so that anyone who reads may understand. I am thoroughly aware that it is difficult to arouse humanity from their accustomed and traditional ways. Therefore, I am not interested in contending with anyone, nor do I care to defend this thesis or myself. Furthermore, I am not interested in anyone's arguments, etc. All the necessary information is included herein, and it is left to the reader to judge as to the merits or demerits of this thesis (p. 15).

In other words: Take it or leave it. I was unwilling to do either, however—at least, not without further examination. I wanted background information. I wanted to know the man behind the work. So I wrote to him at an address provided by Dorothy Starr, of the *Pole Watchers' Newsletter*. His reply, sent in early 1978, stated, "Am familiar with Barber's work. I did not accept his theory. There is a *vast* difference in my theory and his. We had only one thing in common—that there must be a minor orbit in addition to the large orbit. Apart from that we had nothing in common. My work is entirely different. There is no relation to his theory."

So the resemblance I saw wasn't true? I needed to know more. A telephone call one evening put us in better contact. I began by asking about his life. His surprising reply told me this was not going to be an ordinary interview.

Q. Mr. Sepic, may I have some biographical information?

A. I haven't got any. I have no history. I think I went to second grade in school and after that I don't remember whether I went any more or not.

Q. Are you saying you didn't go to high school or college? You have no formal training in science?

A. I have nothing at all—absolutely nothing. I'm just like some of those inventors who never went to school beyond the tenth or eleventh year.

Q. But you obviously learned to read—and to read widely.

A. Oh, definitely. I was inclined that way all my life.

Q. What is your occupation?

A. I've had a number of occupations. I first started out to be a druggist, but I dropped that. Then I turned out to be a machinist. Then I got away from that and turned out to be a carpenter, housebuilder. That was my last occupation.

Q. How old are you?

A. I'll be seventy in a month or so. (This was in September 1978. Therefore, Sepic was about fifty-two when he published *The Imminent Shift of the Earth's Axis* and six or so when he first became interested in the earth's motion.)

Q. When did you first come across the idea of pole shifts?

A. Many years ago. I was interested in the subject all my life. I fooled around with gyroscopes when I was a kid and I had my own idea about gyroscopes. And I was interested in the earth's motions all my life—I don't even know why. Finally I started to put two pieces together and got a pretty good idea of what it's all about. Years later I ran into a man by the name of Barber—he had an ad in the paper—and I didn't agree with him. And he knew it—I told him so. Toward the last he was willing to set his entire theory aside and push my book forward. In other words, he seemed to agree that his theory was all wrong.

Q. Did he state that in writing?

A. To me, yes. He didn't exactly state it in writing. He did concede to me that he accepted my book and was no longer willing to keep selling his or whatever, but he was more interested in pushing my book just before he died.

Q. The last thing he published that I was able to find was his Bulletin #5, which was dated 1957.

A. I never got it. But his theory was all screwed up. You see, he thought that the earth had a magnetic field that was holding the axis in harness about so many millions of miles away under the earth, say at the bottom of the South Pole, and I explained to him how that was not possible, that the earth maintained an inclination of about 23.5° and that it could not have a magnetic field holding the axis in line—no such thing, it was purely gyroscopic. He and I were at total variance.

Q. What other influences, what other references do you have for your own theory?

A. None in the world. I developed it myself—I developed all of it myself. You could call it a minor orbit if you like, which is perfectly correct, I think, and it's more nearly correct than a small orbit. Today when I was measuring the sun at meridian time I found out that the declination was not 23.25 minutes [degrees?] or thereabout—it was only ten minutes of arc. How do you like that for a change? In other words, the oscillations of the earth are approximately, or as far as I've been able to gather this year, are about four times worse than they were ten years ago. Ten minutes of declination is not very much when it should be twenty-three. That amounts to around the oscillation orbit—what I also call the oscillation orbit. It amounts to about fifteen miles of oscillation of the earth's poles.

Q. Have you read the works of Hugh Brown?

A. Oh, yes. I corresponded with Brown. I told him it was not possible to shift the pole by means of the ice. Knowing gyroscopics, [I could see that] it would not shift the earth at all. As far as I'm concerned, that's absolutely a false concept, and I told Brown that. I explained to him every possible way I could, many years ago. I said, "It's just not possible." I said, "Here are the pole shifts every so often and many times the poles have intersected water areas in which the ice would not be accumulated, whereby there could not be pole shifts ever." But it didn't ring a bell with him. He just played the old tune. He wouldn't get away from it—a lifetime musician with one tune. That's the way he was—he wouldn't change.

Q. Have you read Velikovsky or Hapgood?

A. Yes, I read Velikovsky and I read Hapgood.

Q. But you say they played no part in your theory.

A. No, sir. They are absolutely false theories. But the main thing is that the earth does shift and I don't think we have too far to go for another shift. It could be this century, it could be later. I can't pinpoint it down, because it's like trying to pin down a [spinning] two-inch gyroscope to a fraction of a second. It's just not possible.

Q. Did Adam Barber try to convince you that you ought to build a boat, as he suggested?

A. Yes, but I think that's a forlorn hope, because when that water comes charging both ways, there's no boat that's going to withstand that. It just wouldn't amount to a tinker's damn.

The best thing is to be as far away from the ocean shores—
high up, say around Colorado. The only thing I can do is to
watch the pole at meridian time every day, like I've been doing,
and when its oscillations get bad enough, I'll know where to
go.

On page 10 of his treatise, Sepic gives us a partial picture of
things to come.

> There will be many earth oscillations before the shift hap-
> pens. The shift will cause the ocean and other waters to overrun
> the lands at terrific speeds. The waters will come in huge walls
> from all directions as the earth is shifting and finally settling
> on a new axis. The seasons will change times. Most everything
> will be washed away. Our cities will be no more. Glaciers will
> be on the move. A new equator will begin to bulge outwards.
> The old bulge will begin to be pressed inwards and gradually
> disappear. This will cause cracking of the earth's surface, vi-
> olent earthquakes, unheard-of bad weather, and many new vol-
> canoes will come into action. . . . The winds will be violent and
> the atmosphere will become cloudy and it will stay cloudy for
> a long time. The ice caps will be moved out of their places,
> and the melting thereof will cause the waters to rise and remain
> higher for some time to come. Those who are fortunate or
> unfortunate to survive through it all, and there will not be too
> many, will have to start all over again—with only their bare
> hands.

How, precisely, will this occur? Sepic leads off with a section
entitled "Proofs that the Small Orbit Exists," in which he shows,
with useful illustrations, that the minor orbit is akin to the motion
of someone sitting on a Ferris wheel. As the wheel goes around,
so the earth goes around the sun. But "the rider does make an extra
turn each trip of the wheel—the seat rotates once about its axle
for each trip the wheel makes. . . . when the rider was highest [his
head] was farthest from the wheel's axle, and when the rider was
lowest [his head] was closer to the wheel's axle. It is pretty much
the same with the earth" (p. 6). In other words, the major orbit
is nearly circular, but the main orbit produces a motion whose
composite result is the apparently eccentric orbit around the sun
that astronomers claim is the only one.

Every rotating orbiting body travels in a duel orbit, Sepic argues. If there were no small orbit there would be no stability forces

generated. Without knowledge of the existence of the small orbit, he says, the science of astronomy "will never be understood in its true form" (p. 5). Astronomy's contention that the earth's equatorial bulge is what gives planetary stability is false. "Gyro forces create axial stability, and the 'bulge myth' as taught in astronomy books is entirely out of order" (p. 9). "No rotating planet could keep its axis in harness without [continual gyrations], regardless of what the inclination of that particular planet happens to be" (p. 6).

To disprove the "bulge myth," Sepic asks us to suppose that the earth were an eight-inch globe. On that scale the sun would have a seventy-foot diameter and be one and a half miles away. The equatorial bulge would be only as thick as several sheets of paper, Sepic says, and "that is an example of the thing that is given credit for the precessional motion and axial stability." Significant action by the sun on the earth's bulge is not only impossible, Sepic contends, but in addition it adds nothing to axial stability.

> The orbiting and rotating forces and the eccentric path the earth travels in create the small orbit around which the earth gyrates. This small orbit is an imposed orbit around which the earth gyrates, lags and drags. It is not possible for the small orbit to be completed at the same time as the large orbit, and the yearly loss about the small orbit each year represents the amount of ineffective energy required or lost as the earth travels around both orbits. The earth really travels in a dual orbit. Without the small orbit there would be no axial stability, and the earth would weave and tumble throughout its yearly journey. The fact that the small orbit is described makes the weaving and tumbling impossible. The small orbit is spread out over the large orbit, and as the earth travels about the large orbit, both orbits are nearly simultaneously described (p. 5).

The small orbit can be described as an extra "roll around" that the earth makes in each orbit of the sun. Not only does this precess the axis, it creates a wobble, preventing perfect stability in the earth. "The earth partly coasts, skids and skews about the small orbit. And the further the axis points outwards, the larger the imaginary top circles [describing the path of precession] get, and [the] less stabilizing forces are generated" (p. 6).

Eventually, every 7,500 years or so, the earth tumbles. "The axis can shift as much as 90°. Exactly how many degrees the axis will shift is speculative. . . . The most likely period of the year for the shift to happen is in the latter part of June and the first part of July. . . . the North pole should dip toward the sun and at the same time go at right angles with the rotation. The seasons should be entirely upset, and the sun will rise from an apparent new direction. After the shift the earth should rotate, for a time, a little faster than it is now rotating" (p. 7).

Without mentioning their names, Sepic contends with Brown and Hapgood about the trigger mechanism. "It would surprise most everyone to know who the scientists and others were who gave their reasons for large pole displacements as being due to polar ice loads. . . . None of these men even mentioned the possibility that the earth travels in a dual orbit, generates stability forces, and the fact that these forces must become irregular and reduced before a shift could be brought about. Nor have they, to my knowledge, taken into consideration the orbital force" (p. 9). Sepic chides "the scientists and others" for "a lack of knowledge," and shows mathematically that the eccentric throw of the polar ice caps "is of no importance in causing polar shifts" (p. 9). ". . . any imbalances caused by ice caps," Sepic declares, "are reduced to 'nothing' compared with the gyro forces, both normal and abnormal. There is no power in the Antarctic ice to even initiate a shift or even move the so-called crust. When the forces become right for a shift, not even the 27 mile equatorial bulge will be of any help in keeping the earth in harness" (p. 9).

Like Barber, Sepic argues that the tides are also due to the gyro forces, rather than to gravitation. "There is positively no logic to the moon myth. . . . In spite of all the ado made by conventionalism as to how the moon controls the tides, they have yet to explain why at or about the polar areas we have one tide every 24 hours! . . . Some areas, even other than about the poles, have one tide or none at all in a 24 hour period. . . . At the Panama Canal there are two tides every 24 hours on the Pacific side anywhere from 12 to 16 feet high, while on the Atlantic side only 40 miles away there is one tide a foot or so high every same 24 hour period" (pp. 12–13). He concludes his section on "Tides and Currents": "Gyro forces continually change in strength from maximum to zero and vice versa, and they reverse directions twice every 24 hours; these are important factors for the recognized two tides daily for most of the earth, and the one tide in the same 24 hours about the polar areas . . ." (p. 13).

After showing the effects of the dual orbit, Sepic writes briefly about past pole shifts. This is the least developed section of his book, and shows most clearly the danger of dispensing with references. Sepic's stand against documentation, on grounds that it is unimportant for his treatise, is shortsighted and a disservice to his audience. Consider this passage:

> Recent carbon-14 dating of the Antarctic ice indicates that the oldest ice is not quite 10,000 years old. Recent carbon-14 dating of the animals about the Arctic indicates about the same age—10,000 years. Yet despite this, conventional books read that the poles shifted gradually in periods of many millions of years! Books further illustrate numerous pole spots! How could there be pole spots if the movement of the poles were at the rate of a few inches a year? These would be pole lines instead!
> A few past pole areas are here listed: Alaska, California, Wisconsin, Hudson Bay, South America, Sudan Basin, Caspian Sea, China, Siberia, etc. The foregoing [arguments and data] should convince most thinking persons that pole shifts are not gradual, are not millions of years apart, and when they do shift they are quite sudden (p. 7).

I'm sorry to say that neither the arguments nor the data convince me. Consider the "facts" first. We have already seen that the age of the mammoths varies all the way from as long ago as 47,500 years for the famous Berezovka specimen to a mammoth bone found in Mexico and dated only about six centuries before Christ. As for the age of the Antarctic ice, one wishes Sepic had given the source of his information. The same can be said for those "conventional books." He should have named them, especially those books illustrating "numerous pole spots." (Probably he had Brown and Hapgood in mind, but we'll never know. The alleged data remain tantalizing but obscure. When I asked for clarification, Sepic ceased communication, apparently offended that I questioned him.)

This bears directly on the logic of his argument in the passage above. "Conventional books" that concede the reality of pole shifts (as distinguished from those much more numerous conventional sources that merely consider—and usually dismiss—the possibility) are rare indeed. But even those books do not "illustrate numerous pole spots." At least I have found none. Former polar locations are given only by *un*conventional theorists such as Brown, Hapgood and Velikovsky. Sepic's argument incorrectly attributes data to the conventional books, and therefore his conclusion is unjustified, even though the reasoning appears sound.

Despite these shortcomings in *The Imminent Shift of the Earth's Axis,* Sepic's presentation goes—as he himself claims—well beyond Barber's. His illustrations are an especially valuable part of it. And his critique of the official explanation of the earth's orbital and rotational properties certainly ought to be investigated by astronomers and geophysicists. Anyone who could predict in 1960 that there would be more and greater earthquakes, more irregular tides and ocean currents, and sudden and unusual changes in the weather and climate, certainly has some element of truth in his perspective. Sepic makes precisely those forecasts on page 10, and present conditions are proving them to be quite correct.

In closing this chapter, I will let Sepic have the last word. His subject is safety and possible prevention of the imminent shift, and his outlook is most bleak. On page 11 Sepic tells us, "The shift of the axis need not happen. It is my belief that it can be prevented with projects of unheard-of proportions, if the powers that be would listen and act. . . ." However, he says, it is his belief that world leaders will do nothing, because they are basically petty and selfish in their ways of thinking.

> The reader might be wondering how one can be saved from the impending catastrophe—my answer is simple—there is no sure way. Later on when the time of the shift becomes more obvious one may consider such places that are far inland, in the middle of large continents, away from the oceans and other large bodies of water, and the higher the better. As previously mentioned, there are, I believe, remedial measures, but I have reasons to believe that the "powers that are" will not listen. The exit of this civilization will, no doubt, be the same as the exits of numerous past civilizations. Humanity as a rule will not respond to things unless there is an immediate need or necessity.

References and Suggested Reading

SEPIC, EMIL. *The Imminent Shift of the Earth's Axis.* 2218 Buhne Street, Eureka, CA 95501: privately published, 1960.

PRECOGNITION

*What Modern Psychics
Have to Say*

Introduction to the Section

Precognition is the scientific term for paranormal knowledge of future events. It is an aspect of extrasensory perception (ESP), just as telepathy and clairvoyance are. Precognition goes beyond analysis and logic to a direct perception of events unforeseeable through use of reason alone. In other words, precognitive knowledge cannot be inferred from present data by intellectual processes (as they are presently understood).

Interestingly, although precognition is probably the most difficult aspect of parapsychology to understand, surveys show that precognition occurs more often than any kind of ESP, particularly in dreams. Several studies, for example, show that successful businessmen use precognition (often labeled "hunch" or "gut feeling") in making judgments affecting the life and profitability of their companies. Another study—this one of the 1966 Aberfan, Wales, tragedy in which a coal tip demolished a school, killing 128 children—showed that increasing numbers of people dreamed of a tragedy there as the fateful day approached. This suggests that the nearer in time we come to a disastrous event, the more widespread and frequent precognition of it becomes.*

This section will examine predictions of a pole shift from a parapsychological point of view. That is because many of those warnings of impending cataclysm come from modern psychic sources. These sources have predicted changes in the earth's geography and positioning in space—changes that are often specifically dated and that are uniformly perceived as occurring in the very near future.

What is a psychic? The question is as hard to answer as, What is a scientist? People are involved in science all the way from laboratory technicians through professors, engineers and editors to field and laboratory researchers and high-level theoreticians. Much the same can be said about psychics. There is no standard model or role to which a psychic must adhere. A psychic can be as exotic as a robed oracle in the smoky grotto at Delphi or as ordinary as your next-door neighbor. Like those engaged in science, about all they have in common is an interest in acquiring knowledge and a means for doing so. Their ability, training and experience as psychics can range as much as those of people in science.

One highly competent psychic I know has a Ph.D in psychology and is on the teaching and research staffs of a number of major hospitals and universities. Another equally competent psychic acquaintance of mine has not even finished high school and has spent most of her adult life as a housewife and mother. These two people function well as psychics apparently because of two factors: a fortuitous gift for it from birth and a persistent effort to cultivate and train that gift. Like musical ability, everyone probably has some degree of psychic ability, and therefore it is common for people to have uncanny hunches or gut feelings about future events. Déjà vu—the feeling you've already experienced a situation but can't remember when—may be due to precognition. But if most people can play "Chopsticks," virtuoso musicians are rare. A hereditary talent alone is not enough—even extreme talent. Hard, disciplined effort is also necessary to develop that talent. Thus, superpsychics are rare.

*These data come from the chapter on precognition in Edgar D. Mitchell's *Psychic Exploration* (G. P. Putnam's Sons, 1974), which I edited.

And even a superpsychic can be wrong. All too often it is assumed that a genuine psychic's prediction *must* be so, as if the future is rigidly predetermined. However, this is a naïve view. A situation is usually neither black nor white. There are mostly shades of gray, and it must be recognized that psychics face the same kinds of constraints on their perceptions as other "futurecasters," even though they may be seeing farther through time and space than they would if they had used only normal means or logical processes.

Rarely is a psychic 100 percent correct in a prediction, and even the superpsychics, who may predict with pinpoint accuracy, are not right in *every* prediction. Everyone is fallible, including psychics. Each psychic has a certain degree of reliability, and even that varies from day to day, from subject to subject, from year to year. The variables are numerous. They include the psychic's emotional and physical state, the setting in which the psychic is to function, the feeling-tone of the interactions he has both with researchers (if it is an experimental situation) and with the one for whom he is psychically functioning (if it is a psychic reading). They also include long-term learning—i.e., improved performance over time through practice. If symbolism appears to the psychic, his accuracy may be altered by incorrect interpretation of what he sees.

But if psychics are not totally accurate all the time, neither are weather forecasters, stock market analysts and military intelligence specialists. Their judgments are subject to modification as new factors enter the picture. The same applies to psychics. They can have their "off" days, in which they "miss" rather than "hit." Many kinds of influence can creep in.

With regard to prediction, we can say that the more general the topic being examined in the psychic state, the more room there is for error. The scope of earth changes is, of course, far more vast than, say, the health of a particular individual. Moreover, a person's state of health is a present condition, while earth changes are a future condition. For these and other reasons—particularly the concept of thought forms, which will be discussed below—we can expect that when dealing with precognition, rather than present-time clairvoyance, the range of variability for a prediction will increase with the dimensions of the predicted event. A look at the record of professional meteorologists' long-range forecasts tells the story in a nutshell. Or as people say about the weather in New England, "If you don't like it, wait half an hour—it'll change."

That being the case, it should be obvious that we need not regard predictions and prophecies as inevitable. Many psychics themselves—the best ones, in my judgment—honestly point this out because their experiences confirm what all the world's major spiritual traditions tell us, namely, that free will is operative and we can influence the outcome of possible future developments through the application of our physical, mental and spiritual resources. (This will be examined in Part V.)

Thought Forms: The Key to Psychic Predictions?

From the psychic point of view, one of the critical factors that can influence the outcome of a predicted situation is the quality of consciousness in human society. The mechanism by which consciousness modifies a set of circumstances has been described by some psychics as "thought forms." The term and the concept behind it come from esoteric psychology and metaphysics, and both figure importantly in the psychic readings to be examined here. The subject of Chapter 11, for example—Edgar Cayce— offered this statement in Reading 906-3 about the nature of human thoughts and how they affect ongoing experience: "For mind is the builder and that which we think upon may become crimes or miracles. For thoughts are things and as their currents run through the environs of an entity's experience these become barriers or stepping stones, dependent upon the manner in which these are laid as it were."

What are thought forms? The concept posits mental or psychic energy as an intermediate substance between matter and consciousness. From this perspective, as Cayce put it, thoughts are things—real but nonphysical energy configurations, produced by human consciousness, that exist objectively in space outside the human beings who produce them. (I say "nonphysical," meaning "physically unobservable at present.") A thought form is the energetic embodiment of the idea on which a person dwells, consciously or otherwise, and it takes on an existence external to and independent of the thinker. By a process of which official science knows little, our thoughts "take wings."

In other words, when we think—consciously or otherwise— the experience of mentation is not simply electrical activity within the neural pathways of the brain, nor is it confined to the limits of the cranium. Telepathy (mind-to-mind communication) and clairvoyance (remote viewing) tell us this must be so, but what becomes of a thought *after* it has been thought? Does it simply disappear, vanish? Apparently not. From the point of view offered by superpsychics, thought activity extends beyond the physical body, partaking of a "field of mind" surrounding the planet and extending into space for an unspecified distance. This mind field is composed of the collective experience of the human race—our thoughts, feelings and actions. Untold numbers of thought forms over millions of years have contributed to the planetary field of mind. Thoughts of a similar nature tend to coalesce over time, to gather into what could be called *thought fields*. These thought fields are equivalent to what the psychologist Carl Jung called an archetype (and this is why people everywhere have access to archetypal experience). The totality of thought fields, or archetypes, constitute an "atmosphere" of thought energy coextensive with the earth's physical atmosphere and beyond, and can be understood as what Jung called the collective unconscious.

This psychic atmosphere or field of mind energy is not yet recognized by official science, but evidence for its existence abounds.* Mind energy interacts with the physical energy matrix sustaining the planet in space, and can influence it subtly but directly in either a positive or a negative fashion, depending on the vibratory quality of thought forms arising from the human level. Harmonious, loving mental states produce a stabilizing effect on the planetary matrix; disharmonious, hateful thoughts result in a destabilized matrix.

This mind-matter interaction is a two-way process. People may "receive" from the planetary mind field or collective unconscious, as well as "give." For example, certain universal or primordial images and symbols are perceived by people in dreams, meditation and other altered states of consciousness, regardless of race, sex or culture. As another example, consider how a new idea or discovery often appears simultaneously in several widely separated locations, apparently as "fallout" or "precipitation" from nonphysical levels of reality to the physical.

*My anthology *Future Science* (Garden City, NY: Anchor Books, 1977) presents a large portion of the evidence, along with other insights from frontier investigations and pioneering research into the physics of paranormal phenomena.

Spiritual traditions warn that we shall reap what we sow. Psychic traditions offer an explanation of how and why this must be. The many "crimes against nature" that people are perpetrating—over-population, environmental pollution, wasting of non-renewable resources, nuclear testing—along with "crimes against humanity" such as war, economic exploitation, the imposition of inhumane living conditions, religious persecution, political abridgment of human rights, intolerance and bigotry toward minorities, etc., are all pouring negative thought forms into the planet's energetic foundations. The result will be geophysical cataclysm: earth changes and a pole shift.

From the point of view of esoteric psychology and occult science, then, rather than saying we will be punished *for* our sins, it would be more accurate to say we will be punished *by* our sins.

Dr. Jeffrey Goodman (whom I mentioned in the commentary to Section II), has coined a term for this psychokinetic process by which human thoughts influence the total energy pattern of earthly life. It is *biorelativity,* the interaction of people with their physical environment via psychic, or mind, energy. In *We Are the Earthquake Generation,* his recent examination of psychic predictions about vast changes in the geography of the earth, Goodman notes, "Since energy can neither be created nor destroyed, the energy of thought, psychics say, is still in existence as a sort of atmosphere or field surrounding the planet, recording all the experience of humanity. This is the so-called 'akashic record' which Cayce and other psychics claim to 'read' when they obtain paranormal information about the past" (p. 195).

Walene James, a student of Cayce literature, offered a succinct statement about the nature of psychic prediction that I feel is worth including here. She wrote to me, "When a psychic reads thought form patterns or the akasha, he is reading a *process,* a flow of energy patterns, not a fixed state. This process is actually directed by human will (choice) and is subject to change by that same will. One of the purposes of dire predictions is to change them [the predictions]—and they can be changed! According to the Cayce readings, the physical universe is a reflection of man's consciousness and the actions arising from that consciousness, although it appears the other way around."

According to another psychic, Aron Abrahamsen (whom you will meet in Chapter 12), thought forms or thought fields interact with planetary forces of nonhuman origin, such as electromagnetism. One of his psychic readings declares, "Thought forms have . . . vibrations, and vibrations can either add to a destructive force or . . . [take] away from that which is building which will be destructive, or it will add to that which is already constructive. Now, thought forms when they are sent out by a people or a group of people can add much to avert or avoid destructions inasmuch as their thought forms can be in a more positive way. . . . Forms can also be overruled, depending upon the magnitude and the intensity of these various thought forms."

He adds, "Just because a person is good and just because a person desires to do that which is good, that does not mean that his singular entrance will overcome a magnitude of negativity. It takes more than just one, though avenues may be opened for him from time to time when there can be a quiescence, there can be a quietness within the negative thought forms and thereby gain an entrance into to overrule the other negative thought forms."

Prayer, of course, is considered to be one of the most powerful ways to add constructive thought forms to the total energy matrix. It is no coincidence that psychic and spiritual traditions declare the efficacy of selfless prayer as a psychospiritual tool or resource.

Higher Intelligence and Psychic Predictions

The subtle physics of "the power of positive thinking," as presented by psychics, may seem startling to some. But it is something to be considered in dealing with the possibility of a pole shift. We must now consider an even more startling subject that bears on the subject of psychic predictions: higher-than-human forms of life.

Might there be such in the universe? Look at it this way. First, we know there are levels of life below the human. Microbes go their microbial way with a life-span measured in days, presumably unaware of, say, insects, which can be seen as an example of the next-higher level of intelligent organisms. And just as insects surpass microbes in longevity and organic complexity by many orders of magnitude, so do we humans surpass insects.

Microbes, insects and people coexist, but rarely are aware of the others' activities and domains. How often do we intrude into the life of an ant or a termite? Logically speaking, however, if life exists at the microbial, insect and human levels, why should it end there? It has been evolving on earth for several billion years. But the sun is a relatively young star. There are many others in the universe far older than our home star. These older stars' planetary systems could have had life-favorable conditions for much longer than such conditions have existed here on earth. Life on other planets, outside the solar system, could have evolved far beyond the human level—as the evidence of ufology (the study of UFOs) suggests. Simply because we haven't observed it yet, we should not assume it doesn't exist.

This line of reasoning has scientific support. Exobiologists such as Cornell University astronomer Carl Sagan calculate that many life-forms more highly evolved than *Homo sapiens* probably exist. In fact, Sagan recently stated in his book *Other Worlds,* "It is roughly estimated that there are more than one million technical civilizations in our Milky Way galaxy alone that are more advanced than ours."

Astronomer Robert Jastrow goes even further. "Can you imagine a form of life as far beyond man as man is beyond the worm and jellyfish?" he asked in an essay entitled "The Search for Life in Outer Space," published in *Newsday* (6 August 1979).

> Such creatures must exist on the majority of stars and planets around us, if life is common in the universe.
>
> These extraterrestrials are not like the flower children in *Close Encounters of the Third Kind* or the cowboys of *Star Wars.* They are creatures whom we will judge to be possessed of magical powers when we see them. By our standards, they will be immortal, omniscient and omnipotent—in a sense, god-like creatures (p. 37).

In addition to looking at the question of higher intelligence from the perspective of science, we can consider what various mystics, religions and metaphysical traditions have told us. The essence of it is that there are other planes of existence, other sets of dimensions interpenetrating our own three-dimensional space-time framework, normally unperceived by humans. These other levels of creation have beings native to them—beings that can enter our own level, the physical. The totality of these supersensible realms and nonphysical entities forms a great chain of being leading up to the source of creation, God.

Now, nothing in science absolutely disallows the possibility of intelligences existing in nonphysical, discarnate or spirit forms. Moreover, the slightest consideration of how vast life as we know it is—from viruses to dinosaurs (and perhaps even atomic and subatomic particles, which some scientists now feel can best be understood as protobiological fauna)—should make clear that there are other possibilities well within the range of what various spiritual and occult traditions tells us. And what they tell us is that humanity exists at the confluence of higher and lower worlds inhabited by angels, demons, devas, elementals, thunderbirds, herukas, archangels, dakinis, hungry ghosts, cherubim and seraphim, major and minor deities, and all the other heavenly or hellish mythological creatures generally dismissed as the fantastic foolishness of the superstitious.

Certain psychics—we will meet some in the following pages—speak of receiving information from ultraphysical entities of advanced intelligence whose creation predates the human race, whose life span is almost inconceivably long by human standards, and whose purpose is to benevolently guide us in our evolutionary journey back to God. Often the psychics describe their superhuman or nonhuman sources of information as "spirit guides" and "guardian angels." The picture of these entities' existence that emerges from the descriptions given over time is one occurring on a scale enormously beyond the human. Their existence, it is said, is cosmically entwined with humanity's. They feel a moral duty, from their level of reality, to influence and guide human affairs in a noncompulsory manner.

Like thought forms, these higher intelligences exist cospatially with the physical atmosphere of our planet and beyond, just like water vapor in air. David Spangler, an American mystic associated with the spiritual community of Findhorn, in Scotland, once told me of his contact with those higher intelligences called devas ("radiant or celestial beings"). If the devas could be perceived with physical vision, he said, all that would be seen are shifting patterns of color and form. Likewise, the president of the Theosophical Society in America, Dora Kunz, described to me her clairvoyant perceptions of devas as vast in dimensions—often many miles—and radiant with unearthly hues.

From the point of view of these more highly evolved intelligences, then, the future of the human race *already* exists to some unspecified degree because they can see with overarching vision the possibilities ahead for us, and can influence human society in such a way as to guide us toward desirable ends. Insofar as human affairs are subtly organized by higher intelligence, the outcome is known in advance. Time, then, can be said from this point of view to flow, contrary to what common sense tells us, *from the future to the past*. It is the progressive revelation of the higher intelligences' plan for humanity.

Here again, however, human free will is said by most psychics to be operative. For a variety of reasons—including the influence of malevolent, antievolutionary intelligences such as demons and "fallen" angels—the resource of spirit guidance may be rejected. This is still another factor that, from the psychic point of view, complicates the matter of prediction.

In the following chapters, you will meet a number of psychics who claim to have accurate perception of the future—a future that involves vast reordering of the surface features of our planet and its position in space. Bear in mind these cautionary aspects of precognition and the psychics' view of reality as you examine the predictions. As in the first section, I will present the data and save major discussion of it for Section V, where I will analyze and evaluate the information.

Now hear the psychics speak.

Chapter Eleven

Edgar Cayce: Miracle Man of Virginia Beach

In late 1962, during the Cuban missile crisis, I was sent to the U.S. naval base at Norfolk, Virginia, with orders to report aboard a destroyer as the antisubmarine warfare officer. When I arrived, however, my ship was "on the line" blockading Cuba. So, having some spare time, I began to look around the Norfolk area in order to find an apartment for my family.

While driving along the main street in nearby Virginia Beach, I saw a sign that read: ASSOCIATION FOR RESEARCH AND ENLIGHTENMENT. Curious, I parked my car and walked up to the main entrance of a large wooden building on a knoll overlooking the Atlantic Ocean. Inside, a friendly receptionist answered my questions in general terms, gave me some literature, and took me into a library where more books than I could ever hope to read were inviting me to delve. She also mentioned a fireproof vault in the basement where "the Edgar Cayce readings" were stored. The originals were not accessible, she said, but copies known as "circulating files" were available to members.

I spent ten minutes or so looking around, but since I was on a different mission—and on government time—I resisted the urge to curl up with a good book, and instead returned to the lobby. As I was about to leave, a gray-haired man of middle age walked past. "Oh, just a minute," the receptionist said. "Here's Hugh Lynn." She introduced us and I then found myself informed even further about the nature and work of the A.R.E. by the eldest son of the man who had founded the organization. Hugh Lynn Cayce's cordial and helpful manner, despite an obviously pressing schedule, told me as much about the A.R.E. as his words did. We had only a brief chat—he was late for an appointment—and then he excused himself. But as he walked away, he gestured to a literature shelf by the receptionist's desk. He picked up a paperback book entitled *There Is a River,* by Thomas Sugrue. "This is the best thing you could read to find out about us." Then he was off.

But I was on—on target—through one of those constantly occurring "coincidences" that in retrospect seem far less accidental. I bought the book and returned to the base. And that was how, in the midst of a nuclear confrontation, I first learned of America's greatest psychic.

The story of Edgar Cayce, "the sleeping prophet," is known well enough through dozens of books, so I needn't present it here except in thumbnail sketch. Cayce was born in Hopkinsville, Kentucky, in 1877, the only son in a fundamentalist Christian family of five. He demonstrated psychic ability as a young boy by telling his parents, when he was only six or seven, that he was able to see and talk to "visions" of relatives who had recently died and, once, an angelic "presence." Later he demonstrated a remarkable capacity to memorize every word in his spelling book by sleeping with his head on it, rather than reading it.

These faculties faded, however, and Cayce—a poor student otherwise—never got beyond seventh grade. His childhood was normal in other respects, though, except for a vision he had when he was thirteen. "He was asked [by the figure in his vision] what he wanted to do with his life," the A.R.E. introductory brochure states. "He said simply that he wanted to help others, especially children. He was told his prayers had been answered but it was eleven years before he realized just what lay before him, and even he himself said later that he had no concept of the implications of what was to come."

What was to come was the ability, discovered at age twenty-four when he caught a cold and suddenly lost his voice, to go to sleep and answer questions put to him. The answers contained extraordinary—indeed, paranormal—information.

For some time before, Cayce had been developing a paralysis of his throat muscles. Local doctors were unable to treat it, and when he caught cold, his voice dropped off to a rasping whisper. In desperation, he sought help from a local hypnotist but insisted that he put himself to sleep, and only then could the hypnotist ask questions and make suggestions. Once in self-induced trance, Cayce successfully dealt with his own medical condition, recommending a treatment that restored his voice and cured his throat problem.

Word of this incident spread in the medical community. Hugh Lynn Cayce, in his preface to *Edgar Cayce on Atlantis*, by Hugh's brother, Edgar Evans Cayce, tells what happened next:

> A group of physicians from Hopkinsville and Bowling Green, Kentucky, took advantage of his unique talent to diagnose their own patients. They soon discovered that Cayce only needed to be given the name and address of the patient, and was then able to "tune in" telepathically on that individual's mind and body, wherever he was, as easily as if they were both in the same room. He needed, and was given, no other information regarding any patient (p. 10).

One of the participating doctors described Cayce's ability to a clinical research society. *The New York Times* picked up the story, gave it prominent coverage, and from then on people from all over the world sought help from "the miracle man." Eventually Cayce established the Association for Research and Enlightenment in Virginia Beach*—in 1931, to be exact—opened it to public membership, and began the first of many activities that now constitute the work of the A.R.E. These include conducting conferences, seminars and lectures; initiating research on and performing experiments with mental and psychic phenomena; operating a healing clinic and a medical research program; running the excellent A.R.E. library, tape library and bookstore; publishing a journal, books and various other informative materials; and dealing in a

*The address is P. O. Box 595, Virginia Beach, VA 23451.

variety of ways with the Edgar Cayce readings, such as indexing, cross-referencing and—most important of all—disseminating them to the public.

Cayce died of a stroke in 1945. From 1901 until his death, at age sixty-seven, it is estimated he entered his sleeplike state at least sixteen thousand times. The earliest reading in the A.R.E. files dates from 1909, but regular records were not kept on a systematic basis until Cayce's lifetime secretary, Gladys Davis, joined him in 1923.

Almost all the readings given from that time on—a total of 14,256 made for more than eight thousand people—have been carefully cross-indexed and are on file with related correspondence and reports. The massive readings constitute the heart and soul of the A.R.E.—49,135 pages containing many millions of words spoken by Cayce on more than ten thousand major subjects. Most of the subjects were totally unknown to the man in his waking state. But in trance he could do things such as speak foreign languages he'd never learned, diagnose medical conditions of people he'd never met, and prescribe medications and courses of therapy that proved uncannily correct, even in the face of expert medical opinion contrary to the reading.

Nearly nine thousand of the readings—64 percent—deal with the physical or medical condition of people. Another twenty-five hundred deal with the past lives of people, giving information that was sometimes verifiable and that proved to be historical fact. Perhaps the most amazing of all are the readings that make reference to metaphysics and human prehistory. These present a picture of ancient civilizations such as Atlantis, thought by most historians to be only mythical, as literally true. Prehistory, as it emerges from the Cayce readings, is a story that totally overturns conventional concepts of human development and the planet's past. Among the major ideas presented by Cayce is the theme of this book: the earth has undergone cataclysmic shifts of its rotational axis—*and it will again at the end of this century!*

According to the Cayce readings, the pole shift will be preceded by several decades of increasingly severe seismic disturbances on a global scale, resulting in vast changes in the planet's geography. Through earthquakes and volcanic action, through elevation and submergence of land, and through flooding, a new face would be given to the earth. Among the changes predicted by Cayce to occur before the end of the century are these:

• New land will appear in the Atlantic and Pacific oceans.

• Most of Japan will "go into the sea"—i.e., become submerged.

• Northern Europe will be transformed "in the twinkling of an eye" as land submerges and the ocean rolls in.

• The northern Atlantic coast of the United States will undergo many physical changes. New York City, the Connecticut coastline "and the like" will go under the sea.

• Los Angeles and San Francisco will be destroyed before New York City.

• The southern Atlantic coast—Georgia, South Carolina and possibly North Carolina—will disappear beneath the ocean.

• The Great Lakes will empty into the Gulf of Mexico.

• The western coast of North America will be inundated for several hundred miles inland.

• Similar cataclysmic disturbances will occur around the globe— in the Arctic, Antarctic and South Pacific, for example.

These earth change predictions and others have been intelligently examined in *We Are the Earthquake Generation*. Another useful work on the subject is *The Age of Cataclysm* by Alfred L. Webre and Phillip H. Liss. Since these books have adequately covered this aspect of the Cayce readings, I will confine myself here only to those readings that pertain to a pole shift.

But before we look at them, we must consider the source of Cayce's knowledge. Clearly, it was not the conscious mind of the person known as Edgar Cayce. The information in the readings, unorthodox as it may appear, is no less than encyclopedic compared to that which the poorly educated Cayce himself could offer. It is a fact that Cayce, upon waking from trance, claimed to remember nothing of what had been uttered through his vocal apparatus. It is also a fact that the vocabulary, grammar and syntax characteristic of the readings differed markedly from Cayce's not unusual midwestern way of speaking when not in trance. Hugh Lynn Cayce's comment on his father's knowledge of Atlantis in the preface mentioned earlier is instructive:

My brother . . . and I know that Edgar Cayce did not read Plato's material on Atlantis, or books on Atlantis, and that he, so far as we know, had absolutely no knowledge of this subject. If his unconscious fabricated this material or wove it together from existing legends and writings, we believe that it is the most amazing example of a telepathic-clairvoyant scanning of existing legends and stories in print or of the minds of persons dealing with the Atlantis theory (p. 12).

The readings themselves indicate two sources from which the sleeping prophet got his information: the subconscious mind of the person for whom he was giving the reading, and what the readings called "the universal memory of nature." The latter has been compared to Jung's collective unconscious and to what Indian metaphysics calls the akashic records. Western religious thought has also called it the Book of Life and the Recording Angel. When someone asked Cayce during a reading to identify the source of his information, these words were spoken (294–1*):

> Edgar Cayce's mind is amenable to suggestion, the same as all other subconscious minds, but in addition thereto it has the power to interpret to the objective mind of others what it acquires from the subconscious mind of other individuals of the same kind. The subconscious mind forgets nothing. The conscious mind receives the impression from without and transfers all thought to the subconscious, where it remains even though the conscious be destroyed [as in coma or death].

The readings (3744–2) also said:

> The information as given or obtained from this body [E.C.] is gathered from the sources from which the suggestion may derive its information. In this state the conscious mind becomes subjugated to the subconscious, superconscious or soul mind; and may and does communicate with like minds, and the subconscious or soul force becomes universal. From any subconscious mind information may be obtained either from this plane or from the impressions as left by the individuals *that have gone on before,* as we see a mirror reflecting direct that which is before it. It is not the object itself, but that reflected. . . .

Lytle Robinson, in *Edgar Cayce's Story of the Origin and Destiny of Man,* remarks that the preceding passage in the readings,

*Numbers in parentheses in this chapter refer to the A.R.E. system of readings identification. The 294 encodes the name of the person for whom the reading was made. The 1 indicates the first reading for that person.

if true, means that Cayce's unconscious mind was able "to tap the mass of knowledge possessed by millions of other subconscious minds, including those who have passed over to the spiritual, cosmic realms in death. This would be an almost unlimited source of wisdom, since it was universal and Cayce was unhindered in time and space" (p. 218).

Even this comment, however, is not sufficient for explaining how it was that "millions of other subconscious minds" could be tapped and have their information *organized* and *verbalized*. Logically speaking, the situation requires some higher-order intelligence to exercise volition, discrimination, organization and control of the information-gathering system. In the introduction to this part, I showed, from a theoretical point of view, that such life forms may exist, ancient in time and vast in space.

In the final analysis, therefore, it seems to me that the real source of the Cayce readings was not human even in the sense of an extended collective unconscious. The collective unconscious may be likened to the data bank of a cosmic computer, but the *operator* of the computer could only be a form of intelligence from a level of life many orders of magnitude beyond the human. It is perhaps best named with the term chosen by both the Indian mystic Sri Aurobindo and the science fiction writer Arthur C. Clarke (in his *Childhood's End)* to describe an almost unimaginably advanced life form: the Overmind. The perspective of the Overmental level of life when viewing earthly affairs could be described, for all practical purposes, as nearly omniscient.

Notice that I said "nearly omniscient." While it is true that the Cayce readings are internally logical and consistent over more than four decades, and while it is also true that they have perhaps the highest percentage of accuracy ever established over so extended a body of psychic material, it must nevertheless be recognized that the readings had some inaccuracies and misses in the predictions. Forgive the pun, but "Cayce at the bat" did not average 1,000 for his lifetime record. It was exceedingly high, but not perfect.

For example, in 1933 Cayce gave a reading (270–30) that predicted an earthquake for San Francisco three years following. The question was asked, "Will the earth upheavals during 1936 affect San Francisco as it did in 1906?" Cayce replied, "This'll be a baby beside what it'll be in '36." History records that San Francisco experienced no such earthquake.

Likewise, in late 1932, Cayce gave a reading (311–10) that predicted physical changes in the northwestern and extreme southwestern parts of Alabama for 1936–38. The reading said the changes would be gradual and would take the form of "the sinking of portions with the following up of the inundations by this overflow." Again, history records no such earth changes in Alabama.

On the other hand, according to a professional geologist who authored a book entitled *Earth Changes* (a study of the Cayce readings which we will examine later in this chapter), Cayce correctly predicted "an earthquake in California on October 22, 1926; violent wind storms on the 15th and 20th of October, 1926; and the general location of strong earthquakes in California between 1926 and 1950."

To summarize the substance and stature of the Edgar Cayce readings is not easy. But perhaps it is best done in the words of Thomas Sugrue, who described them as presenting a philosophy or system of metaphysics that is "a Christianized version of the mystery religions of ancient Egypt, Chaldea, Persia, India and Greece. It fits the figure of Christ into the tradition of one God for all people, and places Him in His proper place, at the apex of the philosophical structure." The doctrines of karma, reincarnation and thought forms are integral parts of that structure.

What specifically did Cayce say about a radical displacement of the poles? Not much, actually. The few predictions he made are interspersed among various readings over a number of years in the context of a wide spectrum of changes that he said will occur from 1958 to 1998, including geological, climatological, economic, political and spiritual changes. Nevertheless, because Cayce is so well known and widely read, his vision of the future has been fairly well assimilated by those interested in psychic phenomena. It now permeates the psychic community, setting what can be called the "base line" for predictions of earth changes and a pole shift.

The earliest comments on polar shifts in the Cayce readings refer to prehistoric ones. In Reading 364–13, made in 1932, Cayce remarked that 10,500,000 years ago the present polar regions were tropical and semitropical regions, and the oceans were "turned about." I have italicized the relevant portions of his comments.

> In the first, or that known as the beginning, or in the Caucasian and Carpathian, or the Garden of Eden, in that land which lies now much in the desert, yet much in mountain and much in the rolling lands there. The extreme northern portions

were then the southern portions, or *the polar regions were then turned to where they occupied more of the tropical and semi-tropical regions;* hence it would be hard to discern or disseminate [sic—transcription error for discriminate?] the change. The Nile entered into the Atlantic Ocean. What is now the Sahara was an inhabited land and very fertile. What is now the central portion of this country, or the Mississippi basin, was then all in the ocean; only the plateau was existent, or the regions that are now portions of Nevada, Utah, and Arizona formed the greater part of what we know as the United States. That [land] along the Atlantic [sea]board formed the outer portion then, or the lowlands of Atlantis. The Andean or the Pacific coast of South America occupied then the extreme western portion of Lemuria. *The Urals and the northern region of same were turned into a tropical land.* The desert in the Mongolian land was then the fertile portion. This may enable you to form *some* concept of the status of the earth's representations at that time! *The oceans were then turned about....*

In another 1932 reading (364–4; A-1) Cayce said that during the life of the Atlantean culture, many earth changes occurred, including a pole shift. Again, I have added emphasis to the relevant portions.

...the variations [in Atlantis], as we find, extend over a period of some two hundred thousand (200,000) years ... and that there were *many* changes in the surface of what is now called the earth. In the first or greater portion, we find that *now* known as *the southern portions of South America and the Arctic or North regions, while those in what is now Siberia—or that as of Hudson Bay—was rather in that region of the tropics, or that position now occupied by near what would be as the same line would run, of the southern Pacific or central Pacific regions*—and about the same way. Then we find, with this change that came first in that portion, when the first of those peoples used that as prepared *for* the changes in the earth, we stood near in the same position as the earth occupies in the present—as to Capricorn, or the equator, or the poles. Then with that portion, *then* the South Pacific, or Lemuria[?], began its disappearance—even before Atlantis....

Reading 5249–1, made in 1944 during the last months of Cayce's life, spoke of "enormous animals" that threatened society during the Atlantean period. Plans were made by the Atlanteans to deal with the situation, but nature acted—through a pole shift—to exterminate the animals before man did.

> The entity* was then among those who were of that group who gathered to rid the earth of the enormous animals which overran the earth, but ice, the entity found, nature, God, *changed the poles* and the animals were destroyed, though man attempted it in that activity of the meetings.

Speaking of a later period in Atlantean history (about 50,000 B.C.), Cayce said (364–13), "You see, with the changes—when there came the uprisings in the Atlantean land, and the sojourning southward—with *the turning of the axis,* the white and yellow races came more into that portion of Egypt, India, Persia and Arabia."

Thus Cayce identifies spin axis displacement, rather than crustal slippage, as the means of pole shift in the past. Presumably, this would apply to future pole shifts also. With regard to a future pole shift, Cayce's words are few and not very precise. In a 1933 reading (378–16), while speaking of hidden records of the Atlantean race coming to light "when the change was imminent in the earth" (i.e., when Atlantis rises again during this century), Cayce said the change

> . . . begins in '58 and ends when the changes wrought in the upheavals and *the shifting of the poles, as begins then the reign in '98. . . .*

This cryptic phrase, "the shifting of the poles," was clarified during a reading (3976–15) in 1934 that dealt with world affairs. Cayce reviewed the "material changes" that are to come as signs in advance of the reappearance of the Christ.

> As to the changes physical again: The earth will be broken up in the western portion of America. The greater portion of Japan must go into the sea. The upper portion of Europe will be changed as in the twinkling of an eye. Land will appear off the east coast of America. There will be the upheavals in the Arctic and in the Antarctic that will make for eruption of volcanoes in the Torrid areas, and *there will be the shifting then of the poles—so that where there has been those of a frigid or*

*"Entity" means the person for whom the reading was made. In this instance, a previous incarnation of the person is discussed.

the semi-tropical will become the more tropical, and moss and fern will grow. And these will begin in those periods in '58 to '98, when these will be proclaimed as the periods when His light will be seen again in the clouds.

In a 1936 reading (826–8), Cayce was asked:

Q-9. What great change or the beginning of what change, if any, is to take place in the earth in the year 2000 to 2001 A.D.?
A-9. When there is a shifting of the poles. Or a new cycle begins.

Thus, the shifting of the axis as prophesied by Edgar Cayce will result in a dramatic realignment of the earth's position in relation to the sun. The dimensions of the shift are not given with any specificity, but clearly a substantial one is meant. Otherwise, how could areas of the globe that are presently frigid—arctic or subarctic—become semitropical? Webre and Liss, in *The Age of Cataclysm,* offer this interpretation:

... Moss and fern are typically native to the forests of North America, implying that the nature of the change will be at least such as to provide a temperate climate for areas now adjacent to the 70 degree meridian at both the Arctic and Antarctic Circles. This would include parts of northern Canada, the Soviet Union, Greenland, and Alaska in the northern, and the outlying parts of Antarctica in the southern hemisphere. The shift itself is internally consistent with the Cayce predictions of massive earth disturbance in the period 1958–1998, and would be the logical consequence of large-scale crustal displacements during the period (p. 106).

Will the pole shift predicted in the Cayce readings be a sudden cataclysm? Cayce's final comment (in Reading 1602–3) about the nature of the coming event was made in 1939 in context of remarks about Atlantis rising. The question was asked, "Will it cause a sudden convolution and about what year?" Cayce replied:

In 1998 we may find a great deal of the activities as have been wrought by the gradual changes that are coming about . . . [at] the change between the Piscean and the Aquarian age. This is a gradual, not a cataclysmic activity in the experience of the earth in this period.

Paul James, a long-time student of the Cayce legacy, points out in his book *California Superquake 1975–77?* that this statement makes quite clear that the coming polar shift will not be a cataclysmic change but, rather, a relatively gradual one. "This does not mean that there will be no earthquakes and other earth changes," he writes, "but only that they will not be cataclysmic in nature. Many of the world's coastal areas will flood, but over a period of two or more years…" (p. 88).

James interprets the Cayce readings to mean that the pole shift will *begin* in 1998 and still be under way in the year 2000. His understanding of what Cayce meant, he told me in 1978, is that the pole shift "will take place over two or three years—or more. We don't know about that farther end of it. It's supposed to start in 1998 and be going on in the year 2000. So I assume that if it's going on for two or three years, and it's still in the midst of going on at that point, it probably would go on for another two or three years. That's the logic I see in it."

If James is correct, then Cayce's version of the shifting of the poles—as drastic a geological-climatological event as it would be—is far less dramatic and devastating than those scenarios we have thus far seen projected by others.

James's hindsight on the inaccuracy of his central thesis in *California Superquake 1975–77?* is worth considering at this point. In as problematic a field as this, the misses can be as instructive as the hits.

James's interpretation of the Cayce readings, and much other material from psychic and scientific sources, led him to conclude that a "superquake" would hit California no later than the end of 1977. That interpretation is obviously wrong, and James himself was the first to ask the question, "What happened?" In January 1978 he issued a brief "Superquake Update" that gives his answer. He writes:

The great majority of Edgar Cayce interpreters, including myself, felt that when Cayce said, "The early portion will see a change in the physical aspect of the west coast of America," he had in mind the early portion of the 1958–98 earth change period which he had spoken of previously. This would place the big quake before 1978. We must conclude that we were mistaken about this "early portion" interpretation or that Cayce was mistaken.

Discounting this "early portion" consideration and moving on to consider what Cayce may have meant by his use of the term "generation," the great majority of interpreters, including his son, Edgar Evans Cayce, and myself believed Edgar Cayce had in mind the largest numerical definition, which is 33 years. But, if Cayce isn't wrong altogether, he may have had in mind a 40-year generation. . . . In that case, the California superquake would strike around 1982.

By 1982, then, we will have reached the outer limits of variability for the Cayce earth change predictions and will be in a position to more accurately assess their reliability, which in turn will bear directly on the reliability of Cayce's pole shift predictions.

James is not alone in his interpretation of a "gradual" pole shift. His view is shared by the anonymous author of *Earth Changes*, a book published in 1959 by the A.R.E. Press that examines Cayce's geophysical predictions and account of the earth's history against then-current scientific knowledge. The author is described as "a professional geologist, holding the S.B., S.M., and Ph.D. degrees in his field . . . [who] has authored a number of research papers for scientific journals."

In his examination of the possibility of polar shifts, the geologist cites two readings that come as close as I know to pinpointing what, from Cayce's perspective, will be the trigger mechanism for the pole shift. Cayce is puzzlingly silent on this question except for these hints. Reading 3976–10, made in 1932, states:

These [national boundary shifts in Europe] will not come as we find, as broken, before the catastrophes of outside forces to the earth in '36, which will come from the shifting of the equilibrium of the earth itself in space, with those of the consequential effects upon the various portions of the country—or world—affected by same.

Reading 5748–6, also made in 1932, asks:

Q-10. What will be the type and extent of the upheaval in '36?

A-10. The wars, the upheavals in the interior of the earth, and the shifting of same by the differentiation in the axis as respecting the positions from the Polaris center.

The geologist considers these readings to be of "the utmost importance" because they "point to an immediate (terrestrial) mechanism underlying the proposed earth changes—namely, a shift of the earth's axis of rotation, or a shifting of the interior of the earth with respect to the rotational axis" (p. 31).

The phrase "upheavals in the interior of the earth" is taken by the geologist to probably mean convection currents. He points out that the geophysicist Vening Meinesz concluded in a 1958 book, *The Earth and Its Gravity Field,* that previous polar shifts "could not have taken place without currents in the mantle" (Meinesz's words) of the earth's interior. The geologist offers data in support of the phrase "shifting of the equilibrium of the earth itself in space" and then concludes, "It would seem that *upheavals in the interior of the earth* (convection currents?) could, in a few years, bring about many of the rising and sinking land movements described in some of the readings . . ." (p. 32).

That is not to say a pole shift would occur, however. The geologist is undecided on that point. Nevertheless, we have already seen evidence that it might occur, and we shall see still more. Suffice it to say here that Cayce's readings indicate a pole shift due to a repositioning of the planet in space, rather than a crustal displacement, and that the speed with which it will take place— several years or so—is moderate in comparison with the predictions of Brown, Thomas, Barber and Sepic. And as we shall see in the following chapters, it is moderate indeed in comparison with what other psychics have to say on the subject.

References and Suggested Readings

Anonymous. *Earth Changes.* Virginia Beach, VA: A.R.E. Press, 1963.

CAYCE, EDGAR EVANS. *Edgar Cayce on Atlantis.* New York: Paperback Library, 1968

*Circulating File on Earth Changes.*Virginia Beach, VA: A.R.E. Press, 1977. (An annotated research guide to the Edgar Cayce readings on earth changes.)

JAMES, PAUL. *California Superquake 1975–77?* Hicksville, NY: Exposition Press, 1974.

ROBINSON, LYTLE. *Edgar Cayce's Story of the Origin and Destiny of Man*. New York: Berkley, 1972.

SUGRUE, THOMAS. *There Is a River*. New York: Holt, Rinehart and Winston, 1942.

WEBRE, ALFRED L., and LISS, PHILLIP H. *The Age of Cataclysm*. New York: Berkley, 1974.

POLE SHIFT

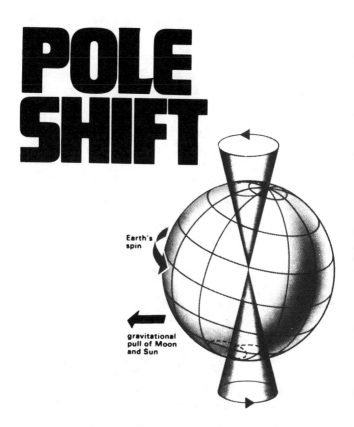

Earth's
spin

gravitational
pull of Moon
and Sun

Precession of the axis. The earth spins at an angle to the plane
of its orbit around the sun; its axis is tilted nearly 23.5°. The
gravitational pull of the sun and moon cause a slow wobble.
The North and South poles take nearly twenty-six thousand
years to complete a circle.

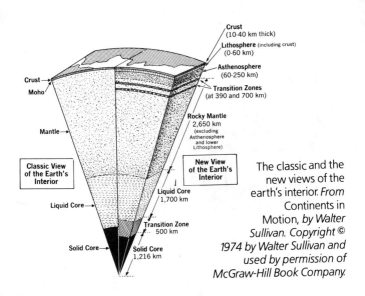

The classic and the new views of the earth's interior. *From Continents in Motion, by Walter Sullivan.* Copyright © 1974 by Walter Sullivan and used by permission of McGraw-Hill Book Company.

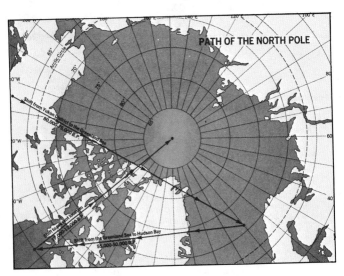

The path of the north pole according to Charles Hapgood. B.P. means "before present." *Courtesy of Charles Hapgood and Chilton Book Company.*

Escape from Atlantis? The beginning scene of a continuous bas-relief frieze discovered in the Yucatan jungle by German archaeologist Teobart Maler shows that the Mayans linked their beginnings to an escape from a cataclysm. At upper left, a pyramidlike structure topples into floodwater, while a volcano erupts and the land sinks. The stylized art also shows heavy rain clouds. The figure in the water suggests that many drowned, although many others escaped by boat. *From The Life & Death of Planet Earth, courtesy of Tom Valentine.*

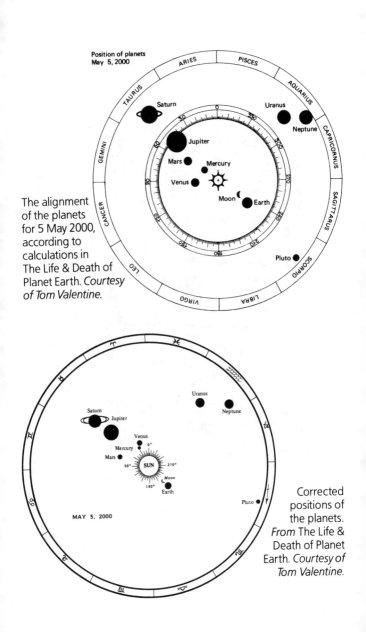

Position of planets
May 5, 2000

ARIES PISCES
TAURUS AQUARIUS
Saturn Uranus CAPRICORNUS
Neptune
GEMINI
Jupiter
Mars Mercury SAGITTARIUS
Venus
Moon Earth
CANCER
Pluto SCORPIO
LEO
VIRGO LIBRA

The alignment
of the planets
for 5 May 2000,
according to
calculations in
The Life & Death of
Planet Earth. *Courtesy
of Tom Valentine.*

Saturn Uranus
Jupiter Neptune
Venus
Mercury 0°
Mars
90° SUN 270°
Moon
180° Earth
Pluto

MAY 5, 2000

Corrected
positions of
the planets.
From The Life &
Death of Planet
Earth. *Courtesy of
Tom Valentine.*

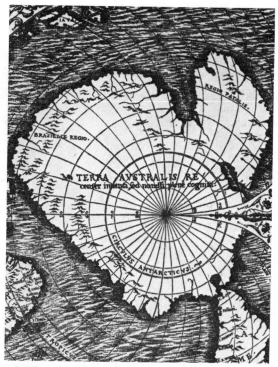

Section of the Oronteus Finaeus map of 1531, showing Antarctica. *Courtesy of Charles Hapgood.*

Antarctica: *right,* the modern map; *left,* Oronteus Finaeus, 1531. *Courtesy of Charles Hapgood.*

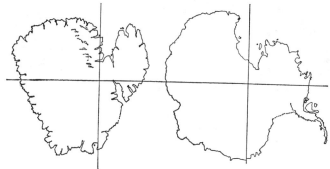

Hopi prophecy. This petroglyph, carved on a rock near Oraibi, Arizona, records coming world events as foreseen by Hopi elders centuries ago. Carving lines are highlighted in lower photo. *Courtesy of Nathan Koenig.*

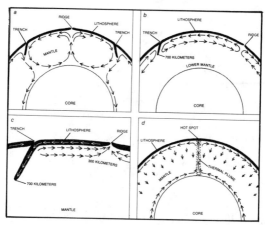

Models of convection have been proposed to explain how activity in the mantle drives the lithospheric plates. In convection, warmer material moves upward and colder material moves downward. One model, (a), holds that convection cells extend through the entire mantle. According to a second model, (b), they are confined to depths above the phase transition from spinel to olivine. A third model, (c), confines movements of the mantle to the asthenosphere. In the thermal-plume model, (d), all upward movement is confined to a few thermal plumes, and the downward flow is accomplished by slow movements of the remainder of the mantle. *From "The Earth's Mantle" by Peter J. Wyllie. Copyright © 1975 by Scientific American, Inc. All rights reserved.*

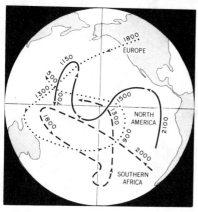

The wandering path of the north magnetic pole, as reconstructed from remanant magnetism in rocks from three continents (age of rocks in millions of years). Differences among pole paths are said to be due to continental drift. *Courtesy of* Earthquake Information Bulletin, *Vol. 5, No. 1, 1973.*

Dima, the baby mammoth found in 1977 in Siberia, being measured by a staff member of the U.S.S.R. Academy of Sciences.
Tass, from Sovfoto.

The restored Berezovka mammoth, in the position in which it was discovered, on display in the Zoological Museum, Leningrad.
Novosti, from Sovfoto.

Chapter Twelve

Aron Abrahamsen: A New Earth and a New Humanity

The most extensive body of earth change prophecies comes from—
or rather, through—the body and personality known as Aron
Abrahamsen, a psychic living in Everett, Washington.* In
1972–73 Aron, as he likes to be called, made a series of readings
that, after transcription, come to more than one hundred fifty
pages. He published them privately in bound mimeographed form
in an edition of several hundred copies. Entitled *Readings on Earth
Changes,* it is a rare volume well worth study, because it shows
in the extreme the difficulties one can experience when working
with psychic information.

---

*Aron Abrahamsen can be reached through the organization he
founded: Association for the Integration of Man, P.O. Box 5008, Everett,
WA 98206. The purpose of A.I.M. is to research and disseminate the
information from the many readings that have come through Aron since
1970.

Consider, for example, this prediction, made in October 1972 about great earth changes that would begin in California:

> Now toward the end of that year in 1977, there will be another disturbance, this time in the area of Sacramento and going south—well, no. It'll go northwest, out toward the coast around Eureka, but the greater disturbance, the greater damage will be in the valley of the Sacramento area and toward the coastline. Here again, some of the areas will be disturbed such as the Bay Area in the San Francisco region, the Salinas Valley, the San Joaquin and Sacramento Valleys but particularly the Sacramento Valley will be disturbed inasmuch as some of the land will be shifted. The hills around the main thoroughfare, the main highway going north will be seen as having big gaps or big crevasses in them, both to the east and to the west. . . .

This statement is obviously incorrect. Does this mean we should disregard Aron Abrahamsen as a source of paranormal information? Not at all. For in addition to documented hits in earthquake forecasting, he has also had some startlingly correct predictions in other fields. These predictions elevate him to the top ranks of contemporary psychics.

With that note of caution, I will now introduce Aron Abrahamsen in terms of his most dramatic "hit": the location of the oldest remains of man in North America.

In 1973, when I was working with astronaut Edgar Mitchell at the Institute of Noetic Sciences in Palo Alto, California, I was contacted from Tucson by a graduate student of geology and archaeology named Jeffrey Goodman. (He has since earned his Ph.D. in anthropology.) He wanted to let the Institute know—because we were investigating psychic abilities—of an important archaeological find he'd made through the services of Aron. Goodman had heard of Aron through friends at the A.R.E. in Virginia Beach, and had contacted him for assistance in his archaeological research. After some initial correspondence and telephone conversation, Aron made a reading for Goodman in 1972 that told him exactly where to find human artifacts more than one hundred thousand years old. Aron was living in Oregon at the time, and from his home he named a location near Flagstaff, Arizona. He had never seen the location, buried in the mountains one thousand miles away. Yet his reading gave an extremely detailed description of what would be found as the digging proceeded. The description, at once archaeological and geological, included the type and the age of the artifacts, and the soil types and rock formations to be encountered at exact depths down to twenty-three feet. At that point, Aron said, the desired evidence would be unearthed. The reading then continued down to fifty feet, at which point a still more extraordinary find would be made—a matter that Goodman has not released to the public yet.

Goodman had the reading evaluated by professional anthropologists, archaeologists and geologists. All agreed that Aron was completely wrong, that there was no chance an excavation would uncover such strange geology, let alone evidence of the presence of humans in North America many tens of thousands of years earlier than signs yet detected by science.

Nevertheless, Goodman went ahead with the excavation in 1973. His faith was founded in part on a precognitive dream that he himself had had in which he clearly saw the valley and the site where he would dig (although he could not locate it until Aron told him). Once there, Goodman found himself experiencing in waking consciousness what he had perceived in the dream state.

The story of Goodman's momentous discovery is told in his recent book *Psychic Archaeology*. Not so incidentally, of Aron's predictions for the Flagstaff site—fifty-eight specific statements about the geology, chronology and artifacts—fifty-two have proved correct, for an 89 percent accuracy rate. Even more astounding are the conclusions Goodman draws from the data. The first Americans, it appears, did not come from Asia but from an altogether unknown culture, possibly Atlantis.

This dramatic situation shows why the Abrahamsen readings on earth changes and a pole shift must be taken seriously. There is another reason, too, already mentioned: Aron's proven accuracy in medical diagnosis through clairvoyance. A brief account of this is given in Goodman's second book, *We Are the Earthquake Generation,* the story of his work with Aron and other psychics on the problem of earthquake prediction. (It was Goodman, incidentally, who proposed the questions to Aron that elicited the responses now published as *Readings on Earth Changes.*) Here is an example of Aron's unusual ability to assist medical science, as reported in *We Are the Earthquake Generation.*

> . . . a group of doctors in a southwestern research clinic with experience in epilepsy recently had a young epileptic whom

they could not help at all. The girl was having 50 to 60 grand mal seizures a day. In desperation the doctors advised the mother to seek the aid of a psychic. Aron's name came up. The mother contacted him, and he gave an emergency reading for the girl which resulted in specific treatments being advised. These treatments were followed and in just two weeks the doctors considered the girl almost totally cured. Her condition had improved to the point where she was having just one or two mild seizures at night in her sleep. The doctors were so impressed that they advised further readings to explore the reasoning behind the unusual treatments (p. 25).

Who is this unusual man? Physically, he appears quite ordinary: nearing sixty, medium height and build, with a slight Scandinavian accent, bald, soft-spoken and shy of publicity. When I first met him and his wife, Doris, in San Francisco in 1974, I was impressed with their humility and desire to be of service to the human race. The two people seemed to function almost as one, yet without losing individuality and the ability to consider critically what each other was saying.

Aron was born in Norway in 1921 of Jewish ancestry. At eighteen he came to the United States and served in the Navy during both World War II and the Korean War. When peace came in 1945, he began to study electrical engineering at California State Polytechnic University. After earning a B.S. degree in 1950, he went into the aerospace industry. During his twenty years' experience in the engineering field, he worked for one of the largest aerospace companies, where he was appointed part of a ten-member study team to determine the feasibility of sending a manned expedition to the moon and back. Many of his concepts, he says, were used in the program that took the Apollo spacecraft to the lunar surface. He also authored several technical papers.

Aron and Doris were married in 1950, seven months after they met on Valentine's Day. Early in their marriage they began their spiritual search for deeper understanding of the religious feelings they shared. In 1955, they tell friends, a very meaningful spiritual experience changed their goals and purpose in life. Many years were spent in Bible study, prayer and meditation. They became active in church life (Baptist), teaching Sunday school and directing youth activities. They also began probing esoteric subjects such as reincarnation and karma. Fluent in Hebrew since childhood, Aron once recorded the Bible in that language for Bible

Voice. He also taught himself Greek in order to go further in his study of the Old and New Testament.

The Abrahamsens found themselves unfolding in a way that felt deeply fulfilling. People began coming to Aron for counseling and, without willing it, he was soon conducting an informal ministry of sorts. Through a spiritual experience in 1963, he learned that he could give personal prophecies, both psychological and medical, for the people he was counseling about life and health.

This gift was radically extended one day in 1969 when Aron had an out-of-body experience that left him with an ability to give past-life readings. Purely for relaxation before counseling, he had begun to meditate, but he soon learned that he received much more than relaxation in his meditations. While in meditation, he found himself at a large hall of records. Two "keepers" of the data there explained that when he needed it, he could have information about a person's previous incarnations—his thoughts, behavior and attainments. This hall of records—undoubtedly what Cayce called the akashic records—is part of the "higher source" to which Aron attunes in order to obtain paranormal information. But the hall of records pertains only to the past. Knowledge of the future also comes from the "higher source." However, what the exact nature of that source is Aron will not, or cannot, say, though it may be simply a logical deduction based on his paranormal insight into past and present—i.e., reading the process of thought form interaction.

Since he had developed this paranormal ability and felt a deep calling to religion and metaphysics, it is understandable that over the years friends began urging Aron to devote himself to this work completely. Although uncertain at first, he agreed to this work when Doris also joined in recommending it. In 1970, therefore, he ended his engineering work and expanded his ministerial activities into a full-time occupation. Thus, as Aron and Doris now describe it, "new and meaningful avenues of service to mankind opened up." Also in 1970, both became ordained ministers in the Universal Christian Church, under the American Ministerial Association.

Eight years earlier, Aron and Doris had begun a spiritual retreat center at Big Bear Lake in southern California, near where they were living at the time. Soon after discovering his life-reading ability, Aron was put to the test by a ministerial friend. The friend was attending a strongly suicidal young woman who had attempted to end her life many times. Her physician and psychiatrists were at a loss to treat her. Heavy sedation was their last resort to keep her from committing suicide.

Aron gave a reading for the woman and mailed the tape to his friend, who used it in his counseling therapy. The woman recognized past life tendencies and mistakes being repeated in her present situation. She saw the need to resolve her problems, rather than run away from them, and began to reorient her life. To everyone's amazement, the woman's desire to commit suicide disappeared.

This encounter helped convince Aron that he had insights of potential merit to offer others from his meditative state, and he began giving readings on a regular basis, assisted by Doris. She presents the names to him once he is in an altered state of consciousness. The readings are done early every morning. Aron reclines on a chair, Doris beside him with a tape recorder. Over the years, Aron has made more than three thousand readings for individuals from all over the world, describing their past lives or present conditions but always with an emphasis upon self-understanding and practical advice for the person's life situation so that he might—as the name of Aron's organization, Association for the Integration of Man, proclaims—become more integrated, whole.

This philosophy extends from the individual to society. Having seen the antagonism that can exist between religions and sects, Aron and Doris do not wear any denominational label but, rather, attempt to "bridge the gap" between religion and metaphysics. "There are truths on both sides," Aron says. "The sad part is, one will not speak to the other. Of the metaphysicians, the religious say, 'You are of the devil—spaced out.' Of the religious, the metaphysicians say, 'You are too narrow—just not with it.' I believe in order for man to be able to get along with himself, he needs the peace of God, and if there is to be peace in the world, it has to start with the individual. We must love one another. This does not mean we need to agree. No one person, denomination or sect has a monopoly on the truth. This is what I mean [when I say] I am endeavoring to bridge the gap."

Although he has not said so explicitly, Aron is also trying to build a bridge between science and religion. His readings often present new data and insights for science, rather than contradict or condemn it. One example of this is given in *We Are the Earth-*

quake Generation: a new forecasting tool for seismologists by which earthquakes can be predicted months in advance. The principles on which this potentially important safety measure rests and the methods by which it can be implemented are well known to science, although Aron—or his higher source—is the first one to bring them together.

Goodman also remarks that Aron performs somewhat differently from the way "sleeping prophets" such as Cayce did. "Unlike them," he tells us, "Aron has recall of what he has said and can respond to questions seeking further details about what was given. Also, he is capable of giving psychic readings that critique his own previous statements. . . ." This would indicate that it is the personality of Aron, not a discarnate source, that is "in charge."

What do the Abrahamsen readings tell us about a pole shift? First, it will take place in 1999 or 2000. Second, it will involve a full 180° "flipping" of the poles that will occur over a period of several days, when the sun will seem to move rapidly through the sky, stand still and then move again. Parts of the world will "go from light to darkness and to light again in a very short time. . . ."

By 1985 the axis will begin to tilt slightly. The North Pole will move in a westerly direction and the South Pole in an easterly direction, when viewed from a position in space above North America. But this movement will be relatively slow and small—taking several years and moving only a few degrees. The axis will wobble back and forth slightly during this time, but this will not be noticed in any particularly unusual way by the public. (Parenthetically speaking, it would certainly have to be noticed by the scientific community and those who depend on exact celestial measurements, such as missile guidance system programmers. And until a stable earth was achieved, long-range missiles could not be fired accurately—an intriguing thought for those, such as the CIA, who predict nuclear warfare in the mid-eighties.)

Let's look in detail at the readings now. In Chapter I, page 5, of *Readings on Earth Changes,** we learn that the axis "will hold itself in a balance of around twenty to twenty-five degrees† for

---

Readings on Earth Changes is not paginated consecutively from start to finish. Each chapter begins with its own page 1. Therefore, I will refer to pages in the book by citing chapter and page. Thus, for example, page 1 of the fifth chapter: p. 5/1.

†Aron is probably referring to the earth's present angle of inclination: 23.45°.

some time. The critical angle is about thirty, thirty degrees. At that point, it will be losing equilibrium and at that point, it will flip rather quickly within a short span. At that point, the sun will be darkened. It will appear darkened for several days."

The reading continues, with questions being asked by Goodman, who is designated as the "Conductor" of the session:

(CONDUCTOR: It will what?) It will darken. The sun will be darkened for the earth has not found its balancing point, its point of equilibrium so it will have the tendency to, in some sections, to shun away from the rays of the sun instead of being attracted to the sun, before it can be spinning up again and start on its new course. For the spinning of the earth on the earth axis will now be done in the same direction as it is now, except it will be opposite. You do not understand this. (CONDUCTOR: No.) As it is now going from east to west, so will it be going in the same direction except it will be going from west to east. In other words, the axis of the earth will be going in the same direction as before after the flip has taken place, after the north and the south pole has then interchanged, so will the axis of the earth, the rotational spin of the spin axis will be in the same rotation as it is presently. This means that the earth will have to be slowed down and be spun up again in the opposite direction so it can go in the same direction as it is now. (CONDUCTOR: How will that happen?) It will happen simply as the earth tips about and the axis or the spinning will slowly stop, so will the attraction between the sun and the moon force the rotation to start again. This will take place in a matter of a few days. (CONDUCTOR: I see.) Is this clear? (CONDUCTOR: Yes.) And then all things will normalize again, except that the climate, the climatic changes will have taken place.

Before this takes place, there is the tremendous earth changes. Now the cataclysmic changes taking place on the western part as well as the eastern part.

(CONDUCTOR: During those few days that the chaotic spinning period will be, will there be earth changes during that time?) Oh yes. But the most of it will have taken place before. By 1990 up to the year 2000, that is when this thing will be taking place. For about ten years, there will be a constant change procedure, so to speak, a constant. Is this clear? (p. 1/5–6).*

In other words, the axis of rotation will be in the same position, with relation to the sun, after the pole shift, and the planet will rotate in the same manner and direction as before. But since the

*See *We Are the Earthquake Generation* for a summary of the earth changes predicted by Aron and others.

world is, in effect, turned upside down, dawn and sunset will be reversed. Boston, for example, instead of being on the east coast of the United States and receiving the sun first will be on the west coast and will receive the sun last.

But even this explanation is not a full description of the event Aron foresees. In a very recent reading, No. 3399, made in March 1978, the following information was given. It adds a new dimension to the pole shift event by revealing not only that the outer crust will slip around the interior of the planet but also that the interior itself will shift by about half the amount of the crustal movement.

> We need here to specify that though the northern and the southern hemispheres will be interchanged . . . that which was on the north will be the south and that which was the south will be the north. However, as far as the poles are concerned, where the ice is formed, so will that not go a complete 180°, but it will go somewhere in between the 180°—somewhere between where it is now and 90°. For that which determines the pole, that is, location of the poles, will now be the rotation of the inner core of the earth. It is here where the magnetic force is also generated. It is like a self-generating magnet—a permanent magnet which also will be thusly oriented only 90° whereas the entire earth will slip and go 180° but the poles will only shift 90° or between where it is now and 90°. Therefore, the areas which are now very cold will then be very warm and temperate. Where it has been somewhat warm, there will be a different climate coming about. . . . But keep in mind that though the earth will flip, the axis—that is, the poles—will not flip in the same degree or at the same distance as the earth itself. There will be somewhat of a displacement such as I mentioned whereas the northern hemisphere will become the southern and yet the north and south poles where the ice caps form will only slip anywhere between the position where they are now and that of 90°. So there is even a slippage within the core of the earth which will not always be synchronized one to the other.

Readings on Earth Changes gives much more information about the conditions to be experienced during the pole shift. This passage describes the effect of the sun "standing still":

> (CONDUCTOR: . . . you indicated that there'd be a darkening of the sun for several days during the flip. Could you give the relationship between what would be observed in particular areas

of the world?) All right, Africa will experience a lot of light. The part, for example, of China which is now China will see the sun as if it were intended either to rise or fall. They aren't sure yet in which direction the sun will come, you see. On the other side, Australia will see something similar. It'll be semi-darkness and there will be other parts of the world, very few though, but other parts such as, when the earth is flipped, such as the United States and even Norway, the Scandinavian countries which will truly exhibit darkness. These are the areas to observe.

(CONDUCTOR: Okay. Will there be any particular sounds or noises or . . . ?) Wind. It'll be almost like a deafening wind which nobody will know what it is for they have not experienced it.

(CONDUCTOR: What speeds will these winds reach?) Several hundred miles. And after the earth has flipped, within twelve to twenty-four hours, the climate is established. . . . (CONDUCTOR: Okay. Are there any specific earth changes to occur during the flip itself?) No (p. 4/11).

Earth changes, no—but climate changes, yes. The readings indicate great changes indeed. The temperate climate of the West Coast will become that of the East Coast. The humidity of the eastern seaboard—the new one—will "suddenly be gone" and the area will develop a more favorable climate, comparable to that of the West Coast today. Both the eastern and southern portions of the country will become somewhat cooler. The new climates, as given above, will establish themselves "within twelve to twenty-four hours" after the planet has tumbled over.

The readings produced by Aron and Goodman explain that there are two factors necessary to bring about a pole shift: interplanetary forces and human thought forms.

Interplanetary forces are described as "the law of the universe, that is maintaining a balance between all the asteroids, planets and also the orbital shape of the planets about the sun and other solar systems." Ordinarily we would think of this description in terms of gravity. But the readings indicate that electromagnetism, not gravity, is the major interplanetary force at work with regard to a pole shift. Like the Solomon readings, which we'll examine in the next chapter, the Abrahamsen readings see the earth's magnetic field as having primary importance here. In fact, a recent reading, No. 2993, made in 1977, tells us that an alignment of the outer planets (date not specified) will "begin to change—weaken—the magnetic force field around the Earth itself. . . ." Simultaneously, "the temperature within the crust of the earth will now begin to rise." The result:

> [So] will there now be that energy also entering into the earth's atmosphere which will now begin to melt the ice cap. And as the ice cap will begin to melt, so will there be a circulation of the heat back into the earth again, which will further increase the temperature. . . . Thus as the ice cap continues to melt—this is now a chain reaction and like a domino effect—and soon one will have reached the critical point whereby the earth will finally flip. This is now that the north pole will take place where the south pole is, and vice versa. And at that point, so will the climate completely have changed; and whereas there has been a very cold climate in the regions of Alaska, that will now become a tropical climate.

Although interplanetary forces are necessary for a pole shift to take place, the readings say, they are not in themselves sufficient. A second force is also necessary—thought forms—and this is the critical element. The interaction of interplanetary forces and thought forms produces a shifting of the poles.

The concept of thought forms was explained in the introduction to this part. Edgar Cayce popularized the esoteric term. The Abrahamsen readings tell us that the earth's axis will flip due to the influence of "the forces about the earth in conjunction with the man's thought forms as well as there would also be interplanetary actions such as comets coming by . . ." (p. 3/10).

Another recent reading, No. 3081, made also in 1977, tells us that a pole shift is not inevitable, due to the potentially saving grace inherent in thought forms, which in turn are dependent upon the quality of human thought and behavior. ". . . there is a hope," the reading states, "that within the time of 1981 till about mid-1982, so is there still an opportunity for mankind to repent and to do good so these changes do not have to take place, but that the earth can continue on its path staying the way it is at the present."

In *Readings on Earth Changes,* beginning on page 3/11, the dates of some previous pole shifts are given, along with the locations of the poles at those times. According to this information, the most *recent* pole shift occurred in 70,000 B.C.—a far cry from what Brown, Hapgood, Velikovsky, Thomas and some others have to say. That pole shift was, as the reading puts it, "a complete flip"—a full 180°, as the coming one will be. At the time of the previous one, the North Pole was about 30° to the west of its present location, when viewed as if in space above North America.

Another pole shift took place about 147,000 years ago, although the axis displacement at that time was only 90° to the east. The North Pole was then between China and India, and the South Pole was on the lost Pacific continent of Lemuria, near the present western coast of South America.

In addition to these, the reading states, "there have been minor shifts, five to ten degrees and there has been also a shift in about the 200,000 [years ago] where there was a 90° shift where the north and the south [poles] was where the equator is."

When Goodman asked Aron in his meditative state whether there had ever been any pole shifts more recent than 70,000 B.C., he replied quite emphatically that there had been none. Then, Goodman inquired, what about the many accounts in the literature of the Egyptians, Mayans, etc., that apparently refer to pole shifts?

> What you are hearing about, for example, in the anecdotes of the Chinese, the philosophies, even in many of the temples in the Himalayas, in Tibet, so will you find accounts of earthquakes, earth shifts and so on, but these are merely retelling what they had heard before. There had been stories carried down, so there hasn't been any real recent, as you call it, recent in your own time (p. 3/13).

The reading also dates certain biblical events (p. 3/14). The flood of Noah, we are told, took place about 20,000 B.C. and the Exodus happened around 7,500 B.C. Again we have tremendous departure in dating from Velikovsky, Brown and Thomas.

The coming pole shift, cataclysmic as it may appear, will not be the end of human history, according to the Abrahamsen readings. Rather, it will be the end of an age and the beginning of another. A new age, a new earth and a new humanity will emerge from the ruins of the old because those who survive will have understanding of universal laws—the laws that inevitably brought on destruction of our civilization because people persisted in violating them.

The Abrahamsen readings portray the recovery period taking place for thirty years or so after the pole shift, during which there will be severe hardship. Eventually, though, the remnants of civilization will develop a solid foundation from which to erect a new world society.

> In that case, mankind will have to learn to live with himself, to love one another, to learn not only to tolerate but learn also what it means to love one another. To live in peace and to live in prosperity without war. That will be the high direction of the new way, of the new direction.
> Mankind will have to learn to understand that as he also begins to find fault with others, he will also find fault with himself. He will have to learn not to blame others for his own intolerance and learn also that all men are equal, all men in the eyes of God are equal. We are coming back to the same message which was spoken from the mountain a long time ago; you shall love your God with all your heart and love your neighbor as yourself (p. 5/17).

This process will be reinforced by the return of the Master Teacher, the Christ, who will appear among the survivors "in about the year 2000." He will teach, as before, that the kingdom must be set up "within each person."

And that is really the whole thrust of the Abrahamsen readings: to awaken people to impending disaster being brought on by their own selfish, unconscious acts of hatred, greed, intolerance, violence and so forth, and to help people to see that their own minds — their thoughts and behavior — are the key to averting catastrophe. If the Great Teacher returns but no one listens and learns, the results will inevitably be as before. But if people awaken to the Teacher within, the Kingdom within, the predictions can be forestalled, even changed altogether.

I have emphasized throughout this book that the intelligent response to claims of psychic prediction must include a test of the accuracy of the source. To accept the Abrahamsen readings — or any others — at face value in every detail would be foolish, since there are obvious differences between them and other prophecies claiming equal validity. I have already indicated that some of Aron's predictions have proved untrue — or at least inaccurate, in the sense that they may still come to pass but not at the indicated time.

One scientist who has made a serious investigation of earthquake predictions is Dr. William Kautz of the Computer Science Laboratory at Stanford Research Institute, in Menlo Park, California. An Sc.D. in electrical engineering, Kautz has been working with a team of psychics to assist his computer-evaluated studies of three topics: crib death, archaeology and the physics of earthquakes. I met Kautz in 1973 when I heard through friends of his interest in the topic I myself was researching: prophecies of earth changes and a pole shift. We quickly became friends, sharing information and resources. Naturally, as soon as Aron and Doris came to my attention, through Goodman, I told Kautz about them. Kautz in turn got in touch with Goodman, later visiting him to inspect the Flagstaff site and to discuss how Goodman worked with psychics. Kautz also began to examine Aron's predictions (I lent him my copy of *Readings on Earth Changes)* along with those of other psychics he was using in his investigations.

It became apparent to Kautz that Aron's predictive ability, impressive as it obviously is, could not be counted on as being totally accurate. (Nor did Kautz expect it to be.) For example, one of the earth-change readings that Aron made for Goodman concerned damage to the Bay Bridge connecting San Francisco with Oakland. The reading states:

> There will be disturbances felt across the bay in about the mid-'73. This will not be so distracting except there will be some buildings will feel the impact. There will be some damage done. We would not call it a real severe earthquake, though it will have its impact. Some of the bridges across the bay will be damaged to a certain extent. The Bay Bridge will not be damaged severely though it will lose some of its moorings on the north side, so it will have a tendency to be a little more flexible than what it should. This will be somewheres in the mid-'73, by July or August, even into the September area.

In a report on earthquake predictions that Kautz prepared in 1975 for personal use only, he noted the following about Aron's Bay Bridge prophecy; "There were the usual number of very small earthquakes around Berkeley that year, although no medium or large ones occurred. I contacted the chief engineer for the Bay Bridge and he informed me that the bridge is inspected regularly and there had been no detectable damage to the moorings during the past several years."

Recently I asked Kautz to give an evaluation of Aron as a psychic. His letter of reply, excerpted below, reminded me that he had become close friends with Aron and Doris since I introduced them, and therefore the objectivity of his evaluation should be readily apparent:

> Aron is one of my best sources on subject matter dealing with people—individuals, groups, present-day persons and figures in history. He has also generated really fine information for me in many technical areas, even some where the subject matter is quite abstract. However, the farther we get away from people's lives, the greater is the variation in reliability. That is, like the others with whom I have been working, his information is sometimes at variance with the consensus. It is conceivable that he is right and the others wrong, or course, though I doubt it. What this all adds up to is: while Aron is generally pretty reliable in technical areas, he is fallible, like all the others, especially on "details": dates, locations, sizes, orientations, particular physical forms and materials (as opposed to functions), separating cause from effect, and identifying the most important of two or more factors. But, then, I do not know of any infallible channels, do you? It is precisely for this reason that I feel that one should never rely on a single source, except possibly for personal guidance. A team approach is required to obtain practically useful technical information from these sources.

In early 1978, when it was clear that several of Aron's predictions for southern California had not come true, I requested Aron to comment on himself. How, I asked him, do you view your own abilities and what can you say to account for the incorrect statements you gave in early 1973? His answer was straightforward and well worth consideration by anyone investigating psychic predictions. It assumes that there is an objective body of data to be given out by higher intelligence channeling it through human minds, and it is the human channel—not the "cosmic data bank" itself—that puts "noise" or misinformation into the system.

> Of course I am not always right, not because the information is incorrect, but because the way it is interpreted through my channel. As you know, it is extremely difficult to pinpoint a date from the other side; we can only hope that we have received the correct interpretation. Other than that, I have no explanation for the misses. It might at the same time also be interesting to

speculate why the right answers came out. There is still so much to learn about the mechanism of this process of obtaining information psychically. Until then we must continue to do our best and give out what we are able to obtain at the time. . . .

In regards to the misses—in one case about a year ago, in a research reading for a friend, several dates were given for tremors in California. The month and day were given. On one of them, the quake fell on the day given, but a month ahead of time. We can only give what we get. Another time the one we expected came six months late. We wish we were able to nail it right down, to the very hour. That would really help the earthquake scientists to check them out. But we aren't there yet. We continually seek to improve our abilities to be of help. It makes us feel terrible when we miss—but we don't condemn ourselves for we know we have done our best for that moment.

Aron and Doris continue to give readings and to work in preparation for earth changes and a pole shift. In August 1978 they sponsored a conference on "Changes to Come." The four-day event, in Livingston, Montana, drew about one hundred people, who participated in lectures, workshops and panel discussions on changes in the earth and in people's lifestyles. Ominously, a serious earthquake struck Santa Barbara, California, at that time, and lesser quakes were recorded in San Diego and Alaska.

The purpose of the conference was, first, to suggest ways to prepare for earth changes if they should come and, second, to explore how inner changes in people ("'quakes' in our own being that shake us out of our lethargy," as Aron and Doris put it) can reduce and perhaps even eliminate earth changes. Specific topics included were nutrition, canning and food preservation, gardening, color and music attunement, working with dreams, alternate energies, healing, massage therapy, Bible study and, of course, earth changes as foreseen in the readings of Aron, Cayce and the man whom you'll meet in the next chapter, Paul Solomon.

Although the conference ended with no plans for a follow-up event, Doris and Aron told me, it produced a lot of love and light—enough so that conferees left well filled with information and inspiration to apply in daily living. That daily application is crucial, according to the Abrahamsen readings. Aron and Doris put it this way in a letter to me in October 1977:

> . . . we have had in many readings that there is still time for a change within the actions of people (there is, of course, always

time for that) but that by 1981 the point of no return will have arrived. Either by that time people will have improved or the momentum will be so great that it will be difficult to induce a change in the events which are to follow. It appears that if these changes in human behavior do not come about by 1981, the world—and in particular the western world—will have to follow in the cycle, just as summer follows spring.

But even if there are tragedies and catastrophes, consider the opportunities for growth; and those who are looking for their teachers—look out, for they will appear, not necessarily in clouds or in shining armor, but in the very experience which is facing them at that very moment.

References and Suggested Readings

ABRAHAMSEN, ARON, and ABRAHAMSEN, DORIS. *Readings on Earth Changes*. Applegate, OR: privately printed, 1973.

GOODMAN, JEFFREY, *Psychic Archaeology*. New York: G. P. Putnam's Sons, 1977.

——. *We Are the Earthquake Generation*. New York: Seaview Books, 1978; Berkley Books, 1979 (paperback edition).

Chapter Thirteen

Paul Solomon: The Prophet and the Pole Shift

> Watch the fifth day of May, 2000 for the time when the planets will be aligned one behind the other across the sky, and the strain of magnetism will shift the surface of the earth until it takes a new shape and form. '84 and 2000, those are the years—a change and a great tribulation, then a change and the dawning of a New Age. These are the times. You have entered the half times.

This prediction of a pole shift was made in 1975 by the Reverend Paul Solomon, founder and pastor of a New Age church called the Fellowship of the Inner Light. Like Edgar Cayce, Solomon functions psychically while in a trance state, giving readings on a wide variety of subjects ranging from diet and health to lost civilizations and coming changes in the geography of the globe.

Solomon's entranced condition has been verified through electroencephalographic (EEG) tests performed in 1975 at Emory

University in Atlanta, Georgia, by an engineer, Don Rhoads, and a psychologist, Dr. Sigfried Worster, who found that Solomon was producing delta brain waves during a reading. This is the brain wave characteristic of deep sleep, and indicates a condition of total unawareness of the external world. Yet Solomon, while apparently in the deepest stage of sleep, hears questions and can respond intelligently—a feat inexplicable in terms of present psychophysiological knowledge.

However, the source of the verbal responses does not seem to be Paul Solomon. When he awakens from a trance session, he invariably needs a few seconds to become aware of where he is and who is present—i.e., to reorient himself to the world that, only a few moments earlier, he had been responding to in an extraordinary fashion.

Most important of all, he says he has no memory of what has just transpired. He knows nothing of what was spoken through his vocal cords and must either listen to the tape recordings that are always made by his associates or else read the transcriptions that are later made.

Who does the listening and speaking, if not Solomon? This question arose in 1972 when Solomon, thirty-three years old, learned of his ability to function psychically. It was asked while Solomon was in trance, and the reply identified itself simply as the Source. People around Solomon now speak of the Paul Solomon Source and distinguish it from the channel or vehicle—the man Solomon—through which it expresses itself.

According to literature prepared by the Fellowship of the Inner Light,* in February 1972 Solomon discovered, while hypnotized by a friend as an experiment, that he had the capacity "to tap the Universal Source." Under hypnosis and speaking in sleepy, incoherent tones, Solomon suddenly jerked as if something had struck him in the solar plexus. The young amateur hypnotist, startled by this unexpected reaction, was doubly surprised by the powerful voice that immediately began to speak through the sleeping but rigid body of Paul Solomon.

"You have not attained sufficient growth or spiritual awareness to understand contact with these records! That which you perform is a foolish experiment, for you attempt to harness powers you do

*The Fellowship of the Inner Light can be reached at 620 14th Street, Virginia Beach, VA 23451.

not understand and to contact sources, records or intelligence you are not familiar with. And how will you try the spirits, should you attain that which you seek? Would you recognize Him whom you do not know, have not been familiar with?"

The hypnotist recovered sufficiently to ask some questions, hoping to gain a clue to the nature or the source of the voice that spoke.

"Are you a familiar spirit, or a spirit guide, and can you give us your name?" he inquired.

The voice answered him in a discourse. "Neither familiar spirit nor guide is needed for the reading of these records. It is rather in yourselves, in your own development, that understanding will come and guidance—not from a lesser spirit, but from the Throne of Grace, that you may come before these records and read what you will for the construction and development of those on your plane who would know more of the Divine. Then go; study with much prayer and meditation; develop the self and spiritual under-standing that you may return, and being familiar with these planes, you will be welcome, and may read and rejoice."

As suddenly as the voice had begun its unexpected discourse, it ceased with a second sudden violent jerk of the body doubling up as if hit in the abdomen and then lapsing into an apparently normal deep sleep.

The excited hypnotist counted Solomon out of the hypnotic state. Solomon awakened dazed and confused, clutching his stom-ach in pain and needing help to stand. Although the pain was short-lived, he remained irritable for some time and certainly didn't share the excitement of the hypnotist for "strange voices" speaking through his sleeping body. His memory of what had taken place was limited to the pain in his stomach. He only remembered drifting off pleasantly and awakening very suddenly with excru-ciating cramps. As far as repeating the experiment was concerned, he wanted no part of it.

It took some time and persuasion to convince Solomon that another attempt should be made but with questions prepared and testing procedures agreed upon—including the presence of a skep-tical friend as witness. From the second reading, Solomon learned that he could avoid the pain associated with the experience by understanding the cause of it. And what was the cause? Solomon, it was said, was leaving his body through astral projection—more technically called out-of-body experience (OBE)—so that the

Source could communicate with minimal interference from the psyche of Solomon.

He also learned that he must follow a rigorous program of mental, physical and spiritual discipline and development to obtain efficient use of this gift for helping himself and others in the attempt to understand a purpose in life and to develop their full potential. A reading soon after that indicated that he should change his name given at birth, William B. Dove, to Paul bar Solomon. He was told that the high standard implied by the combination of names would cause him to aspire to become a new being.

That marked the beginning of the ministry of the Reverend Paul Solomon—a ministry now carried out through the Fellowship of the Inner Light, which offers lectures, courses (called Inner Light Consciousness), tapes and literature drawn from or related to the nine hundred-plus readings that Solomon has made to date. The Fellowship has branches in a dozen cities and six states. It describes itself as "a guided experience in practical mysticism" and claims to have more than ten thousand graduates.

Solomon says that his mission is to heal, prophesy, spread the word of God and be an example for the New Age, which the Source says will begin, as told in the Bible long ago, with a period of turmoil followed by the Second Coming of Christ. The "inner light" to be manifested is inherent in everyone: "the light that lighteth every man that cometh into the world." In the words of the Source, "It is necessary that you express the highest that is within you, creating in this world places of Light to which all men will be drawn for the love that is shared. Then come together and let it be said about you, 'See how they love one another.'" The Fellowship attempts to have people actualize in themselves the same divine consciousness and advanced evolutionary abilities that Christ demonstrated and called upon people to manifest.

Healing was one of Christ's paranormal abilities. Similarly, Solomon has some instances of psychic or faith healing attributed to him. His healing services, which he conducts in full waking consciousness, are often reminiscent of the late Kathryn Kuhlman's. People are laid out, unconscious or nearly so, all over the floor or speaker's platform after having "received the Spirit" from Solomon's touch.

Solomon's clairvoyant medical diagnoses while in trance are equally dramatic. An article by William Beidler in *Fate* magazine (February 1977) gives some accounts of this remarkable faculty.

My friend Carolyn Beckham also had a reading that fall, in September 1973. The information she received was for her husband who had contracted a serious foot infection while serving in the Orient. The fungus (if that's what it was) stubbornly resisted medical treatment. The Source's formula was extremely simple and involved increasing the circulation to his feet, first by plunging them into distilled water as hot as could be endured, then switching to extremely cold water. Within three days his foot condition was relieved and has not returned.

Beidler's second account of Solomon's diagnostic ability is even more dramatic:

In February 1974 Carolyn brought her sister to Atlanta where Paul Solomon was then living. Her sister insisted that Paul not be told the real problem and she pretended that the reading was for herself. Once Paul was in trance, however, they took away the original set of questions from the conductor (the Rev. James Wharton in this case) and substituted a set dealing with the sister's five-year-old son, who was believed to be suffering from muscular dystrophy. Nothing in the questions suggested that the boy even had a health problem but The Source immediately started describing his physical condition.

The child did not have muscular dystrophy, it announced; rather he suffered from a spinal maladjustment, the result of measles having settled in the base of the spine. The Source mentioned the year it happened, which corresponded with the year the lad had contracted measles. It predicted that doctors at Crippled Children's Hospital in Memphis, Ala., where the parents planned to take the boy, would confirm that the problem was not muscular dystrophy.

The mother took the young boy to Crippled Children's Hospital and as the reading had foretold, the doctors there stated the boy did not have muscular dystrophy.

In 1977 I personally witnessed Solomon give a health reading for a man named Albert, who is totally paralyzed from the neck down. It was remarkable to watch Solomon breathe deeply a few times as he lay on a bed, and then exhibit a strong convulsion of his solar plexus, after which he began to speak in a loud, clear voice about Albert's condition. In a rapid-fire delivery that hardly paused for nearly an hour, the voice of Solomon gave forth a detailed and seemingly comprehensive description of the physical, mental and spiritual status of Albert, followed by a medical prescription of half a dozen therapies ranging from drinking two drops per day of a special gold tincture, to the application of a certain shade of blue light along the spine, to hydrotherapy.

I'm unable to report any results in this case because—as I told Solomon over lunch at my home some months later—Albert's doctor refused to implement the treatments. Aside from the medical aspects, there was one small detail of the reading that I was surprised to note: the word "irregardless" was spoken. As a journalist and former teacher of English, I objected mentally to that abuse of language when I heard it. Higher intelligences should know better. I couldn't help wondering, therefore, whether the voice originated with Solomon's unconscious or—as a friend later said in Solomon's favor—the Source was merely using Solomon's body, whose brain happened to include that ungrammatical item in its neural storage of vocabulary. So I can only say: *regardless,* it was a memorable demonstration of someone or something's apparent clairvoyance.

Solomon has not always been considered a prophet of the New Age. Born in Arkansas in 1939, he grew up in Texas. As the son of a minister, he was steeped in the Bible from childhood. Quite naturally, he felt a calling to the ministry and became an ordained Baptist minister after graduating from seminary. He married, had a daughter, and began serving a Baptist church in Texas. But, several years later, a sexual scandal led to Solomon's departure from the church. It also led him into a period of intense, anguished self-searching characterized by theological doubts and disbelief, along with wild explorations of the seamier side of life. As Solomon himself puts it, "If there's something I didn't try, it was because I didn't know of it."

Eventually he found work in an Atlanta gift shop owned by a man who also had a pornographic bookstore. When the man offered him the job of managing the bookstore, he accepted but was there only a few weeks before the police raided it and jailed him overnight.

At that time, also, his wife—from whom he had been living apart—divorced him. The world seemed to be falling upon him, and he entered a period of extreme depression. To help him deal with his emotional problems, he asked his hypnotist-friend to conduct the experiment described above. The ironic result for him was a return to his original, ministerial calling. This time, however, it was to be noninstitutional and nondenominational, with a new sense of relationship to God and with a theology that was at least unorthodox, if not heretical, to the fundamentalist Christian tradition from which he had come. For he proclaimed, as the Fellowship's literature puts it, that "all the acts attributed by historians and scriptural writers to the being known as Jesus Christ are in fact within the realm of man's innate capabilities."

Through lecturing, teaching and giving trance readings, he gathered a following of students and disciples who now constitute the Fellowship of the Inner Light and who consider him to be a living prophet—a man of modern times who listens to God and speaks for Him as did the Old Testament prophets. "The minister who can't ask questions and get answers from a divine source can't do his job properly," Solomon often says.

Certainly Solomon prophesies as of old. His language, his philosophy and his religious organization are decidedly biblical in character, although yoga is a part of the Fellowship's interfaith disciplines. And his trance state utterances contain an enormous number of very specific predictions about the future of society and the planet, especially the earth change readings. These deal with upcoming cataclysms in the earth and the consequent effects upon nations and people, both on the physical and spiritual levels. A recent booklet produced by the Fellowship, *Earth Changes and the New Planet Earth,* gathered many of these readings. Here are some of the most important predictions about geophysical activity and upheaval:

> As to the changes in climate and conditions, in its [England's] entirety will disappear under the cap of ice that will come. This some twenty years in the coming. Could be more, before the final disappearance, but inhabitable then in this time, as the shifting of the poles will move over that area, or toward that area, in periods of change.

This reading, No. 704, was given in 1975. It indicates that the north polar location will move toward or over England. The pole shift, predicted by the Source to begin on 5 May 2000, apparently will involve a dislocation of about 40°. England will be covered with a great sheet of ice after that, although it will be "some twenty years in the coming."

This was amplified in a reading made in December 1977 when Solomon was visiting the Findhorn Community, in Scotland. The reading, No. 989, stated that Findhorn is situated "on the northernmost tip of what once was that Isle of Alta, or that chain, that continent once called Atlantis, the outlying portion thereof."

> And you've settled here, in a place that is, was, destined for violent change. But you have evoked in that place that spirit of the earth itself as proved by its bringing forth. Could it be that the spirit then is healed and this is the demonstration of it? Can it not be that your message to people of the earth is: this is the demonstration of cooperation, the responsibility for the care of the earth? Then would the earth act in revolt under your feet? It will not. It will not in this place.
>
> The climate will change. There will be some changes here, of course. You will be affected not so much in a negative way. Lands about you will be flooded, but there is no concern for moving to a higher place.

Apparently the group consciousness of the Findhorn Community—manifesting love and cooperation among the mineral, vegetable, animal and etheric kingdoms of the planet—has been sufficient to pacify geophysical processes in a localized way, though the greater earth changes are still coming.

Reading No. 193, given in 1973, indicates the precise motion of the earth and its relation to "that red planet" to be experienced during the pole shift:

> . . . that shifting of the poles will come abruptly. And this area will be found in those series of changes or some three slippages of the crust of the earth. As there is the passing close of that red planet, so will the crust of the earth be attracted toward it, as it would be approaching; so then that upper pole or the North Pole will point in the direction of its advance.
>
> "Then as [it] comes close or at its point nearest the earth, so then will the pole point directly upward or in that portion, that direction that now exists or near, and in the passing away of its presence from this planet, so then would the poles shift again a third time toward the opposite direction. So that there will be three separate temperature changes in this area, making near to impossible the survival of any animal species or plants, either in this area or upon the most of the surface of this planet.

The phrase "that red planet" refers to Mars. This is made clear by the Source in Reading No. 850, which states, "Those allusions to the passing near Earth of the planet Mars have, as well, the passing of consciousness, for it has been given that the atmospheres will collide. And as this happens in the physical so, in the consciousness, will the consciousness of that warlike body be at war with Earth and with Venus in that time, until Jupiter intervenes. It is this of which the prophets have spoken, a war among the powers of the air. . . ."

Reading No. 4057, a recent one, also refers to collision between the atmospheres of Earth and Mars. ". . . the willful destruction [by people] of that growing on this planet has caused and will cause and is causing imbalance in the magnetic field about this planet that will hurl the whole planet into the path of that red planet of war, of Mars, upsetting and disturbing the atmospheres. As the atmospheres collide, so will not much be left recognizable upon this ball of earth. . . ."

These readings and many others state clearly that it is almost certain there will not be many people or other life on the earth after the pole shift. In a reading given in 1972, and published in 1974 in *The Paul Solomon Tapes,* the Source stated, "And you will see the earth breaking open as its crust will shift and move. And there will be the noxious gases coming to the surface, and the entire atmosphere will smell of sulphur fumes. And there will be taken much of the plant life from the planet. As a result, the majority of life as you know it will be taken, will be destroyed, will be changed into other forms."

Reading No. 208, given in 1973, contains an important clue to understanding the coming pole shift. As in the Cayce and Abrahamsen readings, the state of human consciousness is noted as the key factor influencing a displacement of the axis. The reading states that the ancient civilization of Lemuria was destroyed because its populace disregarded the balance of nature and tampered with natural conditions of the atmosphere and the planets' energy fields:

. . . there never would have been the shifting of poles upon this planet if there were not the creating of conditions among men that were defiant to the laws of God, but when those conditions of law were subverted among men, and there was the defiance of the Law of God, so there was set in motion upon this planet an energy that created an imbalance between those forces of good and evil or positive and negative. . . . Understand, then, that the movement, the shifting of the poles was caused by the activity of men in defiance of their God.

The Source also has dire predictions of earth changes to occur well before the end of the century. During the period of 1982–84, major catastrophes will affect civilization on a scale far greater than, say, the 1972 Managua, Nicaragua, or the 1906 San Francisco earthquake. However, the severity of this earth change period, although great, will not be as severe as the one at the century's close.

During the 1982–84 period of earth changes, the following events are forecast in the Solomon readings:

• Much of the western coast of North America will disappear totally beneath the sea.

• Japan will also totally disappear beneath the sea.

• New continents will begin to rise in the Atlantic and the Pacific.

• Numerous great earthquakes will strike around the planet.

• Coastal inundation of Georgia will occur, along with northern Florida, turning the Florida Peninsula into an island.

• Land will rise to the east of Florida and later join it to become part of the Atlantic basin continent.

• Atlanta, Georgia will be destroyed by a tidal wave.

• The Great Lakes will empty into the Gulf of Mexico.

• New York City and nearby Long Island will submerge.

• The Connecticut and Massachusetts coastlines will experience inundation through tidal waves (and presumably New Jersey and Rhode Island also).

• The North American continent will be "split in half, down its very center." The eastern half will extend farther into the Atlantic, tilting upward in the southeastern portion and "falling away" in the northeastern.

• The Hawaiian Islands will become the uppermost part of a vast continent, and "cities of splendor will arise from them."

• An ancient temple will rise from the depths of Lake Titicaca in South America.

Such prophecies appear fearsome in the extreme. Yet the Source does not say that every one of them is inevitable. It says there are many factors affecting the events, not the least of which is the human factor. According to the *Earth Changes* booklet, the Source has always said there are two reasons for prophecy: "One is to change prophecy and make sure it never comes true; the other is to prepare for that prophecy so that even should it come true in the manner described, it will not affect you in the way described." Solomon adds, "According to the readings, any prophecy that is fulfilled is a prophecy that failed. The only reason for predicting anything is to prepare for it or to change it, to keep it from happening."

The primary emphasis of the readings, *Earth Changes* states, is one specific thing: spiritual preparation. In line with that emphasis, the Fellowship sponsored the first Earth Changes Congress in November 1977. The weekend gathering was held in Virginia Beach and was attended by about two hundred people, many of whom came as far as five hundred miles.

Solomon's keynote address set the tone for the Congress. "We're going to be looking into history, into the future, and deciding what we can do, how we can respond. ..." An eloquent speaker, Solomon offered a sensible perspective on the possibility of earth changes and a pole shift:

> The readings say that ... Mother Earth is pregnant and about to give birth to a New Age. At an imminent birth there is always a birth pain. There are contractions. There are difficulties. But when a young couple looks forward to a birth in those last few days, they don't begin to say, "Any day now the labor is going to start." They say, "Any day now there's going to be new life." The focus is different. ... Instead of looking at the labor pains, perhaps we can go away from here looking at the birth of a new age. If the changes do come, if the continents are altered, if the weather is altered, if all of these ... do happen, what will be the result? How can we respond to that?

A wide variety of presentations were made over the weekend, with vigorous participation from the audience. One lecture dealt with the evidence of a "supercivilization" prior to our own epoch. Another focused on the "Jupiter Effect" predictions of earthquakes in 1982. A professional geologist spoke on what science knows about our planet's formation and the forces of change at work in it today. My own principal contribution was a lecture on the concept of pole shift, with emphasis on the scientifically oriented theorists and what they feel might be the trigger mechanism.*

Many workshops offered practical information on preparing to survive. Topics included alternative energy use, food growing and food preservation, spiritual farming communities, a nonmonetary economy based on barter, geometric learning devices for spiritual integration and data synthesis, and a comparative study of the earth-change readings from other psychic sources.

The Congress proceedings will be published by the Fellowship and made available to the public. Planning is also underway for Earth Changes Congress II.

Did a consensus or a clarified, focused point of view emerge from Earth Changes I? No. To some degree the Congress illustrated the saying, "If you believe, no explanation is necessary; if you don't believe, no explanation is possible." Many people obviously came to the Congress as true believers. Being already convinced, they were not there to probe the predictions and prophecies but to update their information about when it all would happen and to increase their range of survival skills. Others, curious but skeptical, felt they'd had their minds opened a bit. Still others flatly rejected all predictions and prophecies of an immediate pole shift.

One of the latter group was the professional geologist mentioned above, Dr. William Ryan, of the Lamont-Doherty Geological Observatory, at Columbia University. He is also chief scientist on the oceanographic ship *Glomar Challenger*. We were roommates for the weekend, and I was genuinely pleased to be in his company for the opportunity to raise some acid-test questions about various pole shift theories and predictions.

Ryan's remarks at the closing session of the Congress succinctly expressed his view and offered major opposition to the true believers' position. Someone in the audience asked him to comment on the effects he could foresee from a pole shift. Ryan replied:

> To be honest with you, when I came here I was unaware of the forecasts of changes and they certainly startled me [that the prophecies say] that they're going to be [occurring] tremendously fast. . . . I can't accept that we're going to have a polar shift and I can't accept that there have been polar shifts in the past.
>
> The other day I showed slides of the direction that the earth's

*Copies are available from The Seeker's Quest, P. O. Box 9543, San Jose, CA 95157.

magnetic field has been going through time throughout the period of the ice ages, and there are other techniques that are used to determine the spin axis of the earth. For example, today in the Pacific Ocean, there are [ocean-current] upwellings right across the equator. This is due to a divergence or splitting in the current because currents due to the earth [spinning] rotate in a vortex in one hemisphere and an anti-vortex in the other hemisphere. And there is a belt of silicious productivity [formed from the silica dioxide in marine life such as diatoms, radiolaria, etc.] recorded beneath these sediments on the equator. It's like a chalk line. And as the Pacific [tectonic] plate moves north-ward, these chalky sediments land on a seafloor and are carried northward. So if we drill a hole with the *Glomar Challenger* eight hundred or one thousand miles north of the equator, we find materials that were under the equator. This is the spin part of the earth. This is not determined by the planet's magnetic field 30 million years ago. It is an independent measurement. And so here is a record of exactly what has been the spin direction of the earth [and the equator's location], and this record goes back more than 60 million years. There were no polar flips of the spin axis of the earth in that time.

The thrust of Ryan's remarks is that seabed sediments along the equator show tropical marine fossils laid down there in un-broken fashion for 60 million years—evidence that in his view conclusively rules out the idea of polar ice caps being thrust into equatorial waters. This is a strong example of the sort of data that pole shift theorists must overcome in order to have their hypotheses accepted by the scientific community.

Solomon, like every other psychic, has had "misses." He has made predictions that clearly did not come true as stated. Still others have thus far proved impossible to evaluate on the basis of present data. And although Solomon has a remarkable capacity for medical diagnosis, this does not necessarily lend authority to his predictions in other fields of knowledge—earth changes, for example.

In a reading given on 7 July 1973 and published in *The Paul Solomon Tapes,* Solomon said, "Yet will that capitol of that central city [of Atlantis], or Poseida herself, be seen rising in those waters of Bimini within these next two years and will be identified as that temple" (p. 140). By any standard of evaluation, that statement must be rated a definite miss. No land has risen above the surface of the ocean near the Bimini Islands—a fact that leads me to rate

Edgar Cayce's 1933 prediction of Atlantis rising there in 1968 or 1969 a miss also. (But not a complete miss—see my footnote below.) Moreover, no "capitol" or "central city" has been discovered in the waters there. Last of all, while an ancient megalithic structure has been discovered three fathoms down and half a mile offshore from Paradise Point, on North Bimini Island, there is no evidence that the huge underwater stones were themselves part of a temple—only part of a wall. A principal explorer of the site, Dr. David Zink, notes in his recent book *The Stones of Atlantis* (Prentice-Hall, 1978)* that the site may include a temple (yet to be discovered), because there is evidence of a sacred geometry in the structure thus far discovered. This sacred geometry can also be seen in Stonehenge and other sites rimming the Atlantic basin from the Canary Islands to the pyramids and temples of Central and South America. But the Bimini discovery cannot be said to prove or even support the Solomon prediction, which was made after the Bimini discovery.

So there is a question mark hovering over the pole shift scenario projected by the Paul Solomon Source. This indeterminacy is due to several factors. First, it differs in major ways from those of Cayce and Abrahamsen. Second, the record shows that not all Solomon's predictions thus far verifiable are clear hits and that even some of the hits may have certain enigmatic aspects. Third, the Source itself declares that humanity can influence the future for better or worse, even at the level of geophysical processes. Have the "misses" been flatly wrong, or have they been delayed

---

*Zink points out that this fascinating site was discovered in September 1968 by Dr. Manson Valentine, of Miami, a former professor of zoology at Yale University, and a co-worker, Demetri Rebikoff, President of Rebikoff Institute of Underwater Technology, in Fort Lauderdale. In February 1969, aided by Pino Turolla, a professional underwater photographer and archaeologist, they found another wall and several dozen broken columns. Thus Cayce, who said, "And Poseidia will be among the first portions of Atlantis to rise again. Expect it in '68 or '69. Not so far away!" was precisely right with regard to the time, although an actual elevation of the sea bottom above water there did not occur. Cayce's readings had identified Poseidia as the most important Atlantean island after the continent was broken up by earth changes, and he located it in the area of the Bahama Islands.

in timing, or have they been mitigated by the moral behavior of people? The answer will become more clear in Part IV, where we examine the predictions of The Stelle Group. Therefore I will leave Paul Solomon and the Source for the time being and go on to examine the pole shift predictions of a host of modern sources— a group I describe as "Other Visions, Other Voices."

References and Suggested Readings

BEIDLER, WILLIAM. "Paul Solomon—Another Cayce?" *Fate*, February 1977.

Earth Changes and the New Planet Earth. P. O. Box 206, Virginia Beach, VA 23458: Fellowship of the Inner Light, 1977.

The Paul Solomon Tapes. P. O. Box 444, Virginia Beach, VA 23458: Heritage Publications, 1974.

Chapter Fourteen

Other Visions, Other Voices

When a new idea or a rare experience becomes public knowledge, or at least becomes known within certain circles, there often is a ripple effect. The original experience or presentation generates many offspring—some genuine, some spurious. UFO sightings are a classic example. The first well-publicized UFO sighting in modern history occurred in 1947, when pilot Kenneth Arnold reported seeing "flying saucers." Thereafter, they were reported in geometrically increasing numbers. Additionally new sorts of UFO experiences began to be reported: UFO landings, UFO occupants, then UFO occupants who contacted and communicated with humans, and finally UFO occupants who abducted humans to perform experiments on them.

While the large majority of alleged UFO sightings have been

correctly explained as familiar phenomena—shooting stars, misperceived aircraft or planets, atmospherically generated illusions, etc.—there nevertheless is a core of unexplained reports that indicates a genuine new category of experience for humans: contact with extraterrestrial technology and nonhuman life forms. Perhaps nine out of ten UFO reports can, through competent investigation, be dismissed as cases of mistaken identity, hallucination, fantasy or fraud. But that one remaining case defies conventional explanation and may have genuinely new information to contribute to the worldwide investigation of a mysterious—and undoubtedly momentous—phenomenon.

It is the same with regard to predictions and prophecies of a pole shift. Once the idea got "into the air," it began to generate offspring. I have looked at many of them in the course of this investigation. Some are obviously nothing more than amateurish recastings of, say, the Edgar Cayce readings. Others show more creative embellishment, but nevertheless exhibit certain indications that led me to view them with suspicion and therefore disregard them. (One sad example was a discourse, allegedly from a group of "ascended masters" who speak through an entranced medium in Florida, that quite seriously mentioned the "facts" that Arcturus is presently the pole star and that the earth has several magnetic axes. Ascended masters should know better. In addition, the rest of their data were so vague that they hardly warranted the term "data" at all, consisting largely of metaphysical platitudes. Half a dozen pages of that particular reading were all I could take before throwing it in the wastebasket.)

Some predictions and prophecies, however, appear to contain genuinely new information of a paranormal character. In part these psychic communications recapitulate earlier sources, but they also give new data or insights that amplify the earlier sources or clarify them—or sometimes even contradict them.

In researching this book, I have cast my net far and wide. It is clear, I hope, that the catch of strange creatures I've hauled in has not been accepted indiscriminately. Rather, I have tried to look at each one carefully, as the Buddha advised when he spoke the words noted in Part V.

With that brief prologue to this chapter, I'll now acquaint you with the predictions of a variety of sources that seem to merit further consideration as we attempt to gather information on this important subject.

Ruth Montgomery and the Guides

"The Guides agree with Edgar Cayce that another shift will occur near the close of the century, and that we may as well accustom ourselves to the idea, since there is nothing we can do to prevent it. On the brighter side, they say this awesome event will usher in an era of peace, 'since man will be confronted with enough problems of reconstruction to keep his covetous mind off others' territory.' "

This claim for the inevitability of a pole shift in two decades was made by the Washington-based journalist Ruth Montgomery, who gained fame some years ago as the biographer of psychic-astrologer Jeane Dixon in *A Gift of Prophecy*. Since then Montgomery has herself become a prophetess of sorts, claiming to receive information via mediumship (specifically, automatic writing at her typewriter) about many occult and esoteric aspects of life on earth. The alleged source of this information is a group of spirit guides—entities of great wisdom and compassion who have chosen to watch over Ruth personally and, through her, over the human race. One member of her spirit guide group, she claims, is the famed medium Arthur Ford.

Montgomery has published a number of books containing esoteric knowledge purportedly from the spirit world. Her most recent, *The World Before*, gives the history of the fabled lost worlds of Atlantis and Lemuria. It also makes prognostications for the next few decades. One of the predictions, noted above, is of a pole shift. Not only does Montgomery give us an account of the previous pole shift that destroyed Lemuria, she also tells what will happen the next time, and why.

According to the Guides, "several shifts of the axis have occurred," the last one at approximately 48,000 B.C. Lemuria, the great Pacific continent, went under the sea then. Atlantis, however, was relatively unaffected and continued as the world's foremost civilization for another twenty thousand years, until it destroyed itself through a technological disaster.

The world's topography was altered in the twinkling of an eye [The Guides report], and for a long time afterward there was a haze over the land that had once harbored many millions of human beings and every type of wildlife. Where flat plains had been were now mountain peaks and valleys. Great Peruvian and Mexican cities, built near sea level, were thrust seven to ten thousand feet toward the heavens, leaving only vestigial artifacts of a noble past. Present-day California, then an eastern coastal area of Lemuria, survived the inundation by attaching itself to new lands (our Western United States) which rose from ocean bed. Rivers that had flowed north turned west or east or south, or vice versa, and the Mississippi which had once been coastline, now became a mighty river because of the newly formed land to its west. In Africa the Nile that had emptied into the Atlantic changed course with the tilting of the landmass and found its outlet in the Mediterranean. New land appeared in Europe; and the Sahara, once ocean bottom, now became one of the most fertile spots on earth (p. 118).

The next pole shift, the Guides tell us, will not be as catastrophic as the one that destroyed Lemuria, although great damage will occur. Some lands will disappear under water, while others will emerge and "soon be ripe for cultivation." Outcroppings of older civilizations will appear. "Manhattan and some other sections of the East Coast as far north as Newfoundland will be unaffected until the close of the century, when they will vanish with the shift of the axis, as will Hawaii, Japan, and some other Pacific islands. Florida will also be gravely affected, but Egypt will survive, as will most of the Mediterranean area. Venice, that queen of the Adriatic, will sink from sight, and the Gobi Desert will become fertile and pleasant again."

According to the Guides, the shift of the axis is a natural event in the life of the planet, and is not within the scope of human ability to alter. Immutable divine law is behind it, and therefore it must be regarded as "a soul-cleansing process." However, the Guides remark, our present use of fossil fuels at such a rapid rate "is hastening the shift of the axis" in the late 1990s. They add that our "tinkering with solar and nuclear forces" is an additional factor. The critical one, however, will be "volcanic eruptions." Contrary to those who claim a large meteorite will collide with the earth,* knocking it off balance and shifting the poles, the Guides do not see any such event. A cooling trend in the early 1980s, plus "cyclones of increasing velocity," will herald "the alteration of the earth's position in relation to the sun."

The Guides give this description of the coming pole shift:

---

*The most notable contemporary psychic predicting this is Jeane Dixon, with whom Ruth Montgomery has parted ways.

When the axis shifts it will seem to earthlings as if the sun has not moved across the heavens from horizon to horizon, although the sun will of course continue in its normal revolutions. We here already see this event occurring, so we understand the implications. It will be well for those there to understand that it is not the end of the world, but a process of readjusting sunshine and rain, the sea and the land, so that some areas of the earth are refreshed and others put out to pasture, so to speak. There will be some seas where there is now land, and vice versa, ice caps in new places, and balmy breezes at the poles. When the shift occurs, the souls on earth will be terrified and turn to God in their helpless fear, although some will unfortunately resume their nefarious ways. Yet on the embers of a devastated civilization will arise a better one based on brotherhood, and thereafter the return to earth of him who promised that when he came sinners would be separated from the near saints, and peace would reign for a thousand years. That time is not too far away, in the twenty-first century (pp. 271–72).

Last of all, the Guides explain the cosmic significance of a pole shift. From their perspective, "Those who die will not be wiped out, but returned to spirit with opportunity for renewed spiritual growth. For this reason, we on this side are permitted to tell of the coming event, so that those in physical body will understand the principle and regard it in proper light. The passage from death to spirit, and spirit to physical life is one and the same process, no more to be feared than sleeping and wakening. Those who pass into spirit when the axis shifts will be free from pain and misery, while those who escape death will have an interesting time of it in restoring order and reviving the sweetness of spiritual knowledge" (p. 270).

A Season of Changes Is Coming

The Heritage Store, in Virginia Beach, Virginia,* is a delight to friends of the New Age. It abounds with products for a healthy body, mind and spirit. Among the books it sells is one it published, entitled *Season of Changes, Ways of Response*. The subtitle explains that the book is "A Psychic Interpretation of the Coming

*Located at 317 Laskin Road, with mailing address of P. O. Box 444, Virginia Beach, VA 23458.

Changes in, on and about the Earth and the Corresponding Transformations within Man."

Season of Changes is based on the readings given via the trance
mediumship of a woman who prefers to remain anonymous. It
consists of selections from the readings, plus a supplementary text
written through normal means. The source of the readings is also
unidentified but is considered by the group that "brought through"
the information to be a form of higher intelligence whose perspective on human and planetary affairs is—for all intents and
purposes—omniscient. The supplementary text relates the selected
passages to contemporary affairs such as changing climatic and
economic circumstances, food shortages, social unrest and political
upheavals. The emphasis of the book, however, is on man's capacity to control and transform both personal and planetary circumstances through the application of his spiritual power.

Like many other psychic sources, *Season of Changes* forecasts
calamitous geophysical changes. The climax of the drama, of
course, will be a shifting of the poles. According to the *Season
of Changes* source, " . . . there *are* indications, in various forms,
of the beginning of an *entire* shifting, switching, of poles, polarity.
Now with these come the implications of the various *changes*
which will take place as these begin to come to their greater
manifestation."

The unnamed author of *Season of Changes* comments on this
passage: "The seed of this understanding of a global shifting of
the poles was apparently sown in order to prepare us to receive
the information which was offered during the following reading
session; information indicating that the scope of the approaching
transformations was to be worldwide, and of a magnitude such as
could scarcely be imagined."

The information parallels much of what we have seen in the
Cayce, Solomon and Abrahamsen readings, giving some apparently new data or at least covering previously undisclosed aspects
of the earth changes to come. Among the locations mentioned are
the following:

• Land will rise and become an extension of the coastline from
New York to Massachusetts.

• Many parts of the West Coast will fall away, followed by "land
areas of a more central nature."

• The Great Lakes region will be tipped upward to the north, draining the lakes down the Mississippi until they are emptied. However, much of the water will collect in low spots about a quarter of the way down from the lakes.

• Japan will fall under the sea. So will some parts of the China coast, Arabia and India.

• A great portion of the coastline of Europe will submerge, especially Scandinavia. Nova Scotia will meet a similar fate.

• The Panama Canal Zone will break up.

All these changes will occur by 1998. Shortly thereafter, the pole shift will occur. This prediction was made in a reading that followed soon after the one just summarized. The question was asked: "There has been scientific speculation that the earth's poles are, right now, in a process of gradual shifting. If this is true, how long has such a process been underway, and will it continue to be gradual, or will it be more abrupt?"

The response was specific. "This process has been for the last millennium. Slowly, very slowly at first. But as has been given, within the next 25 or 26 years, there will come again a switch." The text continues:

Question. Most of those scientists who believe in the theory of a cataclysmic shift of the poles also believe that such a shifting occurs all at once, within half a day at the most. Is our understanding correct that the earth changes now occurring will accelerate and will culminate in a sudden shift sometime within the next twenty-five years?

Response. Now, this is true and it shall take a period of time which will be very short in the terms which you count. But the effects of these would be a great deal more lasting. See?

Question. A great deal more lasting than previous changes?

Response. The change in itself, of which you specifically speak, will take a very short duration of a day; and yet the effects of this shall last for a great deal longer than that small period of time. And then after the three-day initial period, the beginning of the quieting and resettling of those forces and natural phenomena that would accompany this. See?

Season of Changes also gives a hint at the trigger mechanism for a pole shift: a pole shift *in the sun*. But even that is not the final cause, for the sun itself will be influenced by celestial bodies and conditions unknown to us at this time. Reading No. 52–5 contains this intriguing statement.

> A shifting [of the sun's poles] is taking place, as is an expansion within the sun itself. As to the effects, these have already affected the earth, and those spots or storms upon the sun are those effects of the expansion and switching of that surface and inner portions, which are taking place.
>
> These affect the earth in those strengths of rays and emanations which are taken into the atmosphere, and bring about, in part, some of the changes in the weather cycles and the greater tendency toward warmth, then. For the orbit of the earth itself, and the expansion of the sun in its gaseous and luminous form becomes more intense to that application in the earth.
>
> Now you feel or think that the changes upon the earth are of such great magnitude, and affect the earth itself. But the *universe,* my friends, is changing. Therefore, all that is within the universe will have its changes in its time and place.
>
> And those influences of a meteor, comet, and atmospheric nature, or a positive and negative flow, are affecting many bodies as well as the sun and the earth. They are as a cleansing, a *purging,* of those things which need to be cleansed.

And then the passage identifies the trigger mechanism:

> And these shall come from what seems to be the voids of space, toward the earth, and envelop such, even the sun, and make, then, such *explosions* of intensity and heat within the sun and its surface that that, *in itself,* will greatly affect the atmosphere of the earth and the earth's reaction of movement, in itself. See?

Lenora Huett and the Path to Illumination

A different cause for the coming pole shift is given by Lenora Huett, coauthor with Wally and Jenny Richardson of *The Path to Illumination:* "An axis change is caused by sudden expulsion of energy from another planet which enters the earth's orbit."

This trigger mechanism is reminiscent of Velikovsky and Solomon. The celestial bodies identified in Velikovsky's account were Venus and Mars. Solomon predicts that the "red planet" will initiate the next pole shift.

Huett, like Solomon, Cayce, Montgomery and the anonymous

author of *Season of Changes,* claims to receive information para-
normally via mediumship, though she describes herself simply as
a "channel." This information is received from a spirit source
whose knowledge—especially foreknowledge—is far greater than
ours. The spirit tells Huett that a pole shift comes about in an
orderly manner:

> There are those planets which course through space in a
> definite pattern amongst the stars and planets. These are the
> police forces of the universe, who bring about a steadying
> influence as planets erode or change. It keeps the planets
> rounded, preventing them from gradually becoming elliptical.
> An occasional change of axis is needed for there are eventually
> those things which must be mutated or removed; and for varying
> levels of the earth to be put to use. The areas which have been
> out of use have been rebuilding their resources. When the time
> comes for change and a newness of beings upon these areas—
> then the change comes. This is one of the ways the eons or
> permutations of time and progressive movement within man
> occur (p. 248).

The pole shift will be quite abrupt, Huett's spirit source tells
us. "... the change which comes about is a very rapid or sudden
thing. It occurs at the entry of a particular planet into the orbit of
your own, and its influence brings about a change with almost
lightning speed. There is no evolving towards this, but a sudden
change at the time it comes about. The catastrophic influences of
the foreign planet often result in such forces and suction being
exerted upon the surface that it causes a sudden rising and falling
of ground levels, in which new and different mountain ranges and
valleys are formed. Thus, these are not always formed through the
ice ages, but often through the pulsating of the earth's crust."

Although freely roaming interstellar planets are unknown to
science, astronomers would presumably be able to see the approach
of such a planet well in advance and thus be able to issue a timely
warning to the world. But this will not be the case, Huett tells us,
because the planet is traveling at such high speed that by the time
it becomes visible it will be too close to do much of anything.
Population resettlement is not possible because "man would not
know which area to be in, for the change causes a sudden jolting
of all areas."

This inevitable event presumably could be given a date in time by Huett's spirit informant; yet when asked to do just this, it refused. Its reason? "... there are many on earth who would take advantage of it and would go to many lengths to cause fear in others because of it. It is necessary that these changes come about in their own timing. If man does not know, he can displace it from his mind and work for his own growth."

It adds, "When man is fully aware that there is neither a now or then, a here or there, he will be less inclined to fear death or that which is beyond what he now knows. It would also be well that he not know his own time of death or sudden change, for it would cause him to close himself off to growth that could come. If he is to be in the spirit, he will be taken. All is well. All is good."

The Ancient and Future Worlds of Michio Kushi

Michio Kushi, founder and president of the East West Foundation in Brookline, Massachusetts, states that the earth has passed through many great changes, particularly axis shifts of a full 90°. Born in Japan some fifty years ago, Kushi became the successor to George Ohsawa, who introduced him to the macrobiotic way of life. Over the past twenty years, Kushi has presented macrobiotics and oriental medicine to the West. The public image of macrobiotics is largely limited to a distorted notion of people eating nothing but brown rice and tea, but this is far from the truth. Kushi, author of the *Book of Macrobiotics,* claims that the ancient "MB" way includes knowledge of ancient and future worlds because the macrobiotic lifestyle is based on the cosmic principles that order the universe.

In his seminars, therefore, Kushi has declared that polar shifts have occurred "thousands of times" in the history of our planet. One such seminar, published in transcript by the East West Foundation as *Acupuncture, Ancient and Future Worlds,* contains Kushi's statement that a pole shift has happened several times within the past quarter million years—most recently only twelve thousand years ago. At that time, "a big event occurred that destroyed world peace or paradise. This was a great catastrophe due to an earth axis shift" (p. 86).

In an illustration accompanying his text, which I am unable to reproduce here, Kushi shows what the world's geography was before the most recent pole shift. The North Pole was in the North

Atlantic, slightly southwest of the British Isles, and the South Pole was in central Australia. (These positions, curiously, are not diametrically opposite.)

This ancient civilization, Kushi says, had memory of people on earth going back a million years, "when their ancestors came from other planets to this earth." These "heavenly visitors" brought the ancestors of present-day mankind knowledge of fire and cultivation, writing, "calendrics," etc. They intermingled with the people of earth, and by this intermingling *Homo sapiens* arose. "The biggest immigration of heavenly visitors," says Kushi, "was 1,700,000 years ago."

Kushi does not identify a trigger mechanism for the pole shift, although he clearly places the primary influence for disaster on human beings who live out of accord with what he terms "the order of the universe." Macrobiotic diet and organic foods are part of the order of the universe, in Kushi's philosophy, but so are displacements of the earth's axis.

I attempted to learn from Kushi what the source of his knowledge is—ancient documents, oral traditions, psychic access to the akashic records, divine revelations, etc.—but I was unsuccessful. Assistants said they didn't have the information and Kushi was inaccessible. It is rumored among Kushi's macrobiotic students that he has predicted a date for the next pole shift, but I could get no confirmation of this through the East West Foundation. All that Kushi has to say on this, so far as I am aware, is the following statement from *Ancient and Future Worlds*, which contains, incidentally, some of the most sensible advice I've heard from anyone dealing with pole shift predictions:

> With the next axis shift, where will the biggest change arise? (No one in class had any idea.) Oh, you will all have to die. Unless you can see where the next axis shift will have effects, your understanding of this axis shift we studied is not complete. And you have to foresee what nations and what continents will sink and where new continents will arise and what places in the future world will become prosperous. This is not difficult to see at all. Please open your map or examine the world globe, and think.
>
> Edgar Cayce said that California would sink, and then Europe will collapse, as well as the western part of Africa.* You

---

*This reference to Africa is an error either by Kushi or in transcription. Cayce spoke of western America, but not Africa.

have to see clearly if his prophecy is correct or not. You know, when they said that California would sink, some people ran away. Then nothing happened. So they ran back again. They are back and forth, leaving every time they think California will sink. We call those people "slaves" because they have no judgment or understanding. [When you do have understanding] you need not be afraid and when the time comes, you can go (p. 124).

Although I could not pin down a date, Kushi is clearly predicting an axis shift in the near future—within our lifetime.

Baird Wallace and the Space Brothers

Still another source of pole shift information is Baird Wallace of Grosse Ile, Michigan. A medium-built man in his fifties, Wallace says that since the 1950s he and a small group of friends have been in contact, through the mediumship or "channeling" of Clyde Trepanier, with higher intelligences referred to in general as Space Brothers. As Wallace reports in his book *The Space Story and the Inner Light*, these entities are not simply flesh-bodied extraterrestrial beings in UFOs. Rather, they are natives of other sets of dimensions—higher physical planes of reality which appear insubstantial or nonphysical to our limited perception. But these planes are real nonetheless, and from them the Space Brothers have a perspective, and therefore knowledge, immensely beyond the human level. Vast cosmic processes involving aeons of time are known to the Space Brothers, Wallace says.

The situation concerning coming catastrophic earth changes, according to the channeled communications, is as follows. The entire solar system is transiting space as the galaxy wheels around its center. (Although this process seems slow from the human point of view, recall from Chapter 1 that the speed of movement through space at our point from the center of the galaxy is hundreds of thousands of miles per hour.) Thus earth, as part of the solar system, is entering a new "sector" of space that has a different vibrational frequency. This higher vibrational rate is affecting every atom of the planet and its inhabitants. The long-term result will be positive changes in our state of consciousness.

The more immediate effects, however, will appear to be disastrous because of a complicating factor. A new sun, larger than our own, is entering our solar system. (This was first reported in Dino Kraspedon's *My Contact with Flying Saucers*. The Trepanier communications confirm this, Wallace says.) More precisely, our solar system is approaching an unknown, unseen star. As it enters our system, it will create a binary star system. This new sun, currently invisible, will begin to be seen as a dull red glow within the next fifteen years. As it does, immense changes will take place within the solar system. A presently undiscovered planet named Vulcan, which has an orbit inside Mercury, will be absorbed into the sun. Pluto will be lost from the solar system. Our moon will leave its orbit and become a planet. And there will be major changes in the paths of other planets. One of the Space Brothers, Oxal, described the coming situation in a communication that took place on 11 October 1975:

> When I mentioned that your solar system is going through a great change, I mean that there are going to be changes in the locations of your planets, stars and satellites. You are going to become a binary sun system and when this new sun moves into your solar system, the magnetic center of your system will change, and so will the locations of your planets. They will have to readjust to the two suns. This is going to make a difference in your solar system. . . .

Another contact, named Yaum, a spiritual teacher, communicated with Trepanier and his group on 29 November 1975. The question, "Is the new sun approaching our system?" was asked. Yaum replied through Trepanier:

> Oh yes, it is on its way, but still a long way off. It has already been seen, but your scientists are not sure what it is. One of your astronauts sighted it on the last trip to the moon and one of them mentioned that it could be a sun. It will not take on its brilliancy until it enters your solar system. Then is when your whole solar system will shift and you will have a new magnetic center of your system. With the influence of the two suns there will be a shift of the planets and other bodies into a new position because the magnetic field will change as everything actually rotates around the magnetic center of your solar system.

Although the Space Brothers do not mention a pole shift by name, it is not difficult to imagine such an event during the process of shifting the solar system from a single-sun center to a binary star system. To anticipate the next section: shades of Nostradamus!

Rodolfo Benavides and the Great Pyramid

In the late nineteenth century, Alan Vaughan relates in *Patterns of Prophecy*, an occult group flourished in England taking inspiration from the Great Pyramid of Egypt. They viewed it as "a symbolic prophecy in stone" and related its architecture to prophecies in the Bible. According to their calculations, the year 2001 was the last date on their prophetic calendar.

Edgar Cayce likewise commented on the subject of pyramid-encoded prophecies for the end of the century. In 1932 his wife conducted him through a reading (5748–5) in which he made this statement in response to her request for "detailed information regarding the origin, purpose and prophecies of the Great Pyramid of Gizeh near Cairo, Egypt":

> This, then, received all the records from the beginning of that given by the priests, Arart, Araaraart and Ra, to that period when there is to be a change in the earth's position and the return of the Great Initiate to that and other lands for the folding up of those prophecies that are depicted there. All changes that came in the religious thought in the world are shown there, in the variations in which the passage through same is reached, from the base to the top—or to the open tomb *and* the top. These are signified by both the layer and the color in what direction the turn is made.

A modern version of occult pyramidology can be found in Rodolfo Benavides' *Dramatic Prophecies of the Great Pyramid*. This book, rampant with dire predictions about the coming years up to 2001, declares itself to be based on exact calculations that unlock the secret wisdom encoded in the design of the great stone structure at Giza. Not surprisingly, the revelations Benavides derives from his study of the pyramid totally confirm what the Bible tells us is to come—or at least what Benavides claims the Bible says. A pole shift, of course, is included.

Dramatic Prophecies of the Great Pyramid was published privately by the author in Mexico, his native country, and later translated into English. The first edition appeared in the 1950s. By 1970 there had been eleven editions, each somewhat crudely updated with new material while leaving intact earlier matter that clearly dates from one or more previous editions. This lack of editorial finesse unintentionally provides a useful means of evaluating the authority of the book. We can examine a fairly instructive example of this on page 312 of the eleventh edition: "About 1975 there will be five tragic months, among them August, in which nuclear energy will incinerate millions of persons. The United States appears to be the country indicated to receive the first demolishing blows ... after that all humanity will be struck!" A somewhat overstated prediction.

There are other weaknesses to the book. Like Chan Thomas, Adam Barber and Emil Sepic, Benavides doesn't believe in footnotes, references or bibliographies. Claims of vast scientific research and exhaustive investigations by the author are made without substantiation, on a take-it-or-leave-it basis.

Despite these shortcomings, I have chosen to include *Dramatic Prophecies of the Great Pyramid* because we won't have long to wait before being able to confirm or deny its pole shift prophecies. That particular event, Benavides declares, will take place sometime between 1977 and 1982!

> In this period the world will suffer a cataclysm and the axis of the earth will change its position with respect to the Sun, coming to a vertical direction with respect to its plane of orbit. This could be a consequence of collision with some errant stellar body (p. 313).

Elsewhere Benavides explains that a pole shift can come about for the 1977–82 period "if not from a direct collision (which could happen), then by a stellar attraction not foreseen at this time. Remember that we have been saying that a gigantic cold planet is approaching" (p. 13). That planet is "nearby" and is named Hercolumbus.

Benavides predicts that 1979 is the year in which the world's climate will change dramatically. It will have its origin in "the uncommon accumulation of snow that is being produced at the South Pole...." He adds that "the polar snows of the north are rapidly lessening and have begun to produce a disequilibrium in the polar axis."

Following the pole shift, there will be deluges as the "polar snows" begin to melt and thus raise the level of the seas. More than 70 percent of humanity will have been destroyed by war and natural disaster. But by the year 2001, a new social order will be established under the "new Messiah"—a man already born and living among us unrecognized. Also, according to Benavides' interpretation of the "symbols of the Great Pyramid," by that time "the Jewish people will no longer be on Earth."

Universal Guidance Through "Eternal Cosmos"

"... it is quite important that all that are present on the earth at this time realize that the changes that come are part of normal activities which, moving the earth forward, tilting it, placing it into a state of side to side, or down and up, becomes part of ancient knowledge which comes from the energy field that is in part your own particular plan on this planet."

The source of this statement identifies itself as Eternal Cosmos. The channel for this spirit teacher is Bella Karish, D.D. An acquaintance whose judgment I respect describes Bella as "the *only* psychic I would go to for counseling." Bella is a codirector of the Fellowship of Universal Guidance,* a nondenominational, nonprofit organization whose goals are, an official brochure states, "to assist those who desire spiritual fulfillment and evolvement to become cleansed, balanced and attuned, physically, emotionally, mentally and spiritually, as well as subconsciously; consciously, blending with their Higher Selves."

Since 1961, Bella and her codirector, Wayne Guthrie (who helped her found the Fellowship), have been heading the Fellowship's multifaceted activities of counseling, publishing, lecturing, teaching, leading prayer groups, and receiving "spirit discourses" from their spiritual guides on a wide range of topics. I learned of Bella in 1977 through Jeffrey Goodman, who worked with her in the course of researching *We Are the Earthquake Generation*. In his Chapter 7, "Bermuda Shorts in Alaska or the Shifting of the Poles," he reports that Bella, so far as he knows, is the first psychic to "put together the range of various phenomena which different psychics foresee for the future" in an intriguing hypothesis concerning how a pole shift might occur. In brief, Bella—speaking as a psychic—said that man's activities, especially nuclear detonations and changes in underground water, are bringing on earthquakes, which in turn will jolt the earth's axis and increase the wobble, eventually leading to a pole shift. "In Bella's model," Goodman writes, "the pole shift predicted for around the year 2000 A.D. is viewed as a natural consequence of accelerating geological activity brought on in part by our own actions" (p. 172).

My request to Bella for access to any information or predictions

*The Fellowship of Universal Guidance is at 1674 Hillhurst Avenue, Los Angeles, CA 90027. It is open to public membership.

she might have about the possibility of a pole shift brought a generous response in the form of a nineteen-page single-spaced unpublished reading from Eternal Cosmos, made in March 1978. The reading offers provocative new data on this extremely complex subject. I have selected highlights, which I'll present without further commentary.

ETERNAL COSMOS: [When the predicted earth changes are fully under way] there will be strange, unusual openings within the seas, the oceans, and as the water rushes in there will come vast pressures, over and above and beyond man's knowledge. These particular activities, moving into areas within the earth that contain a certain amount of molten and of heated areas, will cause steam to rise and there will be many areas on the earth in which these pressures will explode, causing the earth to move outward and upward; in fact, changing the contour of many areas.

QUESTION: Eternal Cosmos, I wish you would explain just exactly how a planet like the earth tilts. Like for instance: an orange. Can you parallel it with an orange or an apple?

ETERNAL COSMOS: We would say this one way, but we prefer to point out that your earth is much more like a top. For you are spinning in space at a rate far greater than you know. So that you are spinning like a top and moving as you know well slowly at the point that is on the ground, but much more quickly at the top. Therefore, Antarctica or the South Pole is totally involved with the slow motion and therefore, when the top moves or tilts to one side or the other, it spins, which is the wobble we have talked about many times. The action of that which is on the pole at the north end becomes then displaced, it does not melt at this point but it becomes loosened, much like the earth we have talked about that crumbles. And it becomes more and more involved with that which is polar ice, moving away from its point above and moving downward at the sides of the top. The spinning movement sends much of this into a different kind of orbit, causing an imbalance as this becomes loosened at the poles, and moves out and away from its particular placement on the earth.

These will become like monstrous icebergs which, because there is no open ocean or sea for them to move, will become impacted, leaving the pole entry (that is what you call rather affectionately "the hole in the pole") to become totally freed of all ice. This then will place a burden on those areas around that central area and because these icebergs weigh vast amounts of weight, the area within these polar regions will suddenly become what you could call collapsed; for within the area there is a central hollow area and these heavy icebergs will move together to form a heavy impact. Tons and tons of ice then will crush and move the earth in its area at the poles inward, which in turn will cause an extremely dangerous pattern in which the earth will become out of balance.

The balance will cause the earth to tilt more and more, but at the same time, like a top that is stopping and spinning to wobble back and forth, this will become the area or period of time which is the most dangerous of all for as it moves in a slow motion, moving through space in a slowed down orbit, the sun, which is part of your particular energy fields, will become intensely involved with melting the ice then. But the ice, impacted and placed into the areas around the pole, crushing it and closing it off, will melt, not outside but inside, moving into the inner earth and causing explosive actions, the ice again meeting this heated area within. It becomes a cataclysm then for the areas that are below the ice form, or the icebergs that are crushing the pole will become totally a blockage that will cause the earth then to turn in its orbit (just as if you have something that has a heavy weight on top and less weight on the bottom turns upside down).

That has been prophesied many times by many prophets of old and we place it before you now. But let us point out again that timing is not within a particular point but the beginning of this will come slowly and will only speed up when the impact of the slippage will start, of the ice then moving inward, flooding into the openings inside of your planet. . . .

No one could live upon the earth as it will be, for it will be completely turned around. It will have what you call the poles in the areas where your heat is, and this in turn will be the area that is the equator; and the equator area will be the area of the poles. However, this tilt will not remain. Therefore life cannot exist for a period of time until the "top," or the earth as I have been calling it, the "top" returns to its stability. You can rewind the top then, place it on its pivot and it will spin until it begins to change and wobble and finally fall down. The pattern then is the "top" (earth), having stopped moving, within its natural, normal orbit will have to be restored to its equilibrium, to make it run again or move again in perfect harmony with the orbiting of the earth. This means that the areas that are tilted or changed or moved through these pressures within the polar area will need to be restored to balance. The ice at that point will have melted in the equator area, into oceans (new ones) which in turn will become part of that which eventually, placed back to this area at the top of your planet, will freeze again, because of the sun's rays changing totally in their axis also.

QUESTION: Are you saying that the poles will warm up and then refreeze afterwards?

ETERNAL COSMOS: That is correct. But they will only refreeze when, through a natural law of gravity which the earth contains, they bounce back to their upper area. In other words, the tilting will come because the ice pack at the top or near the poles will move as water into the earth, into it. Which means it will not be the flooding action which man has expected if the earth tilts or the polar areas tilt. Instead, first, they will crush the end of the area through their weight, then seep down into the openings, meeting with that which you call the molten or heated area: that is, molten rock and other forms that are a protective shield from the outer earth and the inner earth. It is there.

That is what man senses but he does not realize, that it is not in the middle of the earth, it is only a formation between the outer crust and the inner area, which is totally liveable, filled with many different kinds of natural foliage, plants and life; but there are giants that live there, not men of your stature.*

When this occurs then that which is the tilt will come because of the steam pressures which will cause explosive actions and the earth or pole will tilt to an area that you call the equator. And all that is left of the areas of snow and ice will melt then to become part of vast seas that will change the contour of the equator and its land. When this water then has become part of the sea then, through the efforts of the equilibrium, the gravity of your planet will tilt it back to its natural orbit and at that point the action of the seas will become ice again.

*The "hollow earth" theory is considered to be pseudoscientific nonsense by geologists and geophysicists. But see my comment in Part V.

References and Suggested Readings

Anonymous. *Collapse & Comeback.* P. O. Box 747, Franklin, NC 28734: Metascience Corporation, 1979.

Anonymous. *Season of Changes, Ways of Response.* P. O. Box 444, Virginia Beach, VA 23458: Heritage Publications, 1974.

BENAVIDES, RODOLFO. *Dramatic Prophecies of the Great Pyramid*. Apartado 63–226, Mexico 16, D.F.: Editores Mexicanos Unidos, 1969.

HUETT, LENORA; RICHARDSON, WALLY; and RICHARDSON, JENNY. *The Path to Illumination*. Monterey, CA: Angel Press, 1973.

KUSHI, MICHIO. *Acupuncture, Ancient and Future Worlds*. 359 Boylston Street, Boston, MA 02116: East West Foundation, 1974.

MONTGOMERY, RUTH. *The World Before*. Greenwich, CT: Fawcett, 1976.

WALLACE, BAIRD. *The Space Story and the Inner Light*. P. O. Box 158, Grosse Ile, MI 48138: privately printed, 1975.

IV

PROPHECY

*What Ancient Traditions
Have to Say*

Introduction to the Section

This section will focus on five prophetic sources: Native American traditions, the Bible, Nostradamus, an alleged Lemurian tradition and a secret doctrine. All of them either make explicit reference to a pole shift—past or future—or have been responsibly interpreted that way. I have grouped them together primarily because they are ancient sources that many people regard with respect, fashioning their lives about them to varying degrees. As possible source material, they deserve examination.

The distinction between the processes of psychic prediction and prophecy is not great. Both utilize the same mechanism: precognition. However, as described in the preceding section, psychics today tend to regard their visions as information from higher intelligences or information derived from paranormal "data banks," whereas prophets of old interpreted their visions as sacred revelations from God, or the Most High.

The principal difference between psychic prediction and prophecy, then, is not the mechanism but the moral and theological framework in which it is made. A psychic may have many reasons for functioning as he does, and often they are quite mundane (profit, power, prestige, etc.). A prophet, however, has only one reason: to speak the word of God to his people. The prophet might have no special qualities of character as a saint does. He might also have no special qualities of mind as a scholar or genius does. But, in his role as a special instrument with a divine mission, he is fearless and undeterred in delivering his message, whether to a gathering of country dwellers or to a head of state. The Hebrew prophets especially typify this. And the best contemporary psychics carry forward this theme of revelation from on high for the purpose of awakening humanity to divine destiny.

As was pointed out in the preceding section, we need not regard fearful predictions and prophecies as inevitable. They are modifiable to a degree that is directly related to the quality of human thought and behavior, which are themselves reflections of the quality of human consciousness. Moira Timms says in *The Six O'Clock Bus*, "The trouble with prophecy has always been that the laws of prediction have not generally been understood—even by the prophets themselves" (p. 89). However, parapsychology and paraphysics are now "naturalizing" the mind—making rational the laws and mechanisms by which higher-level functions such as precognition occur. Thus we have an emerging picture from these disciplines—supported by what some psychics themselves declare—of prediction and prophecy being subject to change by human consciousness. The practical implications of this, Timms points out, are that we can warn others and prepare ourselves—both of which offer a means in the present for changing the future.

One of the most important modern prophets I know is the Indian yogi-scientist-philosopher Gopi Krishna. Now in his late seventies, this gentle man from Srinagar, Kashmir is remarkable in many ways. His life's purpose is to show humanity the divine purpose behind evolution and the means by which higher life forms appear on earth. The ancient Sanskrit term for the evolutionary power is *kundalini*. This is what Gopi Krishna is demonstrating in terms appropriate to the scientific and intellectual communities. (I have examined the subject of kundalini in my recent anthology *Kundalini, Evolution and Enlightenment*.)

The reason I call Gopi Krishna a prophet is that he, too, has received revelations of the highest sort. He has written them down in *The Shape of Events to Come* and *The Riddle of Consciousness* and has elaborated on them in a dozen other books. They concern the future of the planet and the human race faced with impending nuclear holocaust. They also reveal the means by which disaster might be averted.

"The aim of my writing," he said recently in a letter, "is to show humanity where the error lies and that nature can take recourse to corrective measures when intellectual persuasion fails. What I experience during these spells of absorption is beyond description. If those who are now in power were just to have a fleeting glimpse of the events to come, they would with ashen faces change their habit of thought and behavior at once, thanking heaven for the chance of atonement granted to them."

I had an opportunity to interview Gopi Krishna in 1976 while he was in Zurich, Switzerland. I asked him to describe the process by which revelation is given to him. His reply is noteworthy:

> For me it is a very amazing occurrence. Always when I am in a mood in which these visions and these writings come, it is always a condition so overpowering that I can only express it with tears. As I have said, I feel as if these, my person, my awareness, is not confined to my own body, or to my head, that is, as if I am not body conscious as an individual, but spread in space. I always feel as if there are two personalities existing side by side. One, a titan—spread everywhere, everywhere I. And the other, also a consciousness, an awareness, a part of it, a drop in it, but also one with it. . . .
>
> It seems that this titan, this other intelligence of which I am talking, this universal consciousness, is handing over these thoughts, these ideas, these passages and these verses to me. It is such a breathtaking experience that I feel humble to dust: I am nothing. In this encounter, I am nothing, just a shadow receiving directions and instructions from an intelligence which is spreading everywhere. . . .
>
> And when I start writing, suddenly a paragraph flashes before my vision. It is handed over to me. I can feel it . . . I both hear and see it. It is something that is just coming into my field of consciousness from the other field of which I am always conscious. It is lent to it—it is handed over. And then after some time I start writing. What takes me half an hour to jot down, is transmitted maybe in five seconds sometimes. It just comes and enters my field of knowledge, my field of awareness.

Although Gopi Krishna is quite independent of the psychic tradition, his experience of the process by which prophecy is manifested parallels the psychic view. I asked him if the events he prophesies—man-made and natural cataclysms—are inevitable. He answered:

> You see, they are all inevitable. At least, as the present position is concerned, they are inevitable. But if the position changes, they are not inevitable. But that is unlikely. Therefore I say that they are inevitable.... What I mean to say is that these unnatural and unhealthy actions of human beings can have no other end except that nature takes recourse to drastic methods to change the direction of human life. Then it becomes inevitable.

In other words, if the human race changes its own behavior, the course of the natural world, including geophysical events, will change.

I have dwelt on Gopi Krishna in this introduction to the prophecy section for several reasons. First, he epitomizes the nature of the prophet. (I have also observed him to have erudition, a saintly character and profound wisdom as well.) Second, his prophecies have aroused interest in certain Native American spiritual elders and keepers of their prophecies, due to the strong parallels they see between Gopi Krishna's modern revelations (dating from the 1950s and 1960s) and their own very ancient ones, especially the Hopi, which are presented here. Last of all, I personally received from him what I consider to be a deep insight into the nature of possible earth changes. It happened this way.

One evening during our four-day meeting, I began to describe to Gopi Krishna some of the psychic predictions about geophysical changes. He had just finished his own description of the atomic conflagration revealed to him. If present conditions don't alter radically, he said, in the 1980s millions of people will lose their lives in atomic war and millions more will be injured.

"If the predictions and prophecies of earth changes and a pole shift come to pass," I said, "the situation will be compounded and that figure will be raised to a much higher level. Some predictions declare that 90 percent or more of the human race will be killed."

Gopi Krishna stunned me with his response: "No, the effect will be just the opposite. It will reduce the level."

He didn't elaborate on his comment, but I sensed his meaning from the character of his prophetic visions. Since the central theme of them is that of a divinely created and divinely ordered cosmos in which the human race plays a meaningful part, from Gopi Krishna's point of view those apparently menacing geophysical cataclysms would be instrumentalities by which the Creator keeps his creation in balance.

Imagine the human race struggling in the midst of nuclear war. And then imagine—as I had not until that moment—that the very ground beneath the armies begins to lurch and crumble. Whether such an event is interpreted as supernatural intervention or the outcome of some natural process touched off by atomic explosions, the result could only be one thing: the cessation of hostilities. What military forces would continue to fight—or could—when both their homelands are being shaken apart by seismic action and the battlefields are too unstable for travel? There would be nothing to conquer and nothing called home to which they could return.

Such a development, from Gopi Krishna's perspective, would be divine intervention through natural means in order to return the human race to its proper evolutionary course: growth to a higher state of being.

Gopi Krishna gives an authentic modern restatement of what many ancient prophetic traditions maintain, namely, that there will be a time of great destruction but there will also be a portion of the population saved so that a better, more spiritual society might be constructed. The prophetic traditions presented here all contain this theme.*

In his book *Revelation: Birth of a New Age*, David Spangler points out that the true prophet is a false prophet because, by uttering dire prophetic words, he sets people into a new course of action, a new mode of behavior that eventually either defuses the disaster in the making or else prepares the people sufficiently far in advance so that death and destruction are minimized. The prophet galvanizes people into action, and eventually their actions

---

*I was recently told of an intriguing prophecy by Mohammed in the Koran, but I have been unable to identify it scripturally. As my informant described it, Mohammed prophesied that "the earth will vomit forth great wealth" for Arab lands just before the end of the world. If this is accurate, the recent Near Eastern fortunes built upon oil "vomited" from the earth may be still another prophetic signpost that a pole shift is near.

prove the prophecy wrong—which is exactly what the prophet wanted in the first place! This is an important point to bear in mind as you enter the realm of prophecy.

Another point to ponder is that which Paul Solomon makes, namely, that the significance of earth changes is dependent upon your state of consciousness. The earth is like a pregnant mother, he says—pregnant with the opportunity for a new birth of consciousness. Just as the parents of a child accept labor pains as part of the process by which new life comes into the world, Solomon preaches, so should people mentally relate to earth changes and a pole shift. The preparations and labor pains are short and necessary for us to experience the joy of new life and growth. Even so will the earth go through geological changes as part of the birthing of a higher humanity. Attune yourself to the cosmic processes at work, and then earth changes, though they may occur, will mean something radically different for your existence.

That attunement, however, does not mean—as is often naïvely supposed—that one must sacrifice the intellect in the name of higher consciousness. Abandonment of the intellect is not necessary. The intellect is one of the most important tools we have for determining truth and the nature of reality. What must be abandoned is a false enthronement of intellect. It is a good servant but a bad master, since the spectrum of human intelligence goes far beyond intellect. In other words, the intellect should be used intelligently, properly. It should be used as an aid for clear perception of reality, along with other channels of knowing such as psychic ability, intuition, emotional knowing and the various forms that intelligence takes in human consciousness.

An example of the intelligent use of intellect can be seen in the scholarship supporting Alan Vaughan's *Patterns of Prophecy*, which revealed the spurious nature of a famous pole shift prophecy. About the time that Nostradamus was prophesying in France, an English woman named Ursula Shipton made many predictions that also proved startlingly accurate. Regarded as a witch, she became known as Mother Shipton and is perhaps the most famous seeress in English history. Among her predictions that some allege as proven were Drake's defeat of the Spanish Armada, the discovery of gold in Australia before Australia itself had been discovered, the great fire of London, and the coming of the black death.

Mother Shipton put her predictions into rhyme. One of them reads:

> In 1936 build houses light with straw and sticks.
> For then shall mighty wars be planned
> And fire and sword sweep the land
> But those who live the century through
> In fear and trembling this shall do.
> Flee for the mountains and the dens
> To bogs and forests and wild fens
> For storms shall rage and oceans roar
> When Gabriel stands on sea and shore
> And as he blows his wondrous horn
> Old worlds shall die and new be born.

Some pole watchers understandably point to this passage as still another prophetic vision of an axis shift at the end of this century. Robert Olson comments on it in *This Day Is Ending:* "In 1936 Hitler was planning the conquests which led to World War Two. Anyone familiar with the houses of southern California will agree that her first line is true advice" (p. 37).

Taken at face value, these verses would appear to be remarkable precognition by Mother Shipton. Unfortunately for pole watchers, they are not. Alan Vaughan investigated the subject, studying original editions at the British Museum Library, and learned that "Mother Shipton, a medieval prophetess famous for predicting events of the twentieth century, has actually never predicted anything beyond her own lifetime. A nineteenth century editor has attributed prophecies to her, and these were altered again by writers in the 1960s—an intricate web of hoaxing for fun and profit" (p. 20).

I think this example is doubly instructive because Vaughan not only demonstrates the use of reason to test the validity of prophecy, he also is himself a seer of high caliber whose precognitions, properly documented and registered, have produced some notable hits. As I advised when introducing the section on psychics, bear this example and other cautions in mind when listening to the pole shift prophecies from ancient tradition.

Now hear the prophets speak.

References and Suggested Readings

KRISHNA, GOPI. *Kundalini, The Evolutionary Energy in Man.* Berkeley, CA: Shambhala, 1971.

——. *The Riddle of Consciousness*. 475 Fifth Avenue, NY 10016: Kundalini Research Foundation, 1977.

——. *The Shape of Events to Come*. 475 Fifth Avenue, NY 10016: Kundalini Research Foundation, 1979.

OLSON, ROBERT. *This Day Is Ending*. Privately printed: no information given.

TIMMS, MOIRA. *The Six O'Clock Bus: A Guide to Armageddon and the New Age*. London: Turnstone Press, 1979.

VAUGHAN, ALAN. *Patterns of Prophecy*. New York: Dell, 1976.

Chapter Fifteen

Native American Prophecies: The Great Purification

"It is sealed in the spirit world." The words came from Sun Bear, a Chippewa medicine man and founder of the first interracial tribe, the Bear Tribe Medicine Society. He was speaking to me of the probability he saw in the ancient prophecies of various Native American tribes that foretell a time of terrible tribulation to be experienced by people everywhere.

According to the prophecies, this period of global suffering and destruction—greater than anything known in history—will be the direct result of man's greed, arrogance and insensitivity to the sacred relationship he has with the planet as a living organism. His unrepentant defilement of what Indians call the Earth Mother will bring on a great cleansing and purification of the planet, so long plundered and raped by technological civilization. Man's violations of nature and his fellow humans will cause social disintegration, war and natural catastrophe. The calamitous climax of this period will be earth changes due to seismic and climatic activity, ending in a pole shift.

"Sealed" means absolutely certain. It *must* come to pass. The only question is the degree of severity. That is still modifiable by human action, enough so that many can survive if they are prepared. But happen it must—within twenty-five to fifty years, according to Sun Bear.

I learned this in October 1977, when I first met Sun Bear and his wife, Wabun (Dawning Wind), a white American of Welsh descent from Newark, New Jersey, who put aside her given name (Marlise James) and profession (journalism) to partake fully of Native American culture. We were all speakers at a three-day conference entitled "Odyssey 2001," an exploration of the next few decades, sponsored by the Harrisburg, Pennsylvania, chapter of Spiritual Frontiers Fellowship.

In 1970, they told me as we talked during pauses in the conference, Sun Bear established the Bear Tribe Medicine Society. He was in his early forties and living in Sacramento, California. Later his medicine visions directed him to buy land near Spokane, Washington.* The Bear Tribe now has several dozen people living on sixty acres. The land—which they consider to be held in trust by them for the Great Spirit, rather than owned—has a large house (the longhouse), a sweat lodge, a cabin, a root cellar, a granary, tepees, storage sheds, several hundred chickens, a cow, some horses, goats, dogs and cats. The tribe preaches and practices self-sufficiency as the means to walking in balance and surviving on the Earth Mother. To reduce its dependency on an economic system that it sees as imbalanced, fatally infected and headed straight to perdition, the tribe barters with other tribes and non-Indian people to obtain what it cannot produce itself. It also publishes a quarterly magazine, *Many Smokes,* and books, such as *The Bear Tribe's Self Reliance Book, Buffalo Hearts* and *At Home in the Wilderness.* A Self Reliance Center, open to the public, offers intensive hands-on training in practical skills and an in-depth experience of Native American philosophy, religion, poetry, legends, prayers, medicine—the culture in general.

Sun Bear and Wabun were at the conference to share the Indian way of life, with particular emphasis on its philosophy and proph-

_____
*Sun Bear, Wabun, and the Bear Tribe can be reached at P. O. Box 9167, Spokane, WA 99209.

ecies. They do this often, traveling around the country in an attempt to awaken the public to their theme: Walk in balance on the Earth Mother. They spoke to us of the Indian sense of oneness with the planet and the Indian sense of responsibility for the land put into their care by the Great Spirit. They spoke of the Indian custom of greeting strangers with friendship and hospitality because they might be the Great Spirit in disguise. The centuries-old prophecies they recite warned of what would happen if people became corrupted and forgot this way of life.

And now, according to Sun Bear and Wabun, many Indians feel the prophecies are coming true and moving rapidly toward fulfillment. A "world" is coming to a close, a cycle of civilization is ending.

Indian myths and legends tell of previous worlds in which humanity reached great heights of achievement but ultimately failed to walk in balance on the Earth Mother according to the Great Spirit's teachings. As a result, their epochs were terminated by the Great Spirit through a variety of means: fire, water, ice. The Hopi tradition tells of three prehistoric worlds. Mesoamerican mythology—that of the Mayas and Nahuas—portrays four such worlds, making the present cycle, which began around 3100 B.C., the fifth. The oral tradition of the Iroquois has preserved the legend of the Seven Worlds of the Seneca.*

I learned of this later in the course of my research along the path opened to me that day by Sun Bear: the path of Native American prophecies. My introduction took place in a seminar room, where other conferees and I listened to Sun Bear during an evening session.

Prophecies, visions and dreams are important ways of securing information. They are the way that down through history people are told or warned or the future is conveyed to them.

My medicine and my work have identified me as an Earth

*Frank Waters notes in *Mexico Mystique:* "The mythological creation and destruction of four previous worlds is not a unique conception of the ancient Nahuas and Mayas, or of contemporary Pueblos and Navajos. The belief was common to Hindu and Tibetan Buddhism; to Zoroasterism, the religion of ancient Persia; to the Chinese; and it is found in the myths of Iceland and the Polynesian Islands. Heraclitus and Aristarchus both taught that the earth was destroyed periodically; and Hesiod, the Greek historian, recounted the destruction of four previous worlds" (p. 104).

prophet. I am able through visions and sometimes through other ways to sense changes coming on the Earth Mother. These are things that we rely on. We accept them. If we know something and we believe it, then we feel it is our responsibility to act on it as well. And this is what has come to us at different times through visions and through medicine.

We rely on these things to direct our lives. And in the past this was also true. Many of the old-timers recorded their various visions and prophecies. Some of them were recorded on pictographs, that is, birchbark scrolls. Others were recorded in other ways—on rock, buffalo hide, things like this. Some of these have come down to us. Some of them we can translate, some we can't.

What, specifically, are those prophecies? There are many—too many for a book such as this. Therefore, I will limit this chapter to those of the Hopi, who have perhaps the most extensive and detailed information among Native Americans about future events in the life of our planet.

Hopi means "peaceful people," "one who follows the peaceful path." The Hopi have never been to war with the white man. Nor do they acknowledge the white man's claim upon their land. They consider themselves an independent, sovereign nation and have not signed a treaty with the United States. The land they inhabit—the Black Mesa area of northern Arizona—has been described by Moira Timms in *The Six O'Clock Bus* as "an island in the sky of over two million acres which rises 3300 feet from the desert floor." Timms writes that the Hopi have inhabited Black Mesa "since the beginning," to use their term—many ages before the mesa became known as the "Four Corners area," where the white man later designated the common borders of Utah, Colorado, New Mexico and Arizona. They consider it to be the spiritual center of the continent. The land has been theirs for at least forty-five hundred years, as proved by radiocarbon tests of artifacts, and perhaps as long as ten thousand years, as claimed by the Hopi. Their centuries-old boundary posts are still in place, and they continue to live in the traditional way, farming the desert land and grazing sheep and goats as their forefathers did, resisting incursions of the dominant white culture.*

*For more detailed information on Hopi history and prophecies, especially as they relate to contemporary social, political and economic events, contact: Friends of the Hopi, Box 1852, Flagstaff, AZ 86002.

According to Hopi mythology, the universe was created by Taiowa, the infinite creative force. In the beginning, the individual named Sotuknang (The Finite) was brought forth from Taiowa (The Infinite). Taiowa commanded him to create lesser beings, and so the twins, Poquanghoya and Palongawhoya, were also brought forth. These twins were then sent, one to the North Pole (Poquanghoya) and one to the South Pole (Palongawhoya), where they were jointly commanded to keep the world properly rotating.

At the dawn of time—the First World—things were attuned to the infinite. Eventually, however, people began to lose sight of their origin. In *The Bear Tribe's Self Reliance Book,* which recounts Hopi legend as told by Hopi elders, we learn that the people "lost the use of the vibratory center on the top of the head (*kopave*), and the soft spot that was the doorway between the body and the spirit began to harden" (p. 10).

> Taiowa decided that would never do, and so he ordered Sotuknang to destroy the world, but to save a few people from destruction. He led them into the center of the world, where they were received by the Ant People. The Ant People fed them so well that they, themselves, began to grow thin. It is said that this is the reason why today the Ant People have such thin waists. As the people stayed underground, the volcanoes on the surface of the First World erupted, and the whole world caught fire. After the fires subsided, the people came up from their shelter and began to move to the Second World that had been prepared for them (p. 10).

In the Hopi myth of the Second World, we see the clearest indications of a pole shift. *The Bear Tribe's Self Reliance Book* tells us:

> Here, again, the people lived until they forgot their origin and grew cold and hard to the ways of the Good Life. And so, once again, Sotuknang was ordered to destroy the world. This time he ordered the Twins, Poquanghoya and Palongawhoya, to leave their stations at the North and South poles and let the world be destroyed. They did this, after the people had again hidden with the Ant People underground. After the Twins left their stations, the world's stability was removed and so it flipped end over end and everything on it was destroyed by ice. The world froze over completely. In this legend we see evidence of two ideas now held by scientists: polar reversals and the

glaciation of the last Ice Age. After the ice had melted enough to make the world inhabitable, the people came up from their shelter and began to move into the Third World (pp. 10–11).

A personal communication to me from Frank Waters, author of *Book of the Hopi*, points out that this myth has other specific details appropriate to a pole shift. After the twins were commanded to leave their posts, his letter states, "the world then teetered off balance, spun around crazily, and rolled over twice. This seems to indicate a change in the world axis."

The world just prior to ours, the Hopi say, was destroyed by flood. Thus the Hopi concept of an antediluvian civilization corresponds with those of many cultures that preserve a myth of the Great Deluge. Legend says that in the Third World people traveled in flying machines. But as the people turned evil and did such things as using their flying machines for war, Taiowa destroyed the Third World by sending waves taller than mountains to sweep over the land—still another hint of a pole shift. The faithful ones who did not forsake the ancient teachings given by the Great Spirit became the Chosen People and were sealed into hollow tubes so they could float upon the water. And thus began the Fourth World.

Chief Dan Katchongva, the late Hopi Sun Clan leader, who died in 1972 at age 109, told a reporter, "The Hopi were survivors of another world that was destroyed. Therefore Hopi were here first and made four migrations, north, south, east and west, claiming all the land for the Great Spirit, as commanded by Massau'u and for the True White Brother who will bring on Purification Day."

Who is the True White Brother? Hopi mythology says that after the destruction of the Third World, the Great Spirit made a set of sacred stone tablets into which he breathed all teachings, instructions, prophecies and warnings as a means of safeguarding the Fourth World. The stone tablets, detailing a Life Plan for harmony and balance, were given to the chief who led the faithful to their new land.

The chief had two sons who took responsibility for the tablets when he died. The set of tablets was then divided by Massau'u. The older brother, with one of the tablets, was sent to the East. Upon reaching his destination, he was to return and look for his younger brother to help bring about the Purification Day, when wickedness would be routed from the land and peaceful balance restored. The younger brother would know his older brother because he would have the matching sacred stone tablet. He would also recognize the older brother because he would be light-complexioned—a *true* white brother—and would be wearing a red cap or red cloak. Whether this indicates a single messiah-like figure or a whole group of people is uncertain.

The place where the Hopi migration ended is known as Old Oraibi, in what is now Arizona. On a rock near the present-day village of Oraibi, in the heart of Hopi land, a petroglyph is carved that records the Hopi prophecy of the Great Spirit's return. Brad Steiger, in his book *Medicine Power,* describes the rock carving:

> In the lower left-hand corner are a bow and an arrow, representing the material tools which the Great Spirit, who stands to the right of the implements, gave to the Hopi.
>
> The Great Spirit points to his path, which is straight up. An upper path to the Great Spirit's right is the white man's way. Two white men and one Hopi—symbolizing the Hopi who forsake the old traditions and adopt other ways—walk this line. A vertical line joins the path of the white man with that of the Hopi, indicating their contact since the Hopi's emergence from the Lower World. The Hopi's path is lower, more spiritual, than the way of the white man.
>
> A large circle represents World War I, another stands for World War II. A third circle symbolizes the Great Purification, which the Hopi feel is fast approaching, according to a timetable that was set centuries ago.
>
> After this transitional period, the Great Spirit returns; food and water is abundant; the world is made well. The white man's path becomes more and more erratic until it is but a series of dots that eventually fade away.
>
> A quartered circle in the lower right-hand corner of the petroglyph is the familiar symbol for the spiritual center of the North American Continent, which the Hopi believe is the Southwestern United States, specifically the area around Oraibi (pp. 200–1).

Moira Timms tells us more about the Hopi prophecies of world wars and—the third circle—the Great Purification that is to come.

> The knowledge of world events has been handed down in secret religious societies of the Hopi. The leaders throughout each generation have especially watched for a series of three world-shaking events, each accompanied by a particular sym-

bol. These symbols—the swastika, the sun and the color red—
are inscribed upon a rock and on the sacred gourd rattle used
in Hopi ceremonies. . . . From their sacred teachings the Hopi
knew that out of the violence and destruction of each world-
shaking and purifying experience, the strongest elements would
emerge with a greater force to produce the next event. (World
War I, championed by Germany and its swastika,* merged later
with Japan, "land of the rising sun.") When these symbols
appeared on the international scene, it was very clear to the
Hopi that a major and final phase of world prophecy was being
fulfilled and they then released for humanity (in 1948) many
of their secret teachings in order to help offset the third and
final "great purification" which on the petroglyph is portrayed
by a red hat and cloak. According to the Hopi elders, "The red
hat and cloak people will have a huge population." They will
invade this land in a single day "like a swarm of locusts" and
the sky will be darkened with them as they come. This prophecy
is similar to the biblical Revelation which speaks of the great
Red Dragon which threatens nations and the invasions [in Revelations 9:16] by "the King of the East" and 200 million horse-
men (pp. 122a–b)!

Frank Waters, in *Mexico Mystique*, adds to our understanding:

> As the Day of Purification approaches, the True White
> Brother will come. He will bring with him sacred stone tablets
> to match those given the Hopis long ago. With him will come
> two helpers. One will carry the swastika, male symbol of purity,
> and a cross, female symbol of purity. The second helper will
> carry the symbol of the sun. These two will shake the earth.
> There will be a massive explosion, perhaps a volcanic eruption
> that will be felt throughout North and South America.
> If these three fulfill their mission, finding a few Hopis who
> steadfastly adhere to their ancient teachings, they will lay out
> a new life plan. The earth will become new as it was in the
> beginning. The people saved will share everything in common,
> speak one tongue, and adopt the religion of the Great Spirit,
> Massau. But if the three sacred beings fail in their mission, and
> all people still remain corrupt, the Great Spirit will send "One"
> from the West. He will be many, many people and unmerciful.
> He will destroy the earth, only the ants being left to inhabit it
> (p. 272).

---

*The swastika was the symbol of Germany in World War II. In World
War I, Germany's symbol was the Iron Cross.

And finally, to close this series of long quotations, an article by Tom Tarbet entitled "The Essence of Hopi Prophecy," in *East West Journal* (October 1977) tells about the Day of Purification:

> The final stage, called the "great day of purification," has also been described as a "mystery egg" in which the forces of the swastika and the sun, plus a third force, symbolized by the color red, culminate in either total rebirth or total annihilation—we don't yet know which, but the choice is ours. War and natural catastrophe may be involved. The degree of violence will be determined by the degree of inequity caused among the peoples of the world and in the balance of nature. In this crisis, rich and poor will be forced to struggle as equals to survive.
>
> That it will be very violent is now almost taken for granted among traditional Hopi, but humans may still lessen the violence by correcting their treatment of nature and fellow humans. Ancient spiritually based communities, such as the Hopi, must especially be preserved and not forced to abandon their wise way of life and the natural resources they have vowed to protect (p. 37).

Certain signs and events were predicted as checkpoints in history to tell of the return of the True White Brother.* The Hopi were told that the white man would bring wagons hooked to each other and pulled by something other than a horse. Railroad trains fulfilled this prophecy. The Hopi were told that there would be "roads in the sky," "cobwebs in the air" and lines across the land. These prophecies were fulfilled, the Hopi say, by airline routes and airplane vapor trails, by electric and telephone wires spanning the desert, and by highways.

The Hopi were told that a "gourd full of ashes" would be invented, which would cause great destruction—boiling rivers, spreading strange incurable diseases, burning the land and preventing anything from growing there for many years. The atomic bomb fulfilled that prophecy.

Still another prophecy said that two brothers will build a ladder to the moon. The prophecy was fulfilled by the U.S.-Russian "moon race" to land men and machines on the lunar surface.

The Hopi prophecies said that the time would come when the white man would put his house in the sky, and that this would be

---

*Certain indications recognized by Native Americans imply that the True White Brother went to Tibet.

one of the last signs to look for. After that, the time of great changes on the Earth Mother would be very near. For the house in the sky would be the last creation allowed to man. After that, because he had tampered with the moon and stars—with the universal system—famine, pestilence, plague and civil war will come upon the planet, causing society and nature to go out of balance. The Hopi believe that Skylab, the fallen American space laboratory, is the "house in the sky."

When it became clear to the spiritual elders of the Hopi that the ancient prophecies were coming to pass, they held a council in 1948 and decided to make public what had been spoken many ages earlier. The ancient knowledge states that the Hopi, the First People, were given the responsibility to care for this continent. It also foretold of a place where the leaders of the world would gather. Hopi prophecy describes it as "a great house of mica standing on the eastern shore of our land." Mica is a glasslike mineral, and the Hopi interpret this prophecy to mean the United Nations building. The Great Spirit told the Hopi to carry this message of walking in balance on the Earth Mother to the assembled nations. Three times they were to go to the house of mica, and if rejected, they were to return to their homeland and await the time of the Great Purification, which must then occur violently as the Fourth World comes to an end.

Twice the Hopis were rejected—once in 1948 and once in 1973—as they attempted to speak to the General Assembly. Finally, in 1976, at the U.N.-sponsored Habitat Conference, in Vancouver, British Columbia, the Hopis were received to speak their message. The spokesman for the Hopi nation was Thomas Banyacya, who was appointed as an interpreter for and by the hereditary *kikmongwis* (spiritual elders) and the religious headmen of the Hopi people. There, in Vancouver, Banyacya delivered an extemporaneous oration that made a deep impression on everyone. He ended with these words:

> This message today is our third and perhaps final attempt to inform the world of the present status of man's existence on our Earth Mother. We are not asking the United Nations for help in a material way. We are, according to Hopi prophecy, simply trying to inform the world of what is going to happen if the destruction of the earth and its original peoples continues as is known by our religious Hopi elders. We do not come before the United Nations in order to join it. We come to fulfill Hopi sacred mission and ancient prophecy in order to find one,

two or three nations who should by now recognize their sacred duty to stop the destruction of the First People's land and life throughout the western hemisphere. We, Hopi, know that there are one, two or three nations here in the United Nations that could listen to our message and understand it so that they can fulfill their spiritual sacred duty in order that land and life shall never be totally destroyed as before. If this is not done, you must understand that the Hopi have done their part to fulfill their mission, and whatever results from your failure to fulfill your sacred responsibilities to stop all of this destruction, genocide, harassment, imprisonment, oppression and lack of respect for Native Brothers will be of your own doing because we, the First Peoples of the western hemisphere, have carried out our sacred duty by bringing this spiritual message of warnings and hope for the future to your attention.

In May 1978 I had an opportunity to speak with Thomas Banyacya about Hopi prophecy. The setting was the awesome Cathedral of St. John the Divine in upper Manhattan. The occasion was a benefit for the national Native American struggle against governmental abrogation of their rights. An audience of about four hundred people listened to half a dozen Indians from around the United States speak of the plight of native people and how that relates to the planetary movement for survival against a nuclear arms race, economic exploitation, an upset balance of nature and other features of our time that threaten the existence of all life.

Banyacya spoke last. He unfolded a large cloth painting of the Oraibi petroglyph and, over the course of nearly an hour, explained its meaning to the attentive audience. However, the evening was getting very late, and there wasn't time for him to present fully what was foreseen centuries ago in Hopi prophecy. Banyacya ended his talk and the meeting concluded.

Afterward, Alex Hladky, a liaison between the American Indian spiritual community and the UN Temple of Understanding, brought Banyacya and me together. Hladky knew I was working on this book and thoughtfully saw that I got a chance to question Banyacya about the prophecies. I'm grateful because I learned new information that, so far as I am aware, has not been given out to the public before. I will let Banyacya tell it in his own words.

Q. Is there anything in Hopi prophecy which is more specific than just saying that the earth will shake or quake during the Great Purification?

A. There's more detail to it [the Oraibi pictograph] but I can't get that far in the short time I talked here. But it does have that because that is also about man, you see. If man keeps himself in balance, the earth will keep itself in balance. The more we disturb ourselves—not being in balance—then the earth probably will shift, you see, and that would cause it.

Q. Is that part of the prophecies for the end of the Fourth World?

A. Yes.

Q. According to Hopi legend, the Second World ended with a pole shift. The twins left the North and South poles. Is that predicted for what's coming now?

A. Yes. If we put our thoughts and minds together and come into balance again with everything around us, our environment, then that thing [the pole shift] will stay [away] or get back in balance. But otherwise, it's going to continue [coming] and then this land might sink again or it may break up or something's going to happen. They [the Hopi elders who gave the prophecies] told a lot of things in detail in there [the pictograph], but I wasn't able to go into it this evening.

Q. Is there flooding predicted?

A. Yes. Great flooding. The rain was supposed to come gentle in a way so it just soaked the ground with no runoff. But the more we disturb the balance of nature, then lightning's going to come—destructiveness. It's going to start hitting people and buildings and things. It may hit power lines and knock them out. And these are some things that might happen if we're not careful. We are already seeing greater earthquakes, flooding, severe winter, severe drought. There'll be severe hot summer days, and the wind is going to be very active now. It'll blow things over, and lightning's going to be much stronger now. I think we're going to see more of that this year. This is a dangerous period that we're going into. We're very concerned in trying to get this across to people in high places.

Q. Do the Hopi elders have any general period of time, any specific number of years, before it's too late and a pole shift is inevitable?

A. There was a time in past history when that happened. It couldn't be straightened out then. But through the spirit of our people, through the ceremonies that we have, and other ways, it [the balance of nature] has come back and is in its right place again. This earth is a living thing, and it loves us just like a mother. So just like a mother, it starts shaking us up real hard if we're not careful.

Q. Within how many years would the pole shift come? Can you say?

A. It depends on what we do from now on. We're rushing into it. The only thing to do now is to stop those things [that upset the balance of nature]. If we stop and realize that we're getting into a dangerous period, we can do it [restore the balance] if we just put our minds to it. Otherwise we're going to get hit from both sides. Nature and some evil-minded person's going to push the button and the earth will just fall over, and the cities and towns will be wiped out in a few seconds. It is *known* that this is coming, and I think it's time that we do something.

So Banyacya sees the *possibility* of a pole shift but not the *inevitability* of it. In accordance with the majority of psychics, Hopi prophecy declares that there is an influence operating—human consciousness—which can determine the ultimate outcome of planetary forces. Because of that, the Hopi—along with other Native Americans, such as Sun Bear, whose personal visions confirm ancient knowledge—will continue to seek restoration of the balance of nature by awakening humanity to the suicidal course it is on.

If they are not successful in delivering their message, they will return to their land to await the Great Purification. Already the Hopi have made significant preparations such as storing quantities of food and other necessary survival items. For if the final event in the prophecies of catastrophe comes to pass, the Hopi were told long ago that where they are now—what they believe to be the spiritual center of the North American continent—will be their safety land. It is sealed in the spirit world.

References and Suggested Readings

STEIGER, BRAD. *Medicine Power*. Garden City, NY: Doubleday, 1976.

———. *Medicine Talk*. Garden City, NY: Doubleday, 1975.

SUN BEAR, WABUN and the BEAR TRIBE. *The Bear Tribe's Self Reliance Book*. P. O. Box 9167, Spokane, WA 99209: Bear Tribe Publishing, 1977.

TIMMS, MOIRA. *The Six O'Clock Bus: A Guide to Armageddon and the New Age*. London: Turnstone Press, 1979.

WATERS, FRANK. *Book of the Hopi*. New York: Ballantine, 1969.

———. *Mexico Mystique*. Chicago: Swallow Press, 1975.

Chapter Sixteen

Does the Bible Prophesy a Pole Shift?

Does the Bible prophesy a pole shift? Does it contain evidence of a previous one?

Some pole watchers think that the story in Genesis of Noah and the flood is based on an event—a global deluge—that could only have occurred due to a sudden tilting of the earth's axis. Moreover, they point to passages in both the Old and New Testament that indicate, from their perspective, that an axis shift will again take place, resulting in a displacement of the oceans from their basins, along with geophysical upheavals that would undoubtedly occur if the planet were to suddenly capsize.

Psalm 46:2–3* describes events that seem to predict a pole shift:

> Therefore will not we fear, though the earth be removed, and though the mountains be carried into the midst of the sea;
> Though the waters thereof roar and be troubled, though the mountains shake with the swelling thereof.

*All quotations are from the King James Version of the Bible.

Likewise the prophet Isaiah, in 13:13, foresees an event of cataclysmic proportions for the entire planet:

> Therefore I will shake the heavens, and the earth shall remove out of her place, in the wrath of the Lord of hosts, and in the day of his fierce anger.

Isaiah's dire prophecies are given even more strongly in 24:1 and 24:18–20:

> Behold, the Lord maketh the earth empty, and maketh it waste, and turneth it upside down, and scattereth abroad the inhabitants thereof....
> ... for the windows from on high are open, and the foundations of the earth do shake.
> The earth is utterly broken down, the earth is clean dissolved, the earth is moved exceedingly.
> The earth shall reel to and fro like a drunkard, and shall be removed like a cottage; and the transgression thereof shall be heavy upon it; and it shall fall, and not rise again.*

*The translation of these verses in the Revised Standard Version of the Bible uses language more appropriate to seismology and the earth sciences:

> Behold, the Lord will lay waste the earth and make it desolate, and he will twist its surface and scatter its inhabitants....
> ... For the windows of heaven are opened, and the foundations of the earth tremble.
> The earth is utterly broken, the earth is rent asunder, the earth is violently shaken.
> The earth staggers like a drunken man, it sways like a hut; its transgression lies heavy upon it, and it falls, and will not rise again.

Likewise, Aramaic scholar George Lamsa's translation, the Holman Bible, suggests greater scientific precision:

> Behold, the Lord shall destroy the earth and lay it waste, and *turn it upside down* and scatter its inhabitants....
> For the windows of heaven are open, and the foundations of the earth tremble. The earth is utterly broken down, the earth is utterly moved, the earth is staggering exceedingly. The earth shall reel to and fro like a drunkard and shall be shaken like a booth ... and it shall fall and not rise again. (Emphasis added)

The New Testament gives even stronger indication of an approaching time that will disturb the whole globe. The central event, of course, will be the Second Coming of Christ descending from the heavens. Prior to that, however, Matthew tells us in 24:6–7, there will be "wars and rumours of wars," that "nation shall rise against nation, and kingdom against kingdom: and there shall be famines, and pestilences, and earthquakes, in divers places." Those are only the initial signs. We learn in Matthew 24:29 what the final moments of history will be like:

> Immediately after the tribulation of those days shall the sun be darkened, and the moon shall not give her light, and the stars shall fall from heaven, and the powers of the heavens shall be shaken.

The third synoptic gospel, Luke, gives a parallel account of the "end time" in language only slightly altered from Matthew. According to Luke 21:25–26:

> And there shall be signs in the sun, and in the moon. and in the stars; and upon the earth distress of nations, with perplexity; the sea and the waves roaring;
> Men's hearts failing them for fear, and for looking after those things which are coming on the earth; for the powers of heaven shall be shaken.

A somewhat different view of "the day of the Lord" is given in II Peter 3:10. The watery mode of destruction is not mentioned. Rather, a fire will be the means of destroying the earth.

> But the day of the Lord will come as a thief in the night; in the which the heavens shall pass away with a great noise, and the elements shall melt with fervent heat, the earth also and the works that are therein shall be burned up.

It is Revelation, however, that foretells most dramatically of a planet-wide seismic catastrophe. At the opening of the sixth seal, in Revelation 6:12–14, John of Patmos's visionary experience reveals the following:

> And I beheld when he had opened the sixth seal, and, lo, there was a great earthquake: and the sun became black as sackcloth of hair, and the moon became as blood;

And the stars of heaven fell unto the earth, even as a fig tree casteth her untimely figs, when she is shaken of a mighty wind.

And the heaven departed as a scroll when it is rolled together; and every mountain and island were moved out of their places.

Later, in Revelation 16:18–20, John's vision of the end of earthly life and the works of man showed him this:

And there were voices, and thunders, and lightnings; and there was a great earthquake, such as was not since men were upon the earth, so mighty an earthquake, and so great.

And the great city was divided into three parts, and the cities of the nations fell. . . .

And every island fled away, and the mountains were not found.

The problem here, as always, is interpretation. If we view the Bible as figurative rather than literal—using symbol, allegory, metaphor and parable to convey abstract psychological, philosophical or metaphysical truths—then our investigation of the pole shift concept comes to a total halt along this path. Any scriptural hint or clue cannot be taken literally as given fact. What may seem to be so is in reality merely indicative of a higher-level meaning. Prophecies of world-shaking events might refer to psychological disturbances within an individual or they might refer to societal disruption as a culture's "world view" takes on a new orientation and values. Understanding the precise meaning would depend upon various factors such as hidden knowledge, mystical insight or esoteric "keys" to Scripture.

Thus, for example, Revelation has been interpreted by many to be a coded text whose surface appearance masks descriptions of actual people and places of early Christian history. It has also been interpreted as a statement of the mystical process termed "the awakening of kundalini," by which initiates in certain mystery schools and hermetic philosophies were guided to higher states of consciousness (see "The Restored New Testament," in my anthology *Kundalini, Evolution and Enlightenment*). In any case, a figurative interpretation of the Bible puts an end to its value for the earth sciences. Concrete data, not vague descriptions and poetic images, are what are necessary for a scientific investigation of the pole shift hypothesis.

Since a figurative interpretation of the Scriptures is useless for our purposes here, let's examine them as being literally true. For the sake of investigation, let's accept the Bible as being divinely dictated or inspired, without error and free from internal differences or contradictions. Our object is to see what answer emerges regarding the question raised in this chapter's title.

That being so, we are immediately faced with an obstacle: the theory of Velikovsky. In his interpretation of the Old Testament Scriptures, the prophecies of Isaiah, Amos and other "servants of God" have *already* been fulfilled. The cosmic catastrophes they foresaw were predictable by them on the basis of their astronomical observations and their knowledge of previous cosmic catastrophes in the second millennium B.C. These events came to pass again in the eighth and seventh centuries B.C., as predicted, and were "brought about by an extraterrestrial agent . . . [which] caused some disturbance in the motion of the earth on its axis and along its orbit" (*Worlds in Collision*, pp. 217–18).

Velikovsky is quite explicit in his interpretation of the Old Testament. Whereas others have considered the prophecies to be "a peculiar kind of poetic metaphor, a flowery manner of expression," he considers them to be literal and exact. However, Velikovsky does not deal with New Testament prophecies. The problem raised for us is this: If we accept the Velikovskyan position, biblical prophecy appears to be partially fulfilled already. The Old and New Testament do not point to a single event—a global disaster still to come—and the unified theological character of the Bible is therefore destroyed.

In order to continue this investigation, we must therefore make an additional assumption—an assumption that for millions today is better described as an act of faith. We must assume not only that the Bible is divinely inspired, without error, but also that the prophecies of both the Old and the New Testament are still unfulfilled.

On that basis, the following seems to be the most probable conclusion: *while there may be evidence in the Bible of a previous pole shift, the biblical predictions of the end of human society—however else they might occur—cannot be based upon the pole shift concept.* Let's see why.

According to proponents of the pole shift theory, a sudden lurching of the planet would produce gigantic tidal waves that would sweep across the land, inundating it for some unspecified distance, possibly entire continents. However, according to a literal interpretation of the Bible, such an event is divinely prevented because God made a convenant with Noah after the flood. Genesis 9:11–13 tells us:

And I will establish my covenant with you; neither shall all flesh be cut off any more by the waters of a flood; neither shall there any more be a flood to destroy the earth.

And God said, This is the token of the covenant which I make between me and you and every living creature that is with you, for perpetual generations:

I do set my bow in the cloud, and it shall be for a token of a covenant between me and the earth.

The rainbow, then, is a visible reminder to humanity of the commitment made by God to Noah and all succeeding generations that the world will never again be destroyed by flood or deluge. This explains why II Peter 3:10 and 13 declares that the day of the Lord will involve fire, not water, as the means by which the planet is cleansed in preparation for "a new earth." It also explains why the predictions of global cataclysms in the final days—from Isaiah to Revelation—do not indicate water as the means of destruction. Telluric activity of unprecedented proportions is clearly described, but the "raging oceans" remain in place. In the verses quoted above, only Isaiah and Luke mention water, and even then the action described involves mountains being carried "into the midst of the sea"; the seas themselves do not leave their basins.

In my judgment, therefore, a literal interpretation of the Bible does not support predictions of a pole shift. It may, however, give indications of a previous one. The account of Noah contains a tantalizing hint that is always overlooked in the popular retelling of the story. That hint lends some credence to the idea that the earth has undergone at least one pole shift within human memory.

As most people understand the Noachian events, rain fell upon the earth for forty days and nights, drowning all life. Movies and cartoons of the Bible story depict people wading in water up to their ankles . . . and then their knees . . . and then their hips . . . pounding on the ark and imploring Noah to let them inside. And the water keeps on rising. . . .

But such artistic license overlooks an important verse, Genesis 7:11. There we learn:

> In the six hundredth year of Noah's life, in the second month, the seventeenth day of the month, *the same day were all the fountains of the great deep broken up,* and the windows of heaven were opened.

I have emphasized the reference to "the fountains of the great deep" because it is an important detail that quite opposes the notion that the flood resulted simply from a forty-day rainfall. Notice that the event in Genesis 7:11 was sudden and cataclysmic. It occurred within a single day and was seismic in character. The ocean bottoms were "broken up." We get no more data than this, but simple calculation of the rate of rainfall—using the heaviest ever recorded—over a forty-day period cannot produce a result even *near* the depth of the flood waters, which the Bible records as fifteen cubits higher than the highest mountains.

A cubit is equal to 1.5 feet. Therefore, the water stood 22.5 feet above the mountaintops. The greatest rainfall recorded in a calendar year was 905.1 inches (75.4 feet), at Cherrapunji, India. If we divide to determine a daily rate of rainfall, and then multiply that figure by 40, we find that about eight feet of rain fell during the great flood. Even if we double or triple the rainfall rate, this is obviously not enough to raise sea level above the mountains of the world! There may have been forty days of rain, but another factor (or factors) is needed to account for the story of Noah and the ark.

That factor is given in Genesis 7:11. If the ocean bottoms were heaved up, several consequences would follow. First, there would be tremendous tidal wave activity. Flooding of the land would naturally follow from that. Second, there would be isostatic rebound of crustal mass, so that the disturbed, raised seabeds would be balanced by subsidence of dry land. The obvious consequence of this, too, would be flooding. And if additional groundwater were released during the seismic disturbance—that is, "the fountains of the great deep" were forced to yield their subsurface stores of water to the oceans—sea level would again be raised (by an unknown amount).

Thus, the Bible does contain the suggestion of a previous pole shift or at least gigantic upheavals of the earth's crust such as one theoretically associates with the pole shift concept. However, the details given are far too sketchy to draw any firm conclusions. Moreover, the lack of any specific reference to a change in "the heavens" or the established pattern of stars is curious, as is the lack of reference to other phenomena that presumably accompany a pole shift, such as tremendous winds and instantaneous changes in climate. These are conspicuous by their absence in the story of Noah and the great flood.

One of the strongest voices among pole watchers who maintain that the Bible records a previous pole shift and also prophesies a coming one is Adam Barber. As noted in Chapter 9, Barber warned in his book that a shift of the planet's axis would result in "a great wave and flood that would destroy civilization." "A few have written that we were wrong," he says in the second edition (p. 4), "as the Bible states the world will not again be destroyed by water, but by fire. This biblical statement, however, has no bearing here, as the coming flood is not to be a destruction of the earth, but merely a washing of a larger part of it." He continues:

We devoutly believe in God, but attach no credence whatever to the story in the Bible that Noah's flood was caused by 40 days and 40 nights of rain. A shift of the axis of the earth is what caused it. I also believe Noah knew of all the discoveries I have made and that he calculated, as I did, when the flood would come, and prepared for it with the ark.

Even though my prediction of a flood and shift is purely *scientific*, and religious beliefs do not enter into it, some people have shown resentment toward it because of the biblical story.

It is *not* in conflict with the Bible, as is shown by the abundant prophecies in that book. . . . The statement in the Bible that the world would not again be destroyed by water, but by fire, has no application here, as that statement is abundantly contradicted by the Bible itself in the quotations on page 5 of this book, and as this coming flood is not to be a destruction of the world, but merely quite a thorough washing of the greater part of it (pp. 4–5).

Since I had concluded before reading Barber that a nonliteral interpretation of the Bible was useless for an investigation of the pole shift concept, I was intrigued by Barber's opening salvo. He would, it seemed, take a nonliteral approach to Scripture and prove that a pole shift is clearly prophesied there.

The quotations referred to above by Barber are quite extensive. A dozen are given in full and another dozen merely referenced. But, I am sorry to report, Barber uses them in a variety of ways that, upon close examination, prove to be specious.

First, many of the quotations are ambiguous with regard to a pole shift, because they speak of events that could be merely local or at most limited to the region of the Holy Land. For example, Barber cites Jeremiah 10:10: "the earth shall tremble." While this clearly indicates seismic disturbance, it hardly constitutes a prophecy of a pole shift. Moreover, the context in which it occurs in Jeremiah also clearly shows that a pole shift is *not* the subject of that section. Rather, the subject is the wrath of God upon the enemies of Israel—the heathen—and those Israelites who adopt heathen customs. The verse is obviously parochial in nature, but Barber overlooks this.

The second misuse of Biblical references that Barber commits is even more serious. Some of the quotations are not simply ambiguous—they are explicit in referring to subjects that cannot be overlooked or taken out of context or interpreted as relating to a pole shift except by deliberate disregard of reason or ethics. Barber cites Isaiah 28:17, ". . . and the waters shall overflow the hiding place," and Isaiah 44:3, "For I will pour . . . floods upon the dry ground."

Those are exact quotations from Barber. That is, he presents these biblical references in support of his contention that the next pole shift will bring flooding and is precisely forecast by the Bible. But when we turn to Isaiah and read the full passages in which these quotations occur, a different picture altogether emerges. Isaiah 28:16–17 reads:

> Therefore thus saith the Lord God, Behold, I lay in Zion for a foundation a stone, a tried stone, a precious corner stone, a sure foundation: he that believeth shall not make haste.
> Judgment also will I lay to the line, and righteousness to the plummet: and the hail shall sweep away the refuge of lies, *and the waters shall overflow the hiding place.*

The concept of a pole shift is not even remotely present in these verses, let alone a specific reference to one. Nor is it in Isaiah 44:3, which Barber also cites. Isaiah 44:2–3 tells us something quite different from what Barber would have us believe:

> Thus saith the Lord that made thee, and formed thee from the womb, which will help thee; Fear not, O Jacob, my servant; and thou, Jesurun, whom I have chosen.
> *For I will pour* water upon him that is thirsty, and *floods upon the dry ground:* I will pour my spirit upon thy seed, and my blessing upon thine offspring.

Barber also tries to persuade us through deception with the passage in Matthew (24:38) where Jesus refers to Noah. Barber's citation reads, "For as in the days that were before the flood . . . until the day that Noe entered into the ark." The very next verse, however—which Barber deliberately leaves out—makes clear that Jesus is speaking about the *timing* of the coming of the Son of Man, not the specific geological events that will attend it.

Despite these affronts to reason and honesty, it must be said that not all of Barber's biblical citations are untrustworthy. He quotes Isaiah 13:13, which I have already noted as one of the most explicit prophecies of a literally world-shaking event. He also cites Amos 8:8–9:

> Shall not the land tremble for this, and every one mourn that dwelleth therein? and it shall rise up wholly as a flood; and it shall be cast out and drowned, as by the flood of Egypt.
> And it shall come to pass in that day, saith the Lord God, that I will cause the sun to go down at noon, and I will darken the earth in the clear day.

I had not been aware of this passage until I found it cited by Barber. A close look at it in context reveals that Amos is indeed prophesying doom for Israel in a manner that clearly extends to other peoples and parts of the world. Barber comments on the quotation (p. 5): "Nothing else but a shift of the axis and earth could cause the sun to go down at noon, and it is clear this passage predicts such a shift [verse 9]. . . . After the shift, as I state, land will rise in places and sink in others [verse 8]."

Barber also cites Haggai 2:6, "For thus saith the Lord of hosts; Yet once, it is a little while, and I will shake the heavens, and the earth, and the sea, and the dry land." Again, examination of the passage in context shows a prophecy of global proportions, for the very next verse reads, "And I will shake all nations. . . ." We must credit Barber with discovery of these valid prophetic references to an event in which the earth's rotation apparently will be affected, resulting in a planet-wide cataclysm.

Nevertheless, in making a final assessment of Barber's evidence, my judgment is that he has not proved his case. If anything, he has disproved it. For the prophecy by Amos indicates, at most, a pause in the earth's rotational spin. This is not, technically speaking, a pole shift. The axis of the planet would remain unaffected in its celestial positioning. True, a perturbation in diurnal spin could produce enormous destruction due to flooding and seismic activity. Velikovsky has demonstrated this already. The consequences of such an event would probably equal in magnitude a true axis displacement, depending on the length of the pause or perturbation and the speed with which it was imposed. As for what might cause such an event, we can speculate upon nuclear warfare, solar flare-ups, a passing celestial body or even the impact from a collision with one. Be that as it may, it is not "splitting hairs" when I say that Barber does not establish a case for biblical prophecy of a pole shift.

Nor does he establish himself as a clear thinker and reliable scholar. The fervent tone of *The Coming Disaster Worse Than the H-Bomb* is such that one cannot doubt the sincerity with which Barber tries to warn the world of impending doom. But in his zeal to do so, he violates both logic and ethics in order to "stack the cards" in his favor. For after citing several dozen biblical quotations in what can only be the most literal interpretation—having previously declared himself a nonliteralist—he then states (p. 6): "It has been told to me by biblical students that many of the above verses are symbolic and refer to something else. If so, then certainly the verses about God promising Noah there would be no more floods are also symbolic and also something else."

Symbolic of what? And what is the "something else"? Barber doesn't say, and his "chop logic" produces nothing but confusion. Upon this he proceeds to build another irrational conclusion: "If the Bible contradicts itself on such important matters, then I want no parts of it, and if God actually made such contradictory statements, then I want no parts [sic] of him either." He concludes this nonsensical argument by saying, "But I do *not* believe He made those promises to Noah, and do *not* believe He made any of the other statements above quoted. I give the quotations from the Bible covering both sides of the controversy, so the reader may form his own opinion." This, after stating only two pages earlier, "We devoutly believe in God, but attach no credence whatever to the story in the Bible that Noah's flood was caused by 40 days and 40 nights of rain. A shift of the axis of the earth is what caused it. I also believe Noah knew of all the discoveries I have made and that he calculated, as I did, when the flood would come, and prepared for it with the ark."

Logically and theologically, Barber is confusing and confused. To top it all, he distorts the story of Noah, which is the very foundation of his argument regarding biblical prophecy of a pole shift. For the story of Noah is not a story of the end of the world— the physical planet—but, rather, of "the end of all flesh" (Gen. 6:13). Quite obviously, the geophysical world was not destroyed in the flood. Civilization and the biosphere were destroyed. But Barber first turns the Noachian events into a "destruction of the earth" and then says that the Bible has no bearing on the coming disaster worse than the H-Bomb because the "coming flood is not to be a destruction of the world but merely quite a thorough washing of the greater part of it." This is merely setting up a straw man to knock down.

It is hopeless to try untangling the grotesque strands of Barber's argument regarding biblical prophecy. Not only does he indulge in sophistry to present what he thinks are both sides of the question, he also argues for and against both sides—and believes both arguments! He picks and chooses among the Scriptures, taking what suits his preconception—oftentimes with wanton distortion of the data—and disregarding that which contradicts him.

At most, therefore, Barber can be credited with the discovery of several intriguing verses that seem to indicate a global catastrophe that will include flooding and geophysical upheavals due to a perturbation in the earth's diurnal spin, although such an event is not technically a pole shift. Pole shift theorist proponents who look to Adam Barber for evidence of biblical prophecy are misdirecting their gaze. To answer the question in this chapter's title: In my judgment, the Bible does not prophesy a pole shift.

Having said that, I must now admit that there is one possibility that could yield an affirmative answer. The theme of the Bible is God's miraculous intervention in human history. The means of intervention vary, but always they are paranormal or supernatural—beyond the reach of reason. Although the thrust of science is to naturalize the miraculous by extending the domain of reason and empirical knowledge, making the supernatural merely supernormal, it seems to me that we are not so scientifically sophisticated that we can absolutely rule out the possibility of a "miraculous" pole shift in which the oceans remain in place.

Judging by what we have seen thus far in this book, such an event seems impossible, considering the mechanics of our planet in motion. However, the very nature of miracles is to do the impossible, to bring about that which cannot be predicted through rational means. Taking the widest possible view of the matter, therefore, I will revise my answer to the chapter title's question to "perhaps."

I say that partly in deference to the judgment of Jeffrey Goodman, who concludes in *We Are the Earthquake Generation* that the Bible does indeed prophesy a pole shift. Goodman begins his argument by considering Christian fundamentalist Hal Lindsey's interpretation of Revelation in his *The Late Great Planet Earth:*

> Lindsey says the Battle of Armageddon is supposed to be the greatest battle of all time. A frightful carnage will result, so great that the human mind cannot conceive of it. In the midst of this battle, in accordance with Revelation 16:18, an enormous earthquake will hit.... The shock wave from this earthquake and/or battle is supposed to produce worldwide destruction. As a result of this earthquake and/or nuclear blasts from the battle, the earth will shift on its axis (p. 212).

Goodman then points out other biblical passages that he feels indicate a pole shift: Zechariah 14:6 and 14:4, Isaiah 19:5–6 and 24:19–23. Then he writes, "Luke 21:25 foretells the sea roaring over the land." This statement, of course, contradicts what I said earlier. Let's look again at the verse in question.

In the King James Version of the Bible, as quoted above, Luke 21:25 reads, "And there shall be signs in the sun, and in the moon, and in the stars; and upon the earth distress of nations, with perplexity; the sea and the waves roaring." The Revised Standard Version translates this verse as, "And there will be signs in sun and moon and stars, and upon the earth distress of nations in perplexity at the roaring of the sea and the waves...."

While these translations do not explicitly describe a global flood or "foretell the sea roaring over the land," it does seem reasonable that such an event could be *implied* by the verse. That is, the situation described could logically be seen from a geophysical perspective as preliminary to a great flooding. Moreover, the other pieces of evidence that Goodman uses to support his argument have considerable geophysical value. So I have to admit that my position is not unassailable.

Moreover, one can take the position that prophecies so diverse in cultural setting as Hebrew and Hopi that nevertheless have so much in common must originate from the same source: God, or the Great Spirit. By implication, then, the Hopi prophecy of a pole shift supports the notion that the Bible predicts a pole shift also.

Might it be that a pole shift could occur of *relatively minor proportions* so that flooding and tidal wave action are restricted to coastal areas of the world? If so, it could be described as a miraculous pole shift—the ultimate disaster tamed. It would also meet the Old Testament requirement that the world not be destroyed primarily by water. But this would conflict with other aspects of the Hopi prophecy and—as we shall see next—with Nostradamus and The Stelle Group. Moreover, Goodman's argument for a pole shift on the basis of a literalist interpretation overlooks God's statement to Noah forbidding another deluge. Strict adherence to the literalist approach would negate any attempt to foretell the sea roaring over the land in Luke 21:25.

Altogether, then, there is an enigmatic dimension to the biblical prophecies. My judgment is that they do not forewarn of a pole shift but, rather, of another type of geophysical cataclysm. However, there are cogent arguments against this conclusion. Those arguments themselves raise other questions, though. So at this point it would perhaps be best to invite theologians and biblical scholars to join efforts with scientists in an examination of this topic, while we move on to examine a singular prophetic figure.

Chapter Seventeen

Nostradamus and "The Great Movement of the Globe"

One of the best-known figures of earlier times to forecast catastrophic changes in the stability of our planet was a French astrologer-physician named Michel de Notredame—known popularly by his Latinized name, Nostradamus. He might more properly be called a psychic or a seer because his predictions were not always prophetic in the manner described in the Introduction to this section. Often they were deliberately vague and deviously worded, in order not to arouse anger or disfavor in powerful circles. And though at times Nostradamus disagreed with both medical and religious authorities, it was not because he had any divinely revealed messages to deliver on those occasions, but only because—as one of the most learned men of his day—he considered the authorities either stupid or dangerous. However, the predictions of a pole shift that Nostradamus made were clearly felt by him to be true prophecies received from God. Because of that, and their antiquity, it seemed appropriate to include him here. The book simply would not be complete without a consideration of the man and the prophecies.

Nostradamus was born at Saint-Rémy, France, in 1503 to a family of Jewish descent that had been forcibly converted to Roman Catholicism by a royal edict that threatened loss of all possessions otherwise. Even as a child, Nostradamus showed great intelligence, and thanks to a devoted tutor he mastered Latin, Greek, Hebrew and mathematics while quite young. By the age of twenty he was also an accomplished astrologer. He then studied philosophy at Avignon and went on to study medicine, a subject he had also begun much earlier under his tutor at the University of Montpellier, the best school of medicine in France.

A portrait of Nostradamus made when he was about thirty, according to Bruce Pennington in his beautifully illustrated *Eschatus,* shows "a very handsome man with dark curly hair and beard, a noble forehead, a long nose, a sensitive mouth, delicate eyebrows, and widely set eyes which are dark, burning with intellect and intensity." In other words, Pennington says, "he looks like a prophet."

After graduation from Montpellier in 1529, Nostradamus stayed on to teach, filling a professorial chair but feeling contemptuous of most of the faculty. He also began medical practice.

By the age of thirty Nostradamus was a respected physician and married to a beautiful woman who had borne him two sons. But the plague, against which he had fought so valiantly during an earlier epidemic, broke out again and took the lives of his wife and sons. Frank H. Stuckert, in his recent study of prophecy, *August 1999,* tells us what happened next.

> This disaster was immediately followed by another one. Nostradamus, a man of tomorrow, was accused of heresy due to a careless remark about the Virgin Mary. He was summoned before the Inquisitor, but took good care not to appear. He became a fugitive ... his wanderings lasted eight years. He visited Milan, Genoa, Venice, Sicily, Lorraine, and the Dauphiné. It was certainly during this period that his own gift of prophecy began to manifest itself. We know that during his Italian journey he once met a young swineherd and fell on his knees before him, calling him His Holiness. The remark caused much laughter among the spectators, but this young man became, in time, Cardinal of Montralto, and became Pope Sixtus in 1585 (p. 75).

In 1544 Nostradamus returned to France to help fight a new outbreak of the plague. This he did with great success, distinguishing himself because of the cures he was able to effect through unorthodox remedies. By 1547 he was famous and wealthy. He married again and settled down to raise another family, eventually fathering six more children. He also continued to practice as a physician, but in a reduced way, giving greater amounts of his leisure to other studies. In time he would become the leading astrologer of his day, knighted by the King of France and sought after by royalty and the wealthy. But when he died of gout at age 63 in 1566, it was to be his prophecies for which history would best remember him.

In 1547 Nostradamus began to prophesy. The prophecies, published in 1555 in a book entitled *Les Vrayes Centuries* (*The True Centuries*) had the form of rhymed quatrains (four-line stanzas) grouped in units of one hundred. Each unit was called a century. A separate prophetic writing, done in prose and called *Epistle to Henri II*, was dedicated to the French monarch. In later editions of *The True Centuries,* the *Epistle* was included as part of the text, and traditionally is placed between the seventh and eighth centuries.

Nostradamus claimed to have compiled his prophecies through divine guidance. Apparently he functioned much like the oracle at Delphi. According to Dr. Joey Jochmans of Lincoln, Nebraska, who has researched the subject extensively, most often Nostradamus would gaze at the still, reflecting surface of water in a brass bowl. This induced a trance state in him. Then, apparently in communication with discarnate intelligences, he would speak the prophecies aloud. Later, he would write them down in Latin, and finally he would translate them into French verses. Another technique he used was to make astrological calculations based upon celestial events or planetary configurations described in the utterances; this was done mostly to date events he had already described, and thus to refine the prophecies. As Charles Ward puts it in his *Oracles of Nostradamus,* "The outward signs of his procedure and methods are palpably magical . . ." (p. 36).

In the preface to *Centuries,* dedicated to his infant son César, Nostradamus defined the purpose and source of prophecy:

> By the grace of God and the good angels, the prophets have had committed to them the spirit of vaticination, by which they

see things at a distance, and are enabled to forecast future events. For there is nothing that can be accomplished without Him, whose power and goodness are so great to all His creatures as long as they put their trust in Him, much as they may be [exposed] or subject to other influences, [yet] on account of their likeness to the nature of their good guardian angel [or genius] that heat and prophetic power draweth nigh to us, as do the rays of the sun which cast their influence alike upon bodies that are elementary and nonelementary. As for ourselves personally who are but human, we can attain to nothing by our own unaided natural knowledge, nor the best of our intelligence, in the way of deciphering the recondite secrets of God the Creator.

By whatever means—magical invocation to the "good angels," "by divine revelation and inspiration," as Nostradamus also described it, or however else it might be explained—Nostradamus did decipher many "recondite secrets of God." Those secrets, however, were most often dire, whether referring to an individual, to a nation or to the whole of human society. As Ward puts it, "He chiefly predicts the *evil* to come; what is good only figures in his pages incidentally, and at long intervals" (p. 36).

Because of that, Nostradamus saw clearly—in the most pragmatic, nonpsychic way—that he would have to conceal his meaning somewhat or run the risk of displeasing people in high places. A prophecy of ill fate for a member of royalty could result in danger for him, since the bearer of bad tidings was sometimes dispatched. Moreover, as Frank Stuckert notes, we must not forget that he was a Jew deep in magic and therefore in constant danger of persecution as a sorcerer by the Church.

In such a situation Nostradamus took a careful middle course of continuing to make predictions but releasing them in veiled language that usually obscured the meaning until after the event. Ward tells us: "Everything in our author is ambiguous: the man, the thought, the style. We stumble at every step in the rough paths of his labyrinth" (p. 37).

Nostradamus chose to speak in riddles. *Centuries* seems to have no chronological order—a deliberate move by Nostradamus—and the double translation was also a deliberate tactic for personal safety. To quote Ward again: "Still he is clearly no prophet in the old and Hebrew sense of the word—like Isaiah, Daniel, David, John—a man who neither respects his own person as regards its safety, nor the person of other men as regards their position" (p. 36).

Yet, indefinite as the prophecies sometimes appeared to be, it must also be said that Nostradamus could be remarkably accurate in his visions of the future. For example, he predicted Napoleon and the French Revolution. Some scholars consider the thirty-third quatrain of the fifth century (which I'll designate as 5:33) to be a prophecy of the infamous *noyades*—drownings—of Nantes, an execution of prisoners in 1793 that was carried out during the French Revolution by the Committee of Public Safety. A revolutionary leader and terrorist, Jean Baptiste Carrier, sought to clear the crowded prisons of Nantes in the swiftest way, so he had large numbers of prisoners put on board vessels with trap doors for bottoms, which then were sunk in the Loire. Nostradamus' prophecy reads:

> *Des principaul de cite rebelle*
> *Qui tiendront fort pour liberte r'avoir,*
> *Detrancher masles, infelice meslee*
> *Cris, hurlemens a Nantes piteux voir.*

This translates as, "The city's leaders in revolt, will in the name of liberty slaughter its [Nantes's] inhabitants without regard to age or sex."

As recently as 1978 a prophecy of Nostradamus appeared to be fulfilled by the death of Pope John Paul I after only thirty-four days in the papal office. According to an Associated Press article entitled "Ancient Soothsayers Foresaw Brief Papacy, Crisis," Nostradamus mentioned a "barefoot pope" whose reign would last "no more than two months" and who would be succeeded by "the last two popes." The death of the barefoot pope would follow the death of a pope who reigned for fifteen years. Pope Paul VI, John Paul's predecessor, reigned exactly that long. With the death of Paul VI, according to interpreters of Nostradamus, which the AP reported, there will be a "two-year period of short pontificates. During that time the church is supposed to undergo a period of grave crisis and even schism," the article states, adding that Nostradamus "wrote cryptically" and "his works are open to various interpretations."

Many of Nostradamus' predictions came true in his own lifetime, including a prediction of his own fame. King Henri II summoned him to the royal court in Paris, where he became the "Seer of the Salon." An example of Nostradamus' cryptic but retrospectively precise quatrains can be seen in 1:35. Published in 1555, it read:

> *Le Lyon jeune le vieux surmontera,*
> *En champ bellique par singulier duelle:*
> *Dans cage d'or les yeux luy crevera,*
> *Deux pluyes une, puis mourir, mort cruelle.*

> The young lion shall overcome the old one,
> In warlike field and in single fight:
> In a case of gold he will pierce his eyes,
> Two wounds one, then die a cruel death.

In the summer of 1559, the king rode in a tournament with such prowess and skill that he won all tilts for three days. On the third day, he rode for final victory against the captain of his Scottish guard, a man named Montgomery. Stuckert tells us:

> Failing to unseat him, the king insisted that the bout should be ridden again. Montgomery would have preferred to call it a draw, but the king insisted they continue until a victor was established. In the second bout they missed each other; in the third Montgomery pierced the king's golden visor with his lance, which cost the king an eye, and later his life. In spite of assurances that no harm should come to him, Montgomery fled to England, where he stayed for fifteen years. Then the Court remembered the quatrain (1:35) of Nostradamus, published four years previously. . . .

Henri II died on July 10, 1559. The queen [Catherine de Médicis] waited fifteen years, at which time Montgomery appeared again in the north of France. . . . [When he did] he was arrested in bed by six followers of the queen and murdered in cold blood. Catherine had hoped Nostradamus might be useful to her, but had she forgotten how good a prophet he really was? Quatrain 3:30, also published many years previously, read:

> *Celuy qu'en luitte fer au fait bellique,*
> *Aura porte plus grande que luy le prix:*
> *De nuit au lit six luy feront la pique,*
> *Nud, sans harnois, subit, sera surprins.*

He who in fight on martial field,
Shall have carried off the prize from one greater than he,
Shall be surprised in bed by six men at night,
Suddenly, naked, and without armour.

These "hits" have lent an air of authority to the more nebulous or still-unfulfilled predictions that lack specific times, places, proper names, etc. One such prediction, from Nostradamus' *Epistle to Henri II,* reads:

There will be a solar eclipse more dark and gloomy than any since the creation of the world, except after the death of Christ. And it shall be in the month of October that a great movement of the globe will happen, and it will be such that one will think the gravity of the earth has lost its natural balance and that it will be plunged into the abyss and perpetual blackness of space. There will be portents and signs in the spring, extreme changes, nations overthrown, and mighty earthquakes.

This quotation comes from a privately published work by Joey Jochmans, *Portents and Promises.** His commentary on it: "The earth rocked by mounting disaster and destruction, will suddenly shift on its axis. Nations will fall, earthquakes will sink mountains and islands, and mankind will believe it is truly the end of the world. This event will take place in October of 1999 or 2000, and will be preceded by a series of eclipses. There will be a solar eclipse on February 16, 1999, an eclipse of the moon on July 29, 1999 and another solar eclipse on August 11, 1999."

In a letter to me, Jochmans remarks that this prophecy is his interpretation of Nostradamus' "prophetic, poetic and cryptic language," taking into account three other related predictions of Nostradamus: 1:56, 2:46 and 1:48. These, Jochmans says, especially point to "a literal axis shift." Here are his translations of the three quatrains, with his commentaries following in parentheses.

You will see a great transformation at the turn of a century,
Extreme horror, a judgment upon the wicked,
The moon inclined at another angle,
The sun will appear higher in its orbit. 1:56

*Interested parties can contact Jochmans at P. O. Box 82863, Lincoln, NE 68501. The book will soon appear in a commercial edition.

(As the result of the earth's shift, near the end of the present century, Nostradamus foresees that the globe will be tilted at a new angle, so that the sun and moon will appear to move in different orbits in the sky.)

> After there is great trouble among mankind, a greater one
>   is prepared,
> The Great Mover of the Universe will renew time,
> Rain, blood, thirst, famine, steel weapons and disease,
> In the heavens a fire seen, lengthening into shooting sparks.
>   2:46.

(The war on earth and in the spirit world will mark the end of the Present Age, and afterward a New Age will dawn, in which the bruised earth will be healed.)

> The grand twentieth year (A.D. 2000) ends, also the position
>   of the moon,
> It will hold a different monarchy in the sky for another 7,000
>   years,
> Then the sun, too, will be tired of its place,
> And at that time will my prophecies for the world be finished
>   and ended. 1:48

(The New Age will last in peace for a thousand years, from A.D. 2000 to A.D. 3000. During this period, the dark forces will be kept in check and mankind will be given the opportunity to develop its Higher Self, to evolve to new levels of spiritual consciousness. At the end of the thousand years, the dark forces will be loosed again, only this time mankind will have learned to use adversity as a tool to further refine its spiritual development. According to Nostradamus, the earth shift and resulting changes in the heavenly orbits of the sun and moon will take place in 1999–2000 will mark the beginning of the New Age evolution of man to a new spiritual level. The seer predicts the next great earth upheaval will mark yet another step in the evolution of man, for by that far-distant future age mankind will finally leave behind the material body, to become total Spirit. The earth will be destroyed because its purpose will be fulfilled, its material sphere no longer required to support the needs of the soul. Only the Man-Spirit will remain, immortal and pure, to drift through new universes and dimensions, forever experiencing, forever learning, forever growing....)

Alan Vaughan points out in *Patterns of Prophecy* that although Nostradamus didn't give the year or specify the consequences of the October "great movement of the globe" in the Epistle, he does elsewhere warn of a "great inundation," or mass floodings, before a "universal conflagration," or world-engulfing fire. Before and after the floods, according to Nostradamus, "burning stones shall fall from heaven" (2:18, 2:46). The most spectacular burning stones that fall from the sky are the Leonid meteor showers. Although they are an annual occurrence, they appear brilliantly once every thirty-three years. The last time they did so was in November 1966; they will return brilliantly again in the apocalyptic year of 1999. And in 10:72 Nostradamus specifically mentions the year 1999—in the month of July—when either the "King of Terror" will be seen descending from the heavens or his sign will appear there, heralding the start of a great war.

Stuckert observes that since Nostradamus used the Julian calendar, his prediction for July 1999 must be corrected to August 1999. The August 11, 1999 eclipse of the sun, Stuckert says, will be accompanied by "a very rare sign": a cross formed by the sun, Saturn, Uranus and Mars.

Recall Nostradamus' prediction in his *Epistle to Henri II:* "There will be a solar eclipse more dark and gloomy than any since the creation of the world, except after the death of Christ. And it shall be in the month of October that a great movement of the globe will happen. . . ."

October 1999, then, appears to be the date Nostradamus prophesies for the pole shift (although Jochmans feels it may also be October 2000). The result of these millennarian catastrophes, according to the preface to *Centuries*, will be as follows: ". . . I find the world before the universal conflagration, such deluges and deep submersion, that there will remain scarcely any land not covered with water, and that for so long a period, that everything will perish except topography and ethnography"—everything but a small amount of land and people.

What will trigger it? Nostradamus does not say exactly. But an interpretation can be made that draws upon natural and human factors. According to Nostradamus, at the end of the century the world will be in the throes of an enormous war, Armageddon. The effect on the planet of atomic explosions could be to accentuate the perturbation in the rotational axis (the Chandler wobble). Then if a celestial event were to introduce additional effects, a pole shift might occur.

What sort of celestial event? Nostradamus hints in 2:18, which Jochmans translates:

A swift and severe rain
Will abruptly halt two armies,
Celestial hail and descending fires will cover the sea with pum-
    ice,
Death on seven continents and seas (the world) sudden. 2:18

(The war of Armageddon will come to a disastrous end,
when the opposing forces are suddenly destroyed by a terrible
storm of hail and fire that will cover the globe. Everywhere the
forces of war and death will be annihilated.)

Such a rain of fire and pumice is reminiscent of the ten plagues
of Egypt, which Velikovsky explains as being due to the near
passage of Venus in its cometary, protoplanetary form. Nostradamus hints at another celestial factor—one we have already heard
described by the Space Brothers via Baird Wallace. It is the appearance of a dark star that enters into a binary relationship with
the sun, thus shifting the pattern of forces impinging on the earth
(and other planets) and creating great changes in their motions.
In quatrain 2:41, as translated by Stuckert, Nostradamus says:

A great star will burn seven days,
A [cosmic] cloud will make the sun appear double.
The great dog will cry all night. . . .

Stuckert interprets "the great dog" to be Sirius, part of a binary
star system. Its ocularly invisible companion, Sirius-B, is a dark
star.
    Alan Vaughan finds evidence of the same theme—a celestial
body from outer space approaching earth—in Nostradamus' 1999
prophecy. He admits that the idea appears to be "sheer nonsense"
from an astronomical point of view. But if such a thing did occur,
he says, it would neatly explain some otherwise puzzling prophecies of Nostradamus, such as quatrain 1:84, which he translates
as follows:

The moon shall be obscured in the deepest darkness,
Her brother [the sun] shall pass being of a ferruginous [blood-red]
    color;
The great one long hidden under the shadows
Shall make his iron lukewarm in the bloody rain.

"That one seems to say," Vaughan writes, "that when the moon and the sun are eclipsed, then another celestial body will become visibly red. Another of his prophecies [2:41] is more specific: 'The great star shall burn for the space of seven days, /A cloud shall make two suns appear...'" (p. 213).

Quatrain 2:41, to which Stuckert and Vaughan refer, is considerably different in translation and meaning in each rendering. This example points out the extreme difficulty in dealing with the prophecies of Nostradamus; certainty slips through our hands like an eel. Something special is needed to grasp the true intent of Nostradamus.

That "something" may soon come to light. According to Stuckert, Nostradamus said he had a "key" to the code he developed with which to encrypt his visions. The code is there in the quatrains but is not yet recognizable. However, Nostradamus predicted that the code would be rediscovered in Germany before the end of the century. So we can hope that the prophecies will soon have their real meaning revealed, including the exact dates that Nostradamus said could be assigned to each quatrain.

In the meantime, let's go on to examine what another ancient tradition has to say about a forthcoming pole shift.

References and Suggested Readings

BROWN, FLORENCE V. *Nostradamus: The Truth about Tomorrow*. New York: Tower, 1970.

JOCHMANS, JOEY. *Portents and Promises*. P. O. Box 82863, Lincoln, NE 68501: privately printed, 1979.

PENNINGTON, BRUCE. *Eschatus*. New York: Simon & Schuster, 1978.

ROBB, STEWART. *Prophecies on World Events by Nostradamus*. New York: Ace, 1961.

STUCKERT, FRANK H. *August 1999*. Hicksville, NY: Exposition Press, 1978.

VAUGHAN, ALAN. *Patterns of Prophecy*. New York: Dell, 1976.

WARD, CHARLES A. *Oracles of Nostradamus*. New York: Modern Library, 1942.

Chapter Eighteen

The Stelle Group: From Lemuria to the Nation of God

If you sincerely believed that Doomsday is going to occur through a pole shift and that you knew the exact time of the occurrence, what would you do? The Stelle Group has an answer to that question.

Stelle, Illinois, is the headquarters of a spiritual community, The Stelle Group, which presently has about two hundred members around the globe. The community of Stelle lies on 240 acres of Illinois flatland about sixty-five miles southwest of Chicago. It is an unincorporated village that was started in 1972 but is still so small that you won't find it on most maps. In fact, it doesn't have its own post office. You write to the Stelle Group residents there—some one hundred at present—via the post office in nearby Cabery.*

---

*The Stelle Group can be reached at: Stelle, IL, Cabery P. O. 60919.

But the beginnings of a modern town are in place. Stelle—a German world meaning "place"—has several dozen homes, and a plastics and woodworking business housed in a 27,500-square-foot factory building. It has a water treatment facility and pumping station that will accommodate one thousand people. A small market is open for sale of community-raised meat and produce, as well as manufactured goods and health foods. A sewer system is in place, along with roads and sidewalks in front of the cooperatively-owned houses where Stelle Group families reside.

No one but Stelle Group members may become residents of the town. There is a special reason for this. Although Stelle, Illinois, is small and recent, The Stelle Group claims quite a history behind its founding—a history going back a long, long way. In fact, it says that its roots go all the way back to the legendary civilization of Lemuria, which began nearly eighty thousand years ago on the lost continent of the same name (but often called simply Mu).

Lemuria, The Stelle Group maintains, was a continent that embraced all of present-day Australia, New Zealand, the Philippines, Oceania, western North America and everything in between. It was destroyed by earthquakes and then submerged twenty-six thousand years ago. When this huge continent went under, all the oceans of the world were drastically lowered as water rushed into the newly formed Pacific Basin. The relatively small islands that had existed in the Atlantic during the time of the Lemurian civilization were left high and dry by the receding ocean, and the newly emerged land joined the Poseid Archipelago, of the Atlantic Ocean, into a large continent.* Man's first civilization arose on Lemuria seventy-eight thousand years ago and lasted for fifty-two thousand years, reaching such heights that our present civilization can barely be considered a civilization when compared to it. Government, religion and science achieved such perfection as to be far beyond our present comprehension.

Because The Stelle Group claims to be linked to this ancient esoteric knowledge, people who seek to join the community must demonstrate character of high integrity. They are therefore carefully screened and subjected to rigid qualifications and require-

---

*This account of human prehistory and geography differs greatly from Cayce's, of course, and from other esoteric traditions such as theosophy (see the next chapter). This is only one of many differences still to be resolved by investigators.

ments based on what The Stelle Group declares are the spiritual laws governing life, handed down in unbroken fashion over millennia through the ultrasecret Brotherhoods, which eventually caused The Stelle Group to be formed.

To get right to the point: The Stelle Group believes it is to be among an intended "remnant" and is following an ancient mandate from higher intelligences (Masters) on other planes of existence, under the direction of the archangel Melchizedek, the Christ, to establish the Nation of God on Lemuria after the coming pole shift brings it above water again. As The Stelle Group sees it, the earth's crust will slide over the mantle. This "geographical reapportionment" is a predictable, natural occurrence foreseen long ago by the Brotherhoods, and they have included it in their plans for human upliftment to a more spiritual state of being. The Stelle Group has been told that the trigger for the shift will be an alignment of our solar system's planets that will have its maximum gravitational effect on 5 May 2000. Moreover, the exact reasons behind the "how" and "why" it will happen have also been revealed.

This prophetic information is available in a book published by The Stelle Group in 1963, *The Ultimate Frontier*, by Eklal Kueshana, pen name for Richard Kieninger, a fifty-two-year-old philosopher-metaphysician-woodworker who founded The Stelle Group with his wife, Gail, in 1963. *The Ultimate Frontier* is the biography of a person named Richard—obviously the author himself, as he acknowledged to me in a recent conversation—that simultaneously gives an account of the philosophy and activities of the ancient Brotherhoods. The Stelle Group says that about two hundred thousand copies have been sold.

According to a 1978 Associated Press feature on Stelle, "Small Town Makes Ready for Doomsday in Year 2000,"

> Since 1975, Stelle has been functioning under a board of trustees with a president, Malcolm Carnahan, 41, who emerged from a power struggle as the leader. Kieninger and his wife, Gail, went separate ways, she with followers to Centuria, Wis., and he to Garland, Tex., to set up a new group. But Carnahan says that Kieninger, the founder of Stelle, visits the community once a month.

Important background information on The Stelle Group can be found in a recent paperback, *The Life & Death of Planet Earth*, by Tom Valentine. An author, editor and journalist for twenty years, Valentine has written many books about his explorations into the paranormal and borderland science.

From 1969 to 1977, Valentine was a member of The Stelle Group. In fact, he edited the Stelle newsletter for five years. I am therefore drawing heavily from his book and other information he supplied, especially his monthly publication, *National Exchange,** which has published several articles on the possibility of a pole shift. In his book, Valentine describes The Stelle Group like this:

> The Stelle Group started as two people, Richard and Gail Kieninger, in 1963 and today we have a small but elegant community with fine homes and a bustling factory located in a cornfield near Kankakee, Illinois. We are working to accomplish two primary goals. First, we are creating an environment that is conducive to individual advancement based upon individual effort, especially with respect to our children. Second, our goal is to bridge the gap that will be left by the prospects of destruction ahead. We intend to survive, not only with our skins, but with technology, so that mankind need not revert again to a stone age existence. We intend to implement the great plan of the Brotherhoods as best we can. Even if the second part of our project is not necessary—and we'd dearly love to learn that all the prophecy of economic disaster, Armageddon and Doom's Day is mere fantasy—we will have a magnificent city peopled by responsible citizens.

This is a noble goal, and actually quite practical-minded. The emphasis is on *values,* not events, but the values will manifest themselves in such a way that calamitous events, if they should occur, will be met in the best way—with courage and farsighted planning—and if they don't occur, nothing need be changed in one's lifestyle and life objectives. As Malcolm Carnahan put it to the AP reporter, ". . . our focus now is not what is going to happen in the year 2000. It is getting a community going."

Let's look at the prophetic aspect of *The Ultimate Frontier:* the predictions of what will come to pass during this century and afterward.

Midway through *The Ultimate Frontier,* Richard's mentor from the Brotherhoods, Dr. White, reveals the future to him in terms that are most dire. Richard is told that economic and political strife

**National Exchange* is available by subscription only. For a sample copy, send $1.50 to P. O. Box 147, Morton Grove, IL 60053.

will "plague the United States and the rest of the world for the two decades prior to the coming inundation."

"First, I'd like to know about this business of the ancient continent rising out of the ocean," the young man said.

"As you might imagine, lad, such an upheaval will produce far-reaching and disastrous consequences throughout the world. The event is of such importance that the Bible is full of prophetic references to it. The event is generally referred to as Doom's Day by scholars and laymen alike, and it is aptly named.

"The mechanics by which it shall occur follow the same pattern as has been the case over the long geologic history of our world. The Earth's crust continually is in the process of rising and falling—becoming alternately ocean and continent—and producing the stratified layers of rock so readily observed in mountainous areas. Our planet's surface rocks, which are about thirty miles thick at most, 'float' upon the denser underlayment. The underlayment is subjected to such extreme heat and pressure that it cannot crystallize; and although it is essentially solid material, it does not have the interlocking crystalline structure to give it dimensional stability. It is very much like a rubbery plastic and comprises the intermediate layer buffering our hard rock surface from the liquid core of the planet.

"You probably already know that Greenland was recently a tropical land even though it is now mostly covered by glacial ice. The last *major* shift of the earth's land masses occurred with the sinking of Mu 26,000 years ago.* The Earth doesn't change its axis because the great inertia of the dense core gives our whirling sphere gyroscopic stability; however, our planet's crust slid several thousand miles to reach a tenable equilibrium which it had lacked before it shifted. The shift was started by vast, ever-widening glaciers of thick polar ice accumulating in a lopsided deposit on the spinning surface of the planet. Because of the inconceivable tonnage of these glaciers, the resultant imbalance was sufficient to cause severe centrifugal aberrations in the Earth's rotation which literally shook the foundations of the continents. Although the ice was the trigger for the big slide

---

*According to Richard Kieninger, in a letter to me, "Actually we believe there were two since then, according to my sources, and the [worse one] was the more recent slide of only about 30 degrees of arc but nevertheless cataclysmic so far as mankind was concerned. These presumably were circa 8500 B.C. and 5000 B.C."

of the Earth's surface, other factors contributed to setting up the condition of surface instability.

"The Earth is constantly undergoing a slight shrinkage of overall size; and since the rigid crust cannot conform to this change in dimension as readily as the liquid core, the crust must adjust itself through a settling process that makes itself known by earthquakes. The faults produced by the settling process often permit rock at a depth of about ten to fifteen miles to liquefy as pressure above is relieved. The molten rock wells up through the fault and can sheet out over large areas to considerable depth. For about a thousand years prior to the sinking of Mu this type of terranean activity was quite prevalent in parts of the world and was particularly extensive along the southeast coastal lands of Mu near where Easter Island is presently located. Continental arches were also established by earthquake action which allowed the crust to rest more commodiously on the plastic intermediate layer. As time wore on, the whole surface structure became shot through with a maze of faults and pockets.

"When the polar ice cap began to slide, the whole surface began to break up and move over the plastic underlayment. The principal movement was completed within about seventy-two hours, but relatively minor adjustments continued to wreak havoc for some centuries afterward. New mountain ranges were thrust up, and continents were depressed or elevated. Oceans were displaced, and almost everything was changed seemingly overnight.

"The same situation is evolving in the world today. The crust is again out of a state of equilibrium; the continental arches are ripe for buckling; and the ice caps, though much smaller, are unbalanced. The trigger this time, however, will be from an outside source. *On May 5th of the year 2000 A.D., the planets of the Solar system will be arrayed in practically a straight line across space, and our planet will be subjected to enough gravitational distortion to tip the delicate balance* [emphasis added]. Although one cannot normally expect mere planetary configurations to have such a spectacular effect upon us, many factors within our Earth are conjoining to produce great surface instability around the turn of the century.

"The developing state of the Earth's surface structure and the effect that the planets will have upon it was long ago carefully measured and analyzed by the Masters who, with the aid of Melchizedek, arranged their program for the establishment of the Kingdom of God* to coincide with this horrendous cat-

*Stelle Group usage has changed this term to "Nation of God."

aclysm. The Book of Revelation is mostly concerned with this period.

"Chapter sixteen of Revelation foretells the nature of the final years of this century. During the period which will begin after 1953, changes in the climate will be noted all over the world. Meteorologic upheavals will give rise to destructive winds, droughts, floods, and a generally high incidence of disruptive atmospheric conditions *(Rev. 16:1–12)*; the final battle of Armageddon is then recounted *(Rev. 16:13–17)*; and the chapter finishes with a description of the cataclysmic earthquakes which shortly follow Armageddon *(Rev. 16:18–21)*. That the great earthquakes are an integral part of Judgment Day is stated in Revelation *(6:12–17)* and in the General Epistle of Jude."

"Would you please explain what Armageddon is?" Richard requested.

"Armageddon refers to the final great battle which will be so violent and destructive that nothing previous will be comparable to it. The viciousness and hatefulness of the combatants will reach new lows. The weapon of Armageddon is being forged at this very moment [Spring 1945] here in Chicago. The period of Armageddon began in the year 1914 and will rise intermittently to a crescendo of more frequent and progressively destructive wars. The battle of Armageddon will be the culmination of this series of violent outbursts. The last war will begin in 1998 and will end in wholesale obliteration in November, 1999.

"A few months later the seismic reapportionment of the world's land masses will come upon most of the survivors of Armageddon as a blessing. After Armageddon and Doom's Day, less than a tenth of the world's population will be alive to see the year 2001 A.D. The intensity of the earthquakes will be greater than has ever been measured by scientists. All the volcanoes of the world will burst forth, and a host of new ones will join them. Vast quantities of heavy gases like carbon dioxide and sulfur dioxide will be hurtled into the atmosphere by the erupting volcanoes. The gases will become supercooled in the outer reaches of the atmosphere and then descend upon the surface of the Earth in convection currents of such magnitude that hurricane winds will howl over the face of the world. The skies will be filled with dust and choking fumes so that even the sun will not be seen directly for months. Walls of water a thousand feet high will roar across the submerging land and sweep away everything before them. Sea and land animals, vegetation, silt, and sand will be shredded into jumbled muck. Where soil is not washed away, it will be covered by boulders and stone; and the newly exposed sea bottom will be worthless for growing crops. The stench of decay and the bleak destruction everywhere will drive many human survivors hopelessly insane. Those who have the strength of their convictions will retain their civilization and rebuild the world. Those people, of course, will comprise the Kingdom of God, and they will be brought through the awful destruction soon to be visited upon the world. Doom's Day will not be without advantages, for it ushers in the Golden Age. After October, 2001 A.D. the Kingdom of God shall be formed" (pp. 104–6).

Dr. White adds that the last half of this century will see "much seismic and volcanic activity which, like the wars and atmospheric disturbances, will increase in frequency and destructiveness as the century draws to a close." After the world has become quiescent again, the Nation of God will flourish with unprecedented peace and prosperity. Ten centuries later, Melchizedek will come again in a physical body, just as he did when he became the first Emperor of Lemuria and as he did much later as the Christ. He will reign for one thousand years—the Millennium—during which time "men will rapidly advance toward mastership," a highly evolved state of human development that is one of the stages in the advancement of the human soul.

As interesting as it would be to explore in depth the evidence for and against the existence of Lemuria and other esoteric topics presented in *The Ultimate Frontier,* our purpose here is to investigate predictions and prophecies of a pole shift. We have seen Richard Kieninger's statement of what is to come. Now let's consider how Tom Valentine elaborates on the Brotherhoods' revelation of the trigger mechanism, a rare planetary alignment.

In *The Life & Death of Planet Earth,* Valentine has a diagram—which I have reproduced in the illustration section—of the conjunction of the larger planets projected for 5 May 2000. It shows the sun at the center of the diagram (i.e., it is heliocentric, not geocentric) and the planets in a line pointing directly into the constellation Taurus. Earth is almost all by itself on one side of the sun, at 224°. Directly opposite Earth, on the far side of the sun, are the biggest planets—Jupiter (42°) and Saturn (42°)—and Mercury (42°). Mars (51°) and Venus (92°), also on the far side, are close to this alignment. Uranus (318°) and Neptune are almost square, that is, almost at right angles to the line of planets. Pluto (232°) is in alignment on Earth's side of the sun, and so is the moon.

Valentine feels that this near-grand conjunction will operate similarly to the much-discussed "Jupiter Effect" of 1982, as proposed in the 1974 book *The Jupiter Effect,* by John Gribben and Stephen Plagemann, but it will be many times more powerful than the effects the earth might experience in 1982. "It is notable," Valentine writes (p. 143), "that the moon will also be in Taurus at that time, meaning there will be an alignment effect consisting of either a push or a pull, or some kind of magnetic, electrical, mysterious thing away from the earth in practically a straight line due to the locations of the moon, sun, Mercury, Mars, Jupiter and Saturn."

What will result? In the Stelle Group's view, a crustal slippage is the most likely outcome of this planetary alignment. Valentine acknowledges *The Ultimate Frontier* as the source of this idea and uses Hapgood and Campbell (in *Earth's Shifting Crust*) as supporting evidence. Their study of the centrifugal effect of the Antarctic ice pack, Valentine says in a *National Exchange* article (March 1977), determined that the off-center ice pack is most likely to be thrown off along the 96° East meridian because that is the direction in which the greatest part of the Antarctic ice cap faces. Aware that Hapgood revised his view on the matter of the ice mass as a trigger mechanism, Valentine nevertheless estimates that the centrifugal force may be enough to trigger a crustal shift of nearly one quarter turn—that is, a pole shift of almost 90°—meaning that Asia north of Rangoon will wind up at the North Pole and Chicago could wind up about where Mexico City is today. As he puts it in his book:

> The bulge of the ice cap at Antarctica is greatest at the 96th meridian, so it stands to reason the thrust will be out along that line. This means the crust will be shoved northward along the meridian that runs through much of the Indian Ocean before contacting a major city, Rangoon, Burma, then the rest of Asia.
>
> Looking at a globe you can see that such a displacement must shift some [tectonic] plates toward the equator and some away from the equator. The bottom of the Indian Ocean will be stretched as the violent thrust northward drags part of what is now called the Australian plate over our planet's equatorial bulge. Far to the north, Siberia will be contracted and compressed and shoved even farther into the chill arctic region. An ice cap will immediately begin to form on the land masses of Siberia, except at such points where that continental plate may submerge beneath ocean waves.

Along the opposite side, we find that our meridian of maximum displacement runs right through Cleveland, Ohio, meaning that North America will move southward toward the equator, being stretched out as well as sloshed by ocean waters. Two segments of the earth's surface, diametrically opposite one another, will be pivotal points and will move the least. One is located in the mid-Pacific near Hawaii if the assumption about the 96th meridian is correct and the other is near the west coast of Europe. Such pivotal points are evidently less subject to serious damage . . . (p. 143).

Since The Stelle Group prophecy is based on the position of the planets, it is possible to verify it by calculating the positions of the planets on 5 May 2000. Valentine himself did this through the assistance of a graduate student in astronomy, who worked out the positions by hand. As part of my "homework," I tested the accuracy of those calculations.

The initial step was to ask my friend Dane Rudhyar, a master astrologer, to check his ephemeris for the planets' positions on that date. Rudhyar's texts were unavailable at the moment, but his practiced eye and keen mind told him there was something wrong with the chart in Valentine's book. He suggested that I have it examined by an astronomer.

My next step, therefore, was to check with the Nautical Almanac Office of the U. S. Naval Observatory, in Washington, D.C. There an official told me that the data weren't available that far in advance. "Why don't you check with an astrologer?" he asked. "Their ephemerides often go that far into the future."

Having begun with an astrologer, this brought me full circle, and so I sought the information through local astrological connections. Eventually I obtained the data independently from two astrologers who have designed computer programs to calculate planetary positions.* When their printouts arrived, they agreed almost exactly in the calculations, although each astrologer had used different sources to design his program.

Meanwhile, I followed through on Rudhyar's suggestion by having two Yale University graduate students in astronomy independently calculate the positions of the planets, as Valentine's

*They are Robert Hand, 217 Rock Harbor Road, Orleans, MA 02654, and Capel McCutcheon, P. O. Box 251, Wethersfield, CT 06109. Inquiries are invited by them.

assistant had done. Again, their results were in complete agreement with each other and with the astrologers'. All of them showed significant miscalculations in most of the planetary positions in Valentine's chart.

Rounded off to the nearest degree, the positions of the planets for 5 May 2000 (heliocentric) were calculated by my sources as follows (with Valentine's given in parentheses for comparison):

Mercury	28°	(42°)
Venus	22°	(92°)
Mars	72°	(51°)
Jupiter	48°	(42°)
Saturn	50°	(42°)
Uranus	318°	(318°)
Neptune	305°	(304°)
Pluto	251°	(232°)

Ironically, the new calculations have the effect of tightening up the alignment of planets, except for Pluto. So the prediction by the Brotherhoods, through The Stelle Group, for the planets to be "arrayed in practically a straight line across space" is correct. Whether the cataclysmic results predicted will occur is a question we will examine in the next part.

First, however, listen to what may be the oldest pole shift prediction of all.

References and Suggested Readings

KUESHANA, EKLAL (RICHARD KIENINGER). *The Ultimate Frontier*. Chicago: privately printed, 1963.

Stelle Newsletter. The Stelle Group, Stelle, IL, Cabery P. O. 60919.

VALENTINE, TOM. *The Life & Death of Planet Earth*. New York: Pinnacle, 1977.

——. "Cataclysm Trigger?" *National Exchange*, February 1977.

——. "Spinning Off-Center Ice Pack May Cause Crustal Shift," *National Exchange*, March 1977.

Chapter Nineteen

The Secret Doctrine of Theosophy

> ... Esoteric Philosophy ... teaches distinctly, that after the first geological disturbance of the Earth's axis, which ended in the sweeping down to the bottom of the seas of the whole second Continent, with its primeval races—of which successive continents, or "Earths," Atlantis was the fourth—there came another disturbance owing to the axis again resuming its previous degree of inclination as rapidly as it had changed it; when the Earth was indeed *raised once more* out of the Waters. ...

This statement, made in 1888, appears on page 85 of Volume 2 of a remarkable work by a remarkable woman. The book is *The Secret Doctrine,** the author is Helena Petrovna Blavatsky. H. P. B., or Madame Blavatsky, as she is often called, founded an occult organization in 1875, the Theosophical Society, now

*All references are to the definitive, six-volume 1971 Adyar edition.

worldwide in scope. Theosophy is the term she coined for the study of "Divine Wisdom, or the aggregate knowledge and wisdom that underlie the universe." It has been defined more specifically as "the formulation in human language of the nature, structure, origin, destiny and operations of the cosmos and of the multitudes of beings which infill it."

Theosophy is a doctrine. It purports to be a modern revival of the most ancient traditions on the planet—traditions that are millions of years old. It is chiefly presented to the public by the Theosophical Society, whose international headquarters are in Adyar, India. The society is dedicated to "the promotion of brotherhood and the encouragement of the study of religion, philosophy and science, to the end that man may better understand himself and his place in the universe," without regard for race, creed, color, caste or sex. The Theosophical Society in America, numbering some five thousand members, has its headquarters in Wheaton, Illinois, where it publishes many books on Theosophy and related subjects, conducts public events and publishes its magazine, *The American Theosophist*. Dora Kunz, whom I mentioned earlier, is currently its president. A drive through the parklike grounds of Theosophical Society headquarters, up to the stately administrative building with its magnificent library and muraled walls depicting the evolution of humanity, is not to be missed.

Theosophy emanates principally from the writings and teaching of its founder, H. P. Blavatsky, who attracted to herself a coterie of learned intellectual supporters, many with psychic gifts and literary talents. Blavatsky's first book, *Isis Unveiled*, was published in 1877 and proved an immediate success. It purported to be, as the subtitle proclaimed, "a master-key to the mysteries of ancient and modern science and theology." The source of the information, Blavatsky said, were her own teachers, or mahatmas, spiritual masters in discarnate form (who occasionally took physical form on the earth) whose initials were M. and K. H., standing for Morya and Koot Hoomi Lal Singh.

Blavatsky's background is richly colorful and controversial. Born in Russia to nobility, she showed high intelligence as a child by becoming a linguist, a fine pianist, an artist and a skilled equestrian. She also demonstrated psychic powers, an official biography claims.

In 1849, at eighteen, she married a man much older than herself, but separated from him after a few months. She then traveled the world for many years, visiting North America, Eastern Europe, Egypt, Greece, India, Japan, Nepal and Tibet. At age twenty, she said she met the master Morya in London, who told her of the work that was in store for her as founder of the Theosophical movement and who undertook to guide her both in her inner development and her outward work for humanity.

In 1852 she attempted for the first time to enter Tibet, where her masters lived, but was unable to do so. Four years later she was able to enter, and she stayed there for three years, undergoing spiritual training from her master. She then journeyed to Europe, Russia and other places until 1867, when she again went to Tibet for the completion of her training in occultism.

In 1873 her master instructed her to proceed to New York City. There, as the official biography puts it, "at the height of her exceptional spiritual, mental and psychic powers," she met many prominent persons who were then investigating the phenomena of spiritualism. She came in touch with a number of people who were interested in what she had to demonstrate and teach, and in 1875 the Theosophical Society was founded. Its objectives were, first, to form a nucleus of the Universal Brotherhood of Humanity; second, to encourage the study of comparative religion, philosophy and science; and, third, to investigate unexplained laws of nature and the powers latent in man.

The years that followed saw the rapid growth of the Theosophical movement—not without setbacks and controversy, however. The history of the Theosophical Society contains accounts of schisms within and attacks from without, the most devastating being charges against Blavatsky of using fraud to produce alleged psychic phenomena. It was also charged that her mahatmas were nonexistent and her works largely fiction or plagiarism. Nevertheless, when Blavatsky died, in 1891, she had nearly one hundred thousand acknowledged followers in all parts of the world.

The masterwork of Theosophy is *The Secret Doctrine*, a huge document some thirteen hundred pages long, published in six volumes, including the index. The first part of *The Secret Doctrine*, "Cosmogenesis," describes the structure and operation of the cosmos; the second part, "Anthropogenesis," gives the history of sentient life on earth, meaning an account of the humanities—plural—that have successively inhabited this planet from its origin.

The perspective of *The Secret Doctrine* maintains that truth and wisdom have largely been lost to modern man. The universe and man are far older, more complex and more elevated than either science or contemporary religion comprehends. As Blavatsky put it in her Introduction to the book, "The Secret Doctrine was the universally diffused religion of the ancient and prehistoric world. Proofs of its diffusion, authentic records of its history, a complete chain of documents, showing its character and presence in every land, together with the teaching of all its great adepts, exist to this day in the secret crypts of libraries belonging to the Occult Fraternity."

Elsewhere in the Introduction she says, "Yet there remains enough, even among such mutilated records [as survive publicly outside the Occult Fraternity] to warrant us in saying that there is in them every possible evidence of the actual existence of a Parent Doctrine. Fragments have survived geological and political cataclysms to tell the story; and every survival shows evidence that the now *Secret* Wisdom was once the one fountain head, the ever-flowing perennial source, at which were fed all its streamlets—the later religions of all nations—from the first down to the last."

What sort of records survive to tell this story? According to Blavatsky, who refers to them as "secret books" and "esoteric records," the main body is found scattered throughout thousands of Sanskrit manuscripts, some translated, most not. Some of the records have been transmitted only in oral form. Others exist in the large and wealthy lamaseries, where subterranean cave-libraries preserve them. But aspects of the Secret Doctrine have been preserved in the wisdom literature of all the world's sacred traditions, hermetic philosophies, occult paths and esoteric sciences such as astrology. Only the highest initiates in occultism have access to these records, however. They are not available for examination by scientists and scholars, so we have only Blavatsky's word—and what she offers in her books—that these records actually exist.

The reality of the Secret Doctrine, Blavatsky says, will be proved in this (the twentieth) century as being older than the Vedas, of India, which are widely regarded by scholars as being perhaps the oldest religious teachings. She adds in a footnote, "This is no pretension to prophecy, but simply a statement based on the knowledge of facts. Every century an attempt is being made to show the world that Occultism is no vain superstition. Once the door is permitted to be kept a little ajar, it will be opened wider with every century."

The vastness of the Secret Doctrine is staggering. With regard to anthropogenesis, we can capsulize it by saying that there have been other races on this planet before our own. Specifically, there have been four, and each has perished in a mighty cataclysm that has changed the face of the globe. These "root races" were the progenitors of each succeeding root race, and according to the Secret Doctrine, there will be two more root races millennia from now, in the long evolutionary ascent back to Godhead. Each race, except for the first, began during the declining years of its predecessor, so that there was overlap among them. The cataclysms were the culminating event in a natural cycle of earth history, itself part of a still-greater cycle by which God "divinizes" the universe. Thus, the cataclysms were natural and predictable events in the life of this planet, as was the end of the races and the beginnings of other racial forms of humanity.

The first root race people were known as Chhayas, or "shadows"; their existence was ethereal, close to being pure spirit. That is, they were largely insubstantial and free from gross matter, existing in a state appropriate to the earth's early condition, before it condensed into material form as we know it now. The Chhayas' "continent," or—as Blavatsky puts it—"the first *terra firma*," is called The Imperishable Sacred Land. It never shared the fate of the succeeding continents, because it is destined to last from the beginning to the end of the great cycle of creation. "Of this mysterious and sacred land," Blavatsky writes, "very little can be said. . . ."

The second continent, the Hyperborean, stretched out its promontories southwestward from the North Pole to receive the second race, evolved from the first. It comprised the whole of what is now known as northern Asia. "It was a real continent, a *bona fide* land, which knew no winter in those early days. . . ."

Lemuria is the name Blavatsky gives to the third continent, a gigantic continent that stretched from the northern edge of the Indian Ocean to Australia, now wholly disappeared beneath the waters of the Pacific. It also extended, horseshoe-shaped, past Madagascar, around what is now South Africa (which was then merely in process of forming) through the Atlantic up to Norway. The inhabitants of Lemuria—the third root race—were giants some twenty-seven feet tall. "They lived and flourished one million years ago . . . ," Blavatsky declares, but first came into being 18 million years ago.

The fourth race, the Atlanteans, perished for the most part

when their continent broke up and sank between 700,000 and 850,000 years ago. However, a small number lived on until as recently as twelve thousand years ago, when the small island remnant of Atlantis submerged. This, Blavatsky says, is what Plato refers to in the *Timaeus*.

We are of the fifth race, the Aryan, and our "continent" is the face of the globe as we see it today. Our race came into being about 200,000 years before the time of "the *semi*-universal deluge known to geology—the First Glacial Period . . . i.e., about 850,-000 years ago." Blavatsky says that the glacial periods are due not only to the cause assigned to them by science, an extreme eccentricity of the earth's orbit, but also another factor: "the shifting of the Earth's axis—a proof of which may be found in the *Book of Enoch* [Chap. 64, sect. 11], if the veiled language of the *Puranas* be not understood." Enoch is quite significant and clear, she says, when speaking of "the great inclination of the Earth . . . [which] is in travail . . ." (Vol. 3, p. 152).

In Volume 3, Blavatsky gives a long statement that summarizes the Theosophical perspective on polar shifts.

> This means again, that our Globe is subject to seven peri-odical and *entire* changes which go *pari passu* with the Races. For the Secret Doctrine teaches that during this Round, there must be seven terrestrial Pralayas,* occasioned by the change at its appointed time, and not at all blindly, as science may think, but in strict accordance and harmony with karmic Law. In Occultism this Inexorable Law is referred to as the "great Adjuster." Science confesses its ignorance of the cause pro-ducing climatic vicissitudes and also the changes in the axial direction, which are always followed by these vicissitudes. In fact, it does not seem at all sure of the axial changes. And being unable to account for them, it is prepared to deny the axial phenomena altogether, rather than admit the intelligent hand of the karmic Law which alone can reasonably explain these sudden changes and their accompanying results. It has tried to account for them by various and more or less fantastic speculations; one of which, as de Boucheport imagined, would be the sudden collision of our Earth with a comet, thus causing all the geological revolutions. But we prefer holding to our esoteric explanation. . . .

*A Sanskrit term—of which there are many in Theosophy—meaning a state of latency or rest between the life cycles that produce the root races.

Thus since Vaivasvata Manu's* humanity appeared on this Earth, there have already been four such axial disturbances. The old continents—save the first—were sucked in by the oceans, other lands appeared, and huge mountain chains arose where there had been none before. The face of the Globe was completely changed each time; the "survival of the fittest" nations and races was secured through timely help; and the unfit ones—the failures—were disposed of by being swept off the Earth. Such sorting and shifting does not happen between sunset and sunrise, as one may think, but requires several thousands of years before the new house is set in order (Vol. 3, pp. 329–30).

In establishing her case for previous axis shifts, Blavatsky often draws heavily upon two ancient sources, the *Book of Enoch* and Egyptian astronomical records. The *Book of Enoch* she declares to be a résumé, a compound of the main features of the history of the third, fourth and fifth races—a "long retrospective, introspective and prophetic summary of universal and quite *historical* events—geological, ethnological, astronomical, and psychic—with a touch of theogony out of the antediluvian records" (Vol. 4, p. 104). Elsewhere she says that although astronomers "may pooh-pooh the idea of a periodical change in the behavior of the Globe's axis, and smile at the conversation given in the *Book of Enoch* between Noah and his 'grandfather' Enoch, the allegory is nevertheless a geological and astronomical fact. There is a secular change in the inclination of the Earth's axis, and its appointed time is recorded in one of the great Secret Cycles" (Vol. 4, p. 294).

An example of pole shift reference in the *Book of Enoch* has already been given. In another, Blavatsky quotes a section in which the angel Uriel tells Enoch, "Behold, I have showed thee all things, O Enoch; and all things have I revealed to thee. Thou seest the sun, the moon, and *those which conduct the stars* of heaven, which cause all their operations, seasons, and arrivals to return. In the days of sinners *the years shall be shortened*.... The moon shall change its laws...."

Blavatsky comments upon this passage:

In those days also, years before the great Deluge that carried

*The generic personification of the race that appeared 18 million years ago.

away the Atlanteans and changed the face of the whole Earth (because "the *Earth* [on its axis] *became inclined*"), Nature, geologically, astronomically, and cosmically in general, could not have been the same, just because the Earth *had inclined*. To quote from Enoch:

> Noah cried with a bitter voice, Hear me; hear me; hear me; three times. And he said.... The earth labours and is violently shaken. Surely, I shall perish with it (Vol. 4, p. 103).

Blavatsky then asks how "the apochryphal author of this powerful vision" knew that the planet occasionally inclined its axis. Has Enoch, she wonders, perchance read the statement in Frederic Klee's work on the Deluge, which states, "The position of the terrestrial globe with reference to the sun has evidently been, in primitive times, different from what it is now; and this difference must have been caused by a displacement of the axis of rotation of the earth." Obviously, Enoch did not read Klee; Blavatsky's point is to ridicule scientists who say the ancients knew nothing of geography, astronomy, etc.

Klee's statement, Blavatsky remarks, reminds her of the statement made by the Egyptian priests to Herodotus that the sun has not always risen where it rises now, and that in former times the ecliptic had cut the equator at right angles. "Occult data show," she declares elsewhere, "that ever since the time of the regular establishment of the zodiacal calculations in Egypt, *the poles have been thrice inverted*" (Vol. 3, p. 352).

Egyptian records penetrate far into antiquity, according to the Secret Doctrine—more than 700,000 years—but do not have the complete story. Since our race appeared on earth, "Twice already has the face of the Globe been changed by fire, and twice by water.... As land needs rest and renovation, new forces, and a change for its soil. so does water. Then arises a periodical redistribution of land and water, change of climates, etc., all brought on by geological revolution, and ending in a final change in the axis of the Earth" (Vol. 4, p. 294).

According to the ancient historian Eusebius, the Egyptians symbolized the cosmos by a large fiery circle, with a serpent with a hawk's head lying across the diameter. Blavatsky quotes from Mackey's *The Mythological Astronomy* to explain the symbolism: "Here we see the pole of the earth within the plane of the ecliptic, attended with all the fiery consequences that must arise from such a state of the heavens: when the whole Zodiac, in 25,000 [-odd] years, must have 'redden'd with the solar blaze'; and *each sign must have been vertical* to the polar region" (Vol. 3, p. 356). She quotes him further in a footnote a few pages later to point out that the Egyptians had various ways of representing the poles, one of which had them appearing as straight rods surmounted with hawks' wings to distinguish north from south, and another of which had the poles in the form of serpents with hawks' heads to make the same distinction.

Blavatsky's point is to show that beneath the range of symbolic representation of the earth and its rotational axis, there is a unifying concept, namely, "the poles *inverted,* in consequence of the great inclination of axis, which each time resulted in the displacement of the oceans, the submersion of the polar lands, and the consequent upheaval of new continents in the equatorial regions, and vice versa" (Vol. 3, p. 359).

Farther on, she comments that "the two famous Egyptian Zodiacs* [support the]... assertion of the Egyptian Priests to Herodotus, that the terrestrial Pole and the Pole of the Ecliptic had formerly coincided..." (Vol. 3, p. 429). Moreover, she adds, Mackey corroborates this when he notes, in *The Mythological Astronomy,* that the poles are represented in the zodiacs in both positions:

> And in that which shows the Poles [polar axes] at right angles, there are marks which prove that it was not the last time they were in that position; *but the first* [—after the Zodiacs had been traced]. Capricorn is represented at the North Pole; and Cancer is divided near its middle, at the South Pole; which is a confirmation that originally they had their winter when the Sun was in Cancer. But the chief characteristics of its being a monument commemorating the *first time* that the Pole had been in that position, are the Lion and the Virgin (Vol. 3, p. 429).

There are many such "dark sayings," Blavatsky notes, scattered throughout the Puranas, the Bible and other mythologies. To a true occultist they divulge two facts:

> (*a*) that the Ancients knew as well as, and perhaps better than, the moderns do, astronomy, geognosy and cosmography

*The Dendera zodiac and the zodiac on the ceiling of the tomb of Senmut.

in general; and (b) that the behavior of the Globe has altered more than once since the primitive state of things. Thus, Xenophantes—on the *blind* faith of *his* "ignorant" religion, which taught that Phaethon, in his desire to learn the *hidden* truth, made the Sun deviate from its usual course—asserts somewhere that, "the Sun turned toward another country"; which is a parallel—slightly more scientific, however, if not as bold—of Joshua stopping the course of the Sun altogether. Yet it may explain the teaching of the Northern mythology that before the *actual order* of things the Sun arose in the South, and its placing the Frigid Zone . . . in the East, whereas it is now in the North (Vol. 4, p. 104).

I have quoted heavily from *The Secret Doctrine* because it is such a rich source of references to ancient traditions that claim to have maintained data about the cataclysmic history of our planet. But what about the future? Does Blavatsky specifically prophesy a pole shift? Is it possible, according to the Secret Doctrine, to fix a date for the next pole shift? And if so, how does that dating compare with those we have seen thus far?

Strictly speaking, Blavatsky is not a prophet, nor would she want to be considered one. From her point of view, true prophecy is simply a rational calculation by a mind attuned to the cosmic processes governing creation and evolution. Prophecy is the lawful operation of a higher mind that has access to esoteric data and knowledge of how to apply them. In other words, prophecy is both reasonable and logical. Conclusions, offered as what the world generally refers to as prophecy, are drawn through careful and exact calculations based upon estoteric science. Official science—what Blavatsky calls "profane science"—if it could agree upon the premises underlying the calculations, would confirm the calculations.

Thus, to answer the questions posed: Blavatsky does indeed state when future polar shifts may be expected, but she does not consider her statements to be prophetic in the ordinary sense. They are, to her, matters of fact, which can be verified through the data she gives. Specifically, she states:

> Let anyone, well acquainted with astronomy and mathematics, throw a retrospective glance into the twilight and shadows of the past. Let him observe and take notes of what he knows of the history of peoples and nations, and collate their respective rises and falls with what is known of astronomical cycles—especially with the *Sidereal Year*, which is equal to

25,868 of our solar years. Then, if the observer is gifted with the faintest intuition, he will find how the weal and woe of nations are intimately connected with the beginning and close of this Sidereal Cycle. True, the non-occultist has the disadvantage that he has no such far distant times to rely upon. He knows nothing, through exact science, of what took place nearly 10,000 years ago, yet he may find consolation in the knowledge of, or—if he so prefers—speculation about, the fate of every one of the modern nations he knows of—some 16,000 years hence. Our meaning is very clear. Every Sidereal Year the tropics recede from the pole *four degrees* in each revolution from the equinoctial points, as the equator turns through the Zodiacal constellations. Now, as every astronomer knows, at present the tropic is only twenty-three degrees and a fraction less than half a degree from the equator. Hence it still has two and a half degrees to run before the end of the Sidereal Year. This gives humanity in general, and our civilized races in particular, a reprieve of about 16,000 years (Vol. 3, p. 330).

Sixteen thousand years until the poles next shift. Quite a bit beyond the end of this century! From the point of view of the Secret Doctrine, which claims to be the most ancient tradition of all, we of the fifth root race have a long time indeed before "the ultimate disaster."

This is not to say, however, that the course of human life on the planet will be free from geophysical threat. Blavatsky comments:

...the *last* remnants of the Fifth Continent [ours] will not disappear until some time after the birth of the *new* Race; when another and *new* dwelling, the Sixth Continent, will have appeared above the *new* waters on the face of the Globe, so as to receive the new stranger. To it also will emigrate and there will settle all those who will be fortunate enough to escape the general disaster. When this shall be...it is not for the writer to know. Only, as Nature no more proceeds by sudden jumps and starts, then man changes suddenly from a child into a mature man, the final cataclysm will be preceded by many smaller submersions and destructions both by wave and volcanic fires. The exultant pulse will beat high in the heart of the race now in the American zone, but there will be no more Americans when the Sixth Race commences; no more, in fact, than Europeans; for they will have now become a *new Race, and many new nations*. Yet the Fifth will not die, but will survive for a while; overlapping the new Race for many hundred thousands of years to come...(Vol. 4, p. 443).

POLE SHIFT AT A GLANCE

SCIENTISTS

PSYCHICS

SOURCE	Date of Last Pole Shift	Date of Next Pole Shift	Previous North Pole Location	Next North Pole Location	Speed of Pole Shift	Trigger Mechanism for Next Shift	Comments
Hugh Auchincloss Brown	5,000 B.C.	Overdue	Sudan Basin	Near Philippines (Alternative: Lake Chad)	One day	South polar ice cap	
Charles H. Hapgood	15,000–10,000 B.C.	?	Hudson Bay		Several thousand years	?	Feels shift may be under way.
Immanuel Velikovsky	700 B.C. 1450 B.C.		Baffin Land				Does not predict another pole shift.
Chan Thomas	5,000 B.C.	30–500 years from 1963	Sudan Basin	Indian Ocean	One day	Ice cap + magnetic null zone	
Adam Barber	7,000 B.C.	Within 25 years			One day	Minor orbit of the Earth	
Emil Sepic	5,500 B.C.	Within 20 years			One day?	Minor orbit of the Earth	
Edgar Cayce	ca. 50,000 B.C.	1999–2000	?		Several years?	Convection currents?	Trigger mechanism will be "upheavals in the interior of the Earth."
Aron Abrahamsen	ca. 70,000 B.C.	1999–2000	30° west of present position	Tropical region	One day	Magnetic-field decrease leads to melted ice caps	Axis shifts 180° but core shifts 90°, so new poles form in tropical region.
Paul Solomon		5 May, 2000		Near England	One day	Mars	Axis remains upright but crust slips.
Ruth Montgomery	48,000 B.C.	1999–2000			One day?	Volcanic eruptions	
Season of Changes		1999–2000			One day?	Sudden change in solar radiation due to extrasolar conditions	
Lenora Huett		Not revealed by the source				Foreign planet	

PROPHECIES

Source	Date (past)	Date (predicted)	Location	Duration	Cause	Human activity / notes
Michio Kushi	10,000 B.C.	?		One day?		Human activity not in accord with the order of the universe is connected with trigger mechanism.
Baird Wallace		Within 15 years			Dark star	Pole shift not explicitly predicted.
Rodolfo Benavides		1977-82		One day	Foreign planet	Planet is named Hercolumbus.
Bella Karish		?		?	Polar ice cap + ground water level changes	
Hopi prophecy	?	Within 25 years	?	One day		Human activity that upsets the balance of nature will produce a trigger mechanism.
The Bible	?	?		One day	Divine intervention (via earthquake?)	
Nostradamus		1999-2000		One day	Dark star	
The Stelle Group	24,000 B.C.	5 May, 2000		One day	Planets in alignment + polar ice caps	
Helena P. Blavatsky	8,000 B.C.	18,000 A.D.	?	One day?	Sun-moon influence?	

? = Mentioned in the source, but data are vague.

The last question we need to ask here is whether the Secret Doctrine identifies a trigger mechanism for axis shifts. The answer is yes: the sun and moon. But the only reference to this I can find is not very specific. It occurs in the context of an explanation of the meaning of certain myths. These myths contain allegorical data about previous shifts of the earth. Blavatsky remarks: ". . . the destruction of the children of Niobe by the children of Latona— Apollo and Diana, the deities of light, wisdom and purity, or the Sun and Moon astronomically, whose influence causes changes in the Earth's axis, deluges and other cosmic cataclysms—is thus very clear" (Vol. 4, p. 340).

Altogether, the Secret Doctrine presents a portrait of polar shifts that is at considerable odds with much of what we have seen thus far. Blavatsky, one of the earliest sources on this subject, and claiming access to the most ancient esoteric wisdom, nevertheless makes no prophecies of imminent disaster. Nor does she allow for the influence of thought forms to affect the workings of the astrogeophysical mechanisms involved (though she apparently does not rule this out either). But she does maintain that polar shifts are inevitable; moreover, that they are necessary events in the evolution of humanity.

The presentation of the data is now complete. We have heard from three independent groups—ancient traditions, contemporary psychics and scientifically-oriented researchers. With all the "exhibits" arrayed before us, it is time to examine them critically and weigh them against current scientific perspectives to see if a coherent and reasonable picture can be drawn from various sources. That is, to put it in a question: can we make a rational *projection* that in any way substantiates nonrational (but not irrational) *prediction?* The next section will attempt to do that.

References and Suggested Reading

BLAVATSKY, H. P. *The Secret Doctrine.* Adyar, India: Theosophical Publishing House, 1971. (Available through the Theosophical Publishing House, Box 270, Wheaton IL 60187.)

PROJECTION

The Outlook for Posthistory

Chapter Twenty

Does Science Support the Predictions and Prophecies?

Do the predictions and prophecies allow us to make a scientific forecast? I have already implied so in the premise of this book. But as we shall see, it is no easy task—with so many variables and sometimes conflicting data! For example, the Paul Solomon Source says that the Hawaiian Islands will rise and become peaks of a great mountain chain on the continent that will rise in the Pacific, but Aron Abrahamsen predicts that the Hawaiian Islands will sink beneath the sea. I pointed out this contradiction to Solomon one day and his response was thought-provoking. Both predictions come from the same source, he felt, and so it probably was the case that Aron's reading referred to the Hawaiians at a different "slice in time" than his own reading. Both were true, in his view, but for different periods of future history.

That is something to consider. On the face of it, though, the readings appear to contradict each other, and so an analysis and evaluation of the data will have to wrestle with thorny problems such as this. At this point, however, let's assume that there is sufficient evidence to convince the hardened skeptic that pole shifts may have occurred in the past, and that we therefore ought to examine the possibility of another one in the near future, despite seeming conflict among the "exhibits" about the nature of the pole shift: how it will happen, when, how fast, etc.

What should be our attitude in doing this? In *The Advancement of Learning*, the father of modern science, Sir Francis Bacon, admonishes us thus: "In contemplation, if a man begin with certainties, he shall end in doubts; but if he will be content to begin with doubts, he will end in certainties." And in *Silva Silvarum: The Phaenomena of the Universe* he advises, ". . . now at length condescend with due Submission and Veneration to approach and peruse the Volume of the Creation; dwell some time upon it; and, bringing to the Work a Mind well purged of Opinions, Idols and false Notions, converse familiarly therein."

It has often been pointed out by enlightened teachers of humanity—as Bacon illustrates here—that intellect is a good servant but a bad master. Science, a supremely intellectual endeavor, has often fallen prey to misguided intellect, which attempts to impose upon nature its own arbitrary constructs. When this occurs, "reason" and "logic" are enthroned in their worst forms—as masks for intellectual arrogance and ego glorification.

The only corrective for this, as T. H. Huxley wrote, is to "Sit down before fact like a little child, and be prepared to give up every preconceived notion, follow humbly wherever and to whatever abysses nature leads, or you shall learn nothing."

To sit down, childlike, before data such as these is a difficult task indeed. If ever there was a subject of which we should be skeptical, it is predictions and prophecies of the ultimate disaster. There have been so many! Yet skepticism born of intellectual arrogance is just as bad as credulity born of intellectual poverty. We must navigate between the rocks of blind belief and the shoals of blind disbelief. To put it in more colloquial terms: we walk a narrow path between having an open mind and having a hole in the head. We must be rigorous but yielding, willing to believe (for investigative purposes) yet doubting of all beliefs, skeptical but not cynical. We must recognize that private conviction is not the same as public demonstration. And we must keep in mind the need for discrimination, rigorous logic, careful scholarship and the most scrutinizing analysis. One of the Buddha's maxims seems especially appropriate for an investigation of the possibility of a pole shift: "Believe nothing which is unreasonable and reject nothing as unreasonable without proper examination."

In doing all this, we should bear in mind four principles drawn from observation in many fields of human experience:

1. The acquisition of knowledge is unending, and interpretation of that knowledge changes from time to time.

2. No one has a monopoly on truth.

3. A person's position may be right for the wrong reasons.

4. It sometimes takes an amateur or an outsider to see a complex situation clearly.

In other words, first, every answer we may get about pole shifts can raise a dozen new questions, and, second, while we should not rule out the possibility that one pole shift theory may be fully correct and all others incorrect, it is unlikely that such will prove to be the case. Moreover, third, a person may sometimes have an intuitive insight, psychic perception or revelatory experience that he fails to communicate properly. His presentation of it—the logical exposition, mathematical calculations, analogies or whatever else is used in communicating nonrational, nonverbal experience in rational, verbal terms—may be an inaccurate reflection of it that has the right conclusion but imprecise or incorrect development of the terms leading to it. Last of all, professionals often are conditioned by prevailing perspectives and technical considerations that tend to block a holistic view and fresh insights.

For these reasons, and as I noted in the Preface, we should not expect to find final answers here. Nevertheless, sufficiently strong indications will emerge to warrant a prompt and concerted scientific investigation.

To begin, we will look at the range of factors that have been identified by various sources as the trigger mechanism that might—individually or in conjunction with others—have sufficient influence on the planet or its crust to destabilize it and initiate a change in the location of the poles. The proposed trigger mechanisms can be divided into two groups of two categories each, as I have shown in the following chart: astrophysical, geophysical-climatological, socioeconomic and biorelative. We will examine the full range, including some that have not been presented before now because they came to my attention not in the context of predictions but as an aspect of other perspectives being offered by certain lines of consciousness research.

POSSIBLE TRIGGER MECHANISMS AND FACTORS CONTRIBUTING TO A POLE SHIFT

NATURAL

I. ASTROPHYSICAL
    A. Planetary alignment
    B. Minor orbit of the earth
    C. Unusual sun-moon relation
    D. Passing celestial body (comet, dark star, planet)
    E. Impact by a celestial body (meteorite, planet, comet)
    F. Magnetic or gravitational null zones
    G. Change in solar radiation

II. GEOPHYSICAL-CLIMATOLOGICAL
    A. Polar ice caps increase and/or decrease
    B. Change in surface mass loading (due to erosion, mountain building, coastline changes, isostatic rebound, reduced water tables, glaciation, etc.)
    C. Magnetic field disappears or reverses polarity
    D. Convection currents in the core and/or mantle
    E. Earthquakes and volcanic eruptions

HUMAN

III. SOCIOECONOMIC
    A. Atmospheric pollution and greenhouse effect
    B. Mining, drilling and damming
    C. Nuclear war and nuclear testing

IV. BIORELATIVE
    A. Thought forms
    B. Etherian physics and orgone weather engineering
    C. Psychotronic weapons
    D. Intervention by higher life forms

Trigger Mechanisms—Natural

ASTROPHYSICAL.

The first is a *planetary alignment*, as predicted by The Stelle Group. On 5 May 2000, the prophecy says, "the planets of the solar system will be arrayed in practically a straight line across space, and our planet will be subjected to enough gravitational distortion to tip the delicate balance. Although one cannot normally expect mere planetary configurations to have such a spectacular effect upon us, many factors within our Earth are conjoining to produce great surface instability around the turn of the century."

Astronomers uniformly reject the notion that the gravitational effects of such an alignment would be sufficient to shift the crust. Grand alignments or near-grand alignments have occurred before without noticeable physical effects, so this seems to be good reason—from the astronomical point of view—for presuming that another such planetary conjunction would likewise have little effect.

There are other reasons, too, why scientists might disregard the Stelle prophecy. For one thing, the geological knowledge of the Brotherhoods is not wholly compatible with contemporary earth-science perspectives. Alert readers may have noted mention that the "Earth is constantly undergoing a slight shrinkage of overall size. . . ." This statement was held as a concept by science at the time *The Ultimate Frontier* was published but has since been discarded as incorrect, nonfactual. Moreover, the ice cap trigger mechanism, named by the Brotherhoods as the critical factor initiating crustal slippage, was already identified in print: Hapgood's *Earth's Shifting Crust* and, before that, Brown's media exposure and monograph. And as we have seen, Hapgood later dropped advocacy of the ice cap concept, admitting that it was insufficient to move the crust.

For all these reasons, a case might be made that the Stelle prophecy is merely a fantastic fabrication by Kieninger. I am not saying this is so; yet even if it were proved true, remember that someone can be right for the wrong reasons. Notice this sentence from the prophecy: "Although one cannot normally expect mere planetary configurations to have such a spectacular effect upon us, many factors within our Earth are conjoining to produce great surface instability around the turn of the century." Some might view this as an "escape clause." On the other hand, it may be that

the Brotherhoods and/or Kieninger have identified a critical con-
catenation of influences, nuclear warfare being the next-to-final
link in the chain of events that shifts the crust. The prophecy also
mentions climatic changes. These conceivably could add to the
instability building in the earth, perhaps by increasing the mass
of the polar ice caps. We will examine these possibilities later,
but for now let us not be hasty in dismissing this astrophysical
trigger mechanism.

Adam Barber and Emil Sepic claim that a *minor orbit of the
earth,* when in a critical synchrony with the major orbit, will tilt
the planet 90°. As noted earlier, there is no recognition of this by
science. I hope the Barber-Sepic claims will be examined, but at
the moment nothing more can be said. Evidence other than their
theoretical arguments is entirely lacking.

The same must be said about the *unusual sun-moon relation*
indicated by H. P. Blavatsky. Because her comment on the trigger
mechanism is so vague, and because her prediction for the time
of the next pole shift is so distant, we need not concern ourselves
with it here.

A *passing celestial body* has been identified by Velikovsky as
the trigger mechanism for previous polar shifts. Specifically, a
comet of great size—the protoplanet Venus—and a fully developed planet—Mars—rocked the earth in space. Because they were
so near, their gravitational influence was far less important than
their electrical charges, which interacted with the earth's electromagnetic field. As a result of electromagnetic coupling, the earth
was shifted on its axis the way one magnet can cause another to
move when brought near it. Velikovsky's insight may also apply
to the planetary alignment in May of 2000.

Could the sum of the planets' electromagnetic fields when in
conjunction be greater than the sum of the individual fields when
not aligned, giving rise to a previously unrecognized effect that
is critical in triggering a pole shift? If so, the Stelle prediction may
hinge on electromagnetism, rather than gravitation. At present,
however, astronomers don't know if all planets in the solar system
have an electromagnetic field. Jupiter and Mercury do, space probe
instruments show. The Moon, Venus and Mars also have fields,
but very weak ones. If the other planets have fields, they must
also be very weak, since instrumentation has not yet detected them.
Electromagnetism as the trigger mechanism in a planetary alignment would therefore seem negligible. But it should be considered.

There is obvious incompatibility between the versions of early history given by *The Ultimate Frontier* and by Velikovsky (and many others). If we put that discrepancy aside, however, a remark by C. S. Sherrerd on "Gyroscopic Precessing and Celestial Axis Displacement," in *Pensée* (Fall 1973) seems appropriate comment on the physics involved:

> Historical records establish that such phenomena [axis shift and changes in the earth's rotation] *have* occurred. This simple model [presented in Sherrerd's article] of course can only suggest orders of magnitude, since many unknown orbital and rotational characteristics of the earth and other celestial effects pertaining at the historical times of interest have been ignored. Furthermore, it is most likely that magnetic, gravitational, and electrostatic forces were *all* involved in concert. Nevertheless, such phenomena are quite plausible within the present knowledge of astrophysics. It is intellectual dishonesty to dismiss the historical records on the basis of alleged astrophysical impossibility. A more fruitful endeavor would rather be to explore what quantitative models are consistent with the historical records and what other logical conclusions can be drawn from such models (p. 33).

That "such phenomena are quite plausible within the present knowledge of astrophysics" is also asserted, with reservations, in a quantitative fashion by Dr. Irving Michelson, professor of mechanics at Illinois Institute of Technology. In a Spring 1974 *Pensée* article, "Mechanics Bears Witness," he declares that Velikovsky's contentions about celestial electromechanics "certainly are not at variance with classical mechanics."

Michelson presents "two new findings, curious and tantalizing . . . [that] connect the kinetic energy of Earth rotation, net electric charge on the Earth, the Earth's magnetic moment, and the interplanetary magnetic field" (p. 15). If, Michelson argues, the earth held an electrical charge of a certain value—a point conceptually asserted by several astrophysicists but not yet empirically demonstrated—it would represent another type of energy not considered in celestial mechanics or in dynamical studies of the earth's rotation. Calculations show that the charge "represents a storage of energy, just sufficient in magnitude that its removal from the Earth could be achieved fully by stopping the rotation of the Earth" (p. 19). This is one of Michelson's findings, and its converse statement is that removing the charge might conceivably stop the rotation. However, no astronomical evidence is cited to suggest that this has ever actually happened (and in correspondence with me, Michelson wonders whether any could be).

Michelson's other finding concerns displacement of the earth's rotation axis. Here two distinct types of energy are involved: mechanical and electrical. Mechanical (kinetic rotational) energy is independent of spatial orientation of the axis; hence a slow shift as much as 180° requires essentially zero additional energy supply. Electrical energy is needed for a 180° axis shift because of earth's magnetic dipole moment and the interplanetary magnetic field—very much less (about one trillion times) than the earth's rotational energy. If this energy is—in some manner unspecified by Michelson—drawn away from the earth's rotational or orbital motion, it would not modify that motion to any significant degree. Thus energy considerations present no obstacle to a 180° axis shift. (But, Michelson points out to me, this fact suggests neither a specific mechanism nor an actual occurrence of such an event.)

The *Pensée* article concludes by noting that the energy necessary for effecting a 180° axis shift of the magnetic dipole corresponds closely to modern estimates of the energy of a single moderately strong geomagnetic storm. "Is it possible that Herodotus' baffling allusions to the 'Earth turned upside down,' as reported in Papyrus Ipuwer also, was triggered by a geomagnetic storm [as Velikovsky asserts]" (p. 20)? Michelson concedes that Velikovsky's declaration of the importance of electromagnetic effects in celestial mechanics "appears to increasing numbers of open-minded and objective scientists to have an entirely valid point" (p. 18).

Subsequently, Michelson analyzed the question with specific reference to Venus's orbit, in what Velikovsky viewed as the truest test of his hypothesis relating to electrical effects. In an invited paper given at McMaster University in 1974 but not published, Michelson examined Velikovsky's assertion that the action of electromagnetic fields explains the near-perfect circularity of Venus's orbit around the sun, a modification achieved in less than fifteen hundred years. Prior to that time, Velikovsky asserts, Venus had wandered through the solar system erratically. How, Michelson wonders, could it enter a stable, nearly circular orbit in such a short time? Would the electromechanics that Velikovsky proposes allow this? Michelson's calculations led to a negative conclusion: Even with the most favorably exaggerated conditions, the forces are billions of times too small. "The mechanism for the circularization of Venus' orbit in a period of the order of thousands of years remains enigmatic. If its resolution is to be found by study of electromagnetic field effects, another mechanism must be sought" (p. 12).

Thus Michelson offers favorable but nevertheless inconclusive energy arguments concerning the earth's rotation and possible changes in its speed and position. However, his examination of possible electromagnetic forces needed to make Venus's orbit circular in historic times—the more significant aspect of Michelson's work—is distinctly unfavorable to Velikovsky's hypothesis. Nevertheless, the cataclysmic effect of a passing celestial body, electrically charged, has been theoretically demonstrated.

Other prophecies and predictions also identify a passing celestial body as the trigger mechanism. These include the Paul Solomon readings, Nostradamus, the Space Brothers and Lenora Huett.

According to the Paul Solomon Source, "that red planet," or Mars, will pass close to the earth, causing crustal slippage three times within a short period. It is, of course, hardly conceivable that Mars will leave its orbit in twenty years or so, unless a cosmic catastrophe occurs to dislodge it. Nevertheless, the reading states that an attractive force from the passing planet will draw the earth's crust toward it as it approaches, and pull the crust along the path of flight as it passes. Internal resistance between the earth's crust and mantle will prevent smooth, continuous slippage. Instead, the crust will lurch after the passing planet in three great heaves. Whether this "pull" is due to gravitation or electromagnetism is not stated, but the latter can be inferred from Reading 208, which mentions an imbalance between positive and negative forces. Gravity is not polarized, as is magnetism.

Nostradamus's prophecy, according to the interpretation followed in this book, indicates that a dark star will enter the solar system, becoming visibly red. Taken at literal value, we cannot equate this with Solomon's "red planet" because Nostradamus, an astrologer, surely knew the difference between a star and a planet. Nevertheless, it may be that the same entity was perceived by the Source and Nostradamus but their modes of expression differed slightly and then were further exaggerated in the case of Nostradamus by inaccurate translation and interpretation.

This is a possibility to be considered, yet the Space Brothers' prediction seems to deny it. The Space Brothers state unequivocally that a dark star, not a planet, will enter the solar system. That being the case, we are forced to assume that either the nature of the passing celestial body is presently indeterminable or else *both* a dark star and a planet will appear (in company?) in the vicinity of the earth at the end of this century. The astronomical probability of either event is, of course, small.

That is not to say nonexistent, however. Our sun may have a companion star! Such, at least, is the view of E. R. Harrison, an astronomer at the University of Massachusetts, who reported in an article of that title in *Nature* (24 November 1977) that "the barycentre of the Solar System is accelerated, possibly because the Sun is a member of a binary system and has a hitherto undetected companion star" (p. 324).

Data from pulsar observations and from the sun, Harrison reports, indicate a wobbly motion characteristic of a star bound to a binary companion. Such a companion could not be an ordinary star because it would be the second-brightest thing in the sky and easily detected. Therefore, if it exists, it must be a dark object. Harrison comments:

> If the companion is a neutron star or black hole produced by a supernova in the early history of the Solar System, then it is difficult to understand how our weakly bound binary system has survived such a violent event. It is more likely that a companion . . . is in a hyperbolic orbit, and the present close encounter is a transitory phenomenon lasting only several thousand years. The companion star is therefore either a faint white or red dwarf in closed orbit around the Sun or a gas-accreting nearby neutron star or black hole in open orbit. Obviously a star in closed orbit about the Sun must lie close to the ecliptic plane so as not to disturb the planets excessively . . . short-term perturbations in the motions of Neptune and Pluto are significant, but as far as I can determine, are not unacceptably large (p. 325).

Harrison concludes his article cautiously. "Has the Sun a companion star? I find it hard to believe that a star so close can exist and yet remain undiscovered. On the other hand, pulsar observations of extraordinary precision imply that it might exist, and therefore a search for a companion star is perhaps worth undertaking" (p. 326).

Some months later, two Dutch astronomers. H. F. Henrichs and R. F. A. Staller, responded to Harrison in *Nature* (11 May 1978) with an article, "Has the Sun Really Got a Companion Star?" They conclude that "Harrison's explanation is probably incorrect" (p. 132).

But the case has not been definitively closed. An astronomer from the University of British Columbia, Serge Pineault, seconded Harrison in a follow-up article, "If the Sun Has a Companion . . ." in *Nature* (26 October 1978). Pineault agrees with Henrichs and Staller that the pulsar data could not be due to the presence of a faint white, red or even black dwarf star in closed orbit around the sun. However, he says, the situation for a neutron star or a black hole is not quite so clear. The best interpretation of the data suggests that a close encounter between the sun and a neutron star or a black hole is indeed possible. Such a dark object could have remained undetected until now, he says, but could now be detected by satellite photography. A search would be worthwhile, and if the proposals made by him and Harrison are correct, "one will have the satisfaction of having observed the (most likely) closest compact companion" (p. 730).

In a personal communication with Harrison in December 1978, he summarized the matter. It is now almost certain, he said, that a companion star in closed orbit—i.e., a permanent companion— does not exist, but a temporary companion in open orbit that brings it close to the solar system is not ruled out.

Not only may we be in a binary star system, the solar system may also be larger than nine planets! There may be others than the ones conventionally recognized. In 1859, for instance, the French astronomer Leverrier, who had correctly predicted the existence of Neptune, told the French Academy of Sciences that he had observed what he believed to be another planet between the sun and Mercury. Leverrier named it Vulcan.

Leverrier's observation was never confirmed by others, but the idea persisted. And in 1971 Dr. Henry C. Courten, of Dowling College, announced that he had discovered what appeared to be an object orbiting the sun about 9 million miles from it. Courten's evidence was mysterious tracks on photographs made during solar eclipses in 1966 and 1970. He postulated an asteroid or, more properly, a planetoid. Courten's observation, like Leverrier's, has not been confirmed, and it is now generally assumed that an intramercurian planet does not exist. But Leverrier's notion lingers.

So does the notion of a transplutonian planet. Since the nineteenth century, astronomers have searched for a planet beyond Pluto. This tenth planet is referred to by astronomers as Planet X, meaning both "Roman numeral ten" and "mysterious." The likelihood of its existence is now greatly increased because of the discovery of a moon around Pluto.

James W. Christy, of the U. S. Naval Observatory, made the discovery in June 1978. This led astronomers to recalculate the size of its host planet. Now it appears that Pluto is much smaller than hitherto believed. This means that certain perturbations in the other outer planets—Uranus and Neptune—can no longer be attributed to Pluto. As Dr. Robert S. Harrington, of the Naval Observatory, put it, according to *National Exchange* (November 1978), astronomers have concluded that "This leaves open the possibility that a new massive object may be found in the solar system, possibly even a new planet" (p. 22).

A telephone call to Christy confirmed this. He told me that preparations are being made to search for a tenth planet, and indications are that it will be either a big one, about the size of Neptune, or many small ones. He referred me to Dr. Thomas Van Flandern, of the Naval Observatory, who will head a theoretical study of the subject. Van Flandern in turn told me that something is very definitely disturbing the outer planets, but it will be several years before any firm conclusions might be reached from the research.

The article quoting Harrington, "Ancients May Have Been Right About Twelfth Planet," is by Zecharia Sitchin, author of *The Twelfth Planet* (Avon Books, 1978). Sitchin brings together Sumerian, Assyrian, Babylonian and biblical texts to argue that these ancient peoples had astronomical knowledge surpassing our own. "I reproduced 5,000 year old depictions of our solar system evidencing knowledge of the three outer planets," Sitchin writes, "whereas in our times they were discovered only in 1781, 1846 and 1930 respectively."

> Counting the Moon as a member of our solar system in its own right, their astronomical texts and maps showed a solar system made up of twelve members: the Sun, Moon and ten planets.
> Accordingly this twelfth member of our solar system has an orbit that brings it to Earth's vicinity every 3,600 years (p. 22).

Sitchin, following the ancients, names the twelfth planet Marduk. It is "many times the size of Earth" and "we have in our possession detailed astronomical data from thousands of years ago, describing the planet and its orbit." Apparently Marduk traverses the entire solar system in a vast cometary orbit that intersects all others. According to Sitchin, Marduk never gets closer to earth than the distance of the asteroid belt, between Mars and Jupiter, during its 3,600-year period.

Neither Christy nor Van Flandern feels that Marduk, if it exists, will behave as Sitchin predicts. Both have been in communication with Sitchin, and both have objections to his contention on the basis of dynamical considerations, i.e., Marduk as proposed violates some basic laws of planetary mechanics. In addition, a recent issue of *Kronos* (Vol. 4, No. 4, 1979) examined Sitchin's claims in detail and offered strong arguments against them. Even allowing that Sitchin's basic assertion is correct, C. Leroy Ellenberger's astronomical analysis shows that Marduk's last passage through the solar system would have been around 200 B.C. Thus the next passage is still some 1,400 years away, and Marduk can therefore be discounted as a trigger mechanism.

Another possibility for triggering a pole shift involves *impact from a celestial body*. The most likely candidates are comets and meteorites. The massive 1908 explosion at Tunguska, Russia, in north central Siberia, which flattened trees over an enormous area and seared them up to twenty miles away, has been attributed to a small comet that fragmented just above the ground. Its total mass, according to Dr. Gerrit L. Verschuur in *Cosmic Catastrophes*, may have been some forty thousand tons.* A larger comet might be sufficient to destabilize the earth if other factors brought it to a critical condition.

So might a meteor. Verschuur notes that the Barringer Meteorite Crater, in the Arizona desert, which fell to earth about twenty-two thousand years ago, dug a hole one hundred and fifty meters deep and more than a kilometer across. This is the equivalent of a three-megaton atomic bomb, or 3 million tons of TNT. "However," he

---

*An alternative explanation, offered by John Baxter and Thomas Atkins in *The Fire Came By* (Doubleday, 1976), is that a spaceship exploded at Tunguska. This view was supported in 1978 by an eminent Soviet scientist, astronomer Felix Zigel, who concluded that the explosion was probably caused by a UFO with a nuclear power source.

writes, "there are even larger craters found on earth, some as much as twenty-four kilometers across, which indicate explosions of up to two thousand megatons of force, larger than any explosion ever set off by humans. Collisions of this magnitude are still expected to occur about once in a million years . . ." (pp. 121–22).

Magnetic field reversals have been attributed to meteoric impact by Drs. Billy P. Glass and Bruce C. Heezen. When both were at the Lamont Geological Observatory, they coauthored a paper in *Scientific American* (July 1967) entitled "Tektites and Geomagnetic Reversals." Tektites are small, jet-black to translucent-green glass bodies two to four centimeters in diameter, and have an extraterrestrial origin. They are formed when a parent body enters the earth's atmosphere and explodes close to the surface or actually slams into the planet, breaking into hot droplets of material that spray out in a gigantic splash and cool into spherical or teardrop form. One tektite bed, or "strewnfield," for example, covers an area four thousand by six thousand miles in extent, from Tasmania to well north of the Philippines and from the East Indies to the eastern coast of Africa.

Glass and Heezen concluded from their studies of tektites that "the earth has been subject to massive cosmic collisions at fairly frequent intervals, and that it may be due for another major encounter" (p. 3). Furthermore, they said, the fact that they were "all laid down in sediment dated to the time of the last geomagnetic reversal immediately points to a common cause, possibly an encounter between the earth and some massive cosmic body" (p. 34). What kind of body? Glass and Heezen could not identify it at the time of the article but suggested asteroids, meteorites or comets. Present thinking is that tektites are meteoric in origin (although a recently advanced theory identified an explosion on the moon as the source).

In any case, the Glass-Heezen position is that the impact of the parent body that produced the tektites also caused the planet's magnetic field to be reversed. "If it can be assumed that the reversal of 700,000 years ago took place during a cosmic encounter, the encounter may have somehow disturbed the earth's magnetohydrodynamic dynamo" (p. 36). Since meteorites large enough to cause such a reversal may arrive every few hundred thousand years, by their estimate, a tektite fall is, statistically speaking, overdue. Glass and Heezen do not claim that *all* reversals are due to such impacts. However, in a weakened-field condition—due perhaps to some sort of internal instability of the earth—a major meteorite hit could conceivably destabilize the planet into *both* a magnetic pole reversal and a geographic pole shift.

Glass's continuing study of the tektite-magnetic reversal possibility confirms his 1967 position. In *Transactions of the American Geophysical Union* (December 1977), he reported still more evidence that "supports a previous suggestion that these tektite falls and geomagnetic reversals might be related."

His most recent statement, according to *Science News* (7 April 1979), indicates that three of the world's four major known strewn-fields are substantially larger than formerly believed. At least two of them "originated at times close to periods of reversal of the earth's magnetic field."

> The North American microtektite layer may not match a known geomagnetic reversal, the researchers say, but it does seem to fit the time associated with the extinction of several species of radiolaria—a phenomenon that has been circumstantially linked with geomagnetic reversals in other studies (p. 232).

The theme of cosmic catastrophe via meteorites was recently carried one step further by geologists Carl K. Seyfert and John G. Murtaugh of Buffalo State University College. In *Science News* (19 November 1977), the two are reported as proposing "nothing less than an extraterrestrial hypothesis for initiation of major events that have shaped the earth. Seyfert and Murtaugh suggest that the impact of very large meteorites (those creating craters at least 1 kilometer across) cause production of hotspots and mantle plumes, and that these in turn may cause continental breakups or changes in the directions and rates of sea-floor spreading" (p. 341). After a summary of the evidence leading Seyfert and Murtaugh to this conclusion, *Science News* commented, "Needless to say, the Seyfert proposal has a long way to go before it gains acceptance. But there is a vacuum in understanding what starts particular episodes of plate motions, and Seyfert's impact hypothesis could help fill the gap."

If comets, asteroids or meteorites are capable of causing a magnetic reversal, breaking up continents and changing the rate and direction of sea-floor spreading, it is not farfetched to see these events as byproducts of a polar shift initiated by impact from such a body.

As discussed in Chapter 8, the *magnetic null zones* proposed by Chan Thomas as the trigger mechanism for pole shifts are without any scientific foundation, so far as I have learned. However, this should not prevent recognition of the possibility that geophysical factors deep within the earth, rather than an astrophysical one in space, might reduce the planet's magnetic strength to the point where the magnetohydrodynamic energy is changed, thereby altering the properties of the mantle and possibly "freeing up" the crust for slippage. As noted already, the earth's magnetic field strength is approaching zero and may reverse as early as A.D. 2030. Even without impact from a meteorite, crustal slippage becomes at least conceivable in that situation.

Interestingly, a recent article in *Catastrophist Geology* (December 1977) by a British physicist, Harold Aspden, suggests an astrophysical phenomenon of galactic proportions that is intriguingly similar to Thomas's proposal and that may explain magnetic field reversals and crustal shifts. However, "Galactic Domains, G Fluctuations and Geomagnetic Reversals" concerns electricity and gravity, not magnetism.

Aspden begins by asking whether G, the constant of gravitation, is in fact constant. Physics assumes it is a universal constant, but a few voices have questioned this. So far, no experimental evidence has been found to indicate otherwise, yet it may be, Aspden states, that gravity decreases with time, as the physicist Dirac once suggested. If the earth is expanding while gravity is decreasing, then a change in G could escape detection.

Moreover, Aspden says, in view of the intimate links between space and time, the value of G could fluctuate within the planet, depending upon the earth's position within the space-time continuum. "Are there perhaps discrete domains in space within which G is universally the same but which have boundaries affecting the Earth's gravitation in its transit from one space domain to an adjacent domain?" (p. 43). Acknowledging that his argument is speculative, Aspden declares:

> Suppose that G can be linked in a unified field theory to electrodynamic interactions and that the parameter σ in the Ginzberg scheme is electrical in character. We may then expect the domain boundary transits to be marked by two phenomena, a transient fluctuation of G and a reversal of polarity of space-dependent electrical effects. The latter would be a reversal lasting until the Earth, for example, countered the next domain

boundary in its path, whereas the former would be of a few seconds' duration.

The new physics providing a domain structure in space . . . [means that Earth] would traverse a boundary in 80 seconds for a normal interception. An oblique crossing would mean that it would sit astride two domains for a rather longer period. A weakening of G between the Earth's matter for such a period would have a violent effect upon the crust. Earth rotation combined with the relaxation of the gravitational force would accentuate the surface nonconformities and cause an east-west shift of the crustal substance. There would possibly be some expansion followed by slow settling over the years before the next space domain was encountered. Yet, having regard to ecological effect, life-forms must, in the main, survive the violence of these few seconds.

This is speculation [but the] space domain hypothesis links two phenomena, the catastrophic upheavals accompanying domain boundary crossings and the simultaneous polarity reversals of electrical character (pp. 43–44).

Aspden then shows that there is solid evidence that "crustal upheaval and reversal of an electrical or magnetic nature occur together in the Earth on a recurrent basis in geological history." However, it seems improbable, given the data, that the earth will encounter another domain boundary for another half million years or so. Aspden concludes: "The idea of G changing suddenly is not easy to assimilate, but if the geological evidence helps, then the basic physics involved is worth rescrutiny. . . . our Earth provides us with . . . food for thought and, until we are satisfied that there is a viable explanation for geomagnetic field reversals and the possibly-related geological periodicities, we must be open to the thought that space itself has structure interacting with our Earth" (pp. 46–47).

Some theorists have proposed that ice ages are due to *changes in solar radiation*.* Hapgood's investigation led him to reject this position, but other scientists still propose it. Most recently, Dr. William C. Livingston of the Kitt Peak National Observatory in Tucson, Arizona, brought up the subject. As reported widely in the press early in 1978, he and co-workers recorded a drop of 11° F. in the sun's surface temperature during 1976. Livingston's

---

*Recall that *Season of Changes* identifies this as the direct cause of the next pole shift.

observations were confirmed by data from weather satellite Nimbus 6, according to the National Oceanic and Atmospheric Administration (NOAA). This decrease began when sunspot activity passed its minimum and began to increase. Livingston assumes that the changes seen are part of a solar cycle that could trigger variations in the earth's climate in the next few years. The 11° drop is a change of only 0.5 percent, Livingston remarked, but he also noted that some climatologists have said that a 2 percent decline over a period of as little as fifty years would be enough to glaciate the entire earth.

"It would be premature to look for climate change right now," he said. "By that, I mean I don't think you can blame the last two winters on what we're seeing on the sun right now. But I do think we can look ahead to some change, whatever it might be."

According to Thomas O'Toole, reporting in the *Washington Post* (13 February 1978), a growing number of scientists believe that changes in the sun trigger changes in the earth's climate. Scientists now believe the eleven-year sunspot cycle has a strong connection with drought on the earth. The drought in southern New Mexico that ended in August 1972 came to an end at the same time that large solar flares appeared on the surface of the sun. A second drought ended two years later, when flares again streaked from the sun's surface. Writes O'Toole, "The most striking evidence came in the 70 years that ended in 1715. In that time, there were almost no sunspots and no aurorae in the Earth's atmosphere. The same period became known on Earth as the Little Ice Age, when the coldest temperatures in the last 1,000 years were recorded in the northern hemisphere."

That seventy-year period is known as the Maunder minimum, so named for its discoverer, who found it through researching historical records. Long considered an anomaly, it was recently shown by A. Eddy, of NOAA, that there have been no fewer than twelve periods of minimum solar activity in the past five thousand years. Thus, according to Kendrick Frazier in *Science News* (22 April 1978), scientists have begun to view these as really quite ordinary. "It now seems quite possible," Eddy says, "that the common 11-year sunspot cycle is but a temporary feature of the most recent solar history or that it gets switched off and on in a program that seems almost random. . . . The new solar physics tells us that the 11-year cycle is but a ripple on an ocean of great sweeping tides" (p. 254).

As if this weren't enough, other scientists studying the sun have come up with a mystery that throws confusion right into the very foundations of stellar physics. As summarized by Dr. James S. Trefil in *Smithsonian* (March 1978), the article's title tells it all: "Missing Particles Cast Doubt on Our Solar Theories." The major point for us here is that astronomers' ideas about the sun may be wrong. The sun, rather than burning steadily through the ages, may be fluctuating irregularly on a scale of hundreds of thousands of years. It may be a variable star, Trefil suggests, whose core fusion reactions, in a typical but unpredictable manner, have recently switched off. This might explain Livingston's observation of surface cooling, and also the historical periods of minimums. Comments Trefil, "Perhaps the sun *is* switched off. Anything that turns off the fusion reaction will also cause a long-term cooling trend in the sun, and this trend will presumably be reflected in the weather on Earth" (p. 82).

Changes in solar radiation are also linked to earthquakes, according to British researcher Guy Playfair in his 1978 book *The Cycles of Heaven*. As we shall see shortly, there is a possible linkage between earthquakes and polar shift. Thus Playfair adds a link to the chain that may bring the ultimate disaster. As he summarizes it:

> Particles from solar flares disturb Earth's atmosphere at points where their relatively weak energy triggers the much stronger latent energy of cyclones and anticyclones. This leads to a redistribution of air around the Earth's atmosphere, creating a loading effect on its centre of mass. The energetic effect of such a process can be enough to cause polar displacements and fluctuations in the rotational velocity of Earth. Thus, increased solar activity increases the potential energy of the atmosphere, leading to a reduction in rotational velocity and a release of kinetic energy—in other words, an earthquake.

This brings our examination back to earth, and we now turn to a consideration of geophysical and climatological factors that may play a part in triggering a pole shift.

GEOPHYSICAL-CLIMATOLOGICAL.

An *increase in the polar ice caps* has long been proposed to explain polar shifts. As noted particularly in Chapter 4, and at other points in the book, a sudden surge in the ice caps is not at all improbable. U.S. satellite maps, according to *National Geographic* (November 1976) show that winter sea-ice buildup each year more than doubles the total expanse of ice in the Antarctic. While the thickness of the winter sea ice is far from glacial in dimension, the present cooling trends apparently seen in world climate could—as they have in the past—touch off huge polar ice marches across the planet's face in a relatively short time of perhaps only several centuries.

In *We Are the Earthquake Generation*, Goodman cites these and other data showing that "the ice age cometh."

• In 1971 the area covered by ice and snow in the Arctic suddenly increased by 12 percent, an increase equal to the combined area of Great Britain, Italy and France. This increase has persisted ever since, and some experts think the ice is on the move again. For example, areas of Baffin Island, in the Canadian Arctic, that were once totally free of snow in the summer are now covered year round.

• The great ice mass of the Antarctic, where average ice and snow cover is always high, is also expanding. In one year, 1966–67, the ice mass grew by 10 percent.

• The North Atlantic is "cooling down about as fast as an ocean can cool," according to an international group of climatic experts who gathered in Bonn, Germany, in May 1974. For example, sea ice has returned to Iceland's coast after more than forty years' absence.

• There are also widening caps of cool air at the poles. There has been a noticeable expansion of the great belts of high-altitude dry polar winds—the so-called circumpolar vortex, which sweeps from west to east around the top and the bottom of the world.

• Glaciers in Alaska and Europe that were retreating until 1940 have begun advancing again.

• The winters of the northern hemisphere are growing longer. Satellite photography shows that in just six years, winters in the northern hemisphere increased by almost three weeks. They averaged 84 days in 1967 and 104 days in 1973.

The approach of a new ice age is debatable on several climatological grounds. In 1977 a panel of the National Academy of

Sciences issued a report on energy and climate forecasting an estimated temperature rise of 5° F. by 2050 as a result of greenhouse effect from carbon dioxide released into the atmosphere from fossil fuels and from a decrease in the world's forests. *Science News* (30 July 1977) reports that the panel's chairman, Roger Revelle, says fossil fuels may be used safely for another twenty or thirty years—a position that was immediately disputed from both sides by ecology advocates and fossil fuel-producing groups.

Columnist Jack Anderson added to the controversy with his 28 August 1978 report on an unreleased U. S. Department of Energy study that concluded that if fossil fuels continue to be burned at present growth rates, the average temperature on the surface of the earth may rise two or three degrees Celsius by the early part of the twenty-first century. This could be the most drastic temperature change to occur on the planet in the past ten thousand years, the study said, because it could create severe environmental, economic and even political problems. Rainfall patterns could shift, thus changing the areas of agricultural productivity. This in turn would lead to "international migrations of the world's people" and a shift of the balance of power among today's have and have-not nations. Furthermore, Anderson notes, "Some reputable scientists fear that a global warm-up could, in the centuries to come, melt large portions of the polar ice caps."

Still another perspective on global climate change is offered in an unpublicized February 1978 report, *Climate Change to the Year 2000*, prepared for the Joint Chiefs of Staff by the National Defense University, the Department of Agriculture and NOAA and issued mostly to Congress and governmental agencies. This survey of opinion among twenty-four climatologists from seven countries constructed five possible climate scenarios for the year 2000, each having a probability of occurrence.

The scenarios showed a broad range of perceptions about possible temperature trends to the end of the century. Collectively, however, they suggest as most likely a climate resembling the average for the past thirty years. The experts tended to anticipate a slight global warming, rather than a cooling. According to the report abstract:

More specifically, their assessments pointed toward only one chance in five that changes in average global temperatures will fall outside the range of $-0.3°$ C to $+0.6°$ C, although any temperature change was generally perceived as being amplified in the higher latitudes of both hemispheres. The respondents also gave fairly strong credence to a 20- to 22-year cycle of drought in the High Plains of the United States but did not agree on its causes.

All these data on climate change serve to highlight the principles I noted at the beginning of this chapter. Every answer we get raises many new questions, and expert opinion is often divided in radical opposition. This also shows clearly the need for a holistic, integrated study of the elements involved, first, in climate change and, second, in the possibility of polar shift. Changes in the core of the sun may lead to changes in the earth's weather and also—insofar as the planet's magnetic field is linked with the sun's—to changes in the earth's core, which in turn could affect the crust.

The situation is further complicated by the X factor called man. Human activity may be causing massive interference with natural processes, aggravating the situation and speeding up certain trends. Imagine, for example, that the earth is now heading into the start of a new ice age. Ordinarily we would regard that as being millennia away. But industrial activity and deforestation might raise the temperature of earth's atmosphere dramatically, especially in the industrialized northern hemisphere. Might it be possible that the south polar ice cap could build while the northern one melts? This would lead to great *changes in surface mass loading*. Coastal areas would be submerged by meltwater from the north and by raised sea levels from the surging south polar ice cap. Isostatic rebound would accelerate in the northern hemisphere, and various other natural processes would be increased to an unknown degree. Such a redistribution of mass would certainly add to the instability of the planet.

A different kind of change in surface mass loading is foreseen by Bella Karish, who says that the crust at the poles will collapse after the ice caps have built up to an unsupportable weight. The Karish scenario is predicated upon the hollow earth theory—something no earth scientist accepts. Let's remember, however, that a person may be right for the wrong reason, and that the factors Karish identifies—which include changes in underground water levels that bring on more earthquakes and greater wobble of the earth's axis—might contribute to global destabilization. For example, a large portion of Arizona is sinking, according to a USGS report issued in late 1978. As the Associated Press described it,

"Declines in groundwater levels have resulted in a large area of Arizona sinking more than seven feet since 1952. . . . much of the sinking has occurred in a 120-square-mile area southeast of Phoenix . . . part of a 4,500-square-mile [area] where the land has sunk and narrow cracks have appeared in the surface. . . . Since 1915 more than 35.5 trillion gallons of water has been withdrawn from the area, much more than could be replaced naturally, officials said."

Walter H. Munk and Gordon MacDonald note in the 1975 revised edition of *The Rotation of the Earth*, "In many regions of the world the ground water table has been lowered by tens of meters as a result of human activity and drought. . . . if it were not for other, larger effects, the exhaustion of ground water would constitute the only case known to us where human activity might produce an observable motion of the pole" (pp. 238–39). And although they say that the problem of polar wandering is unsolved—i.e., neither proved clearly possible nor proved impossible—they feel that "a crust gliding over the mantle does not appear to be a reasonable model for polar wandering" (p. 285).

However, we have seen that opinion is divided on that point, and therefore we can ask: If these events synchronized with a *decrease of the earth's magnetic field* strength to zero, changing the magnetohydrodynamic energy of the mantle, what might happen? We have discussed magnetic field changes already, but here it is important to note an intriguing new study by Thomas G. Barnes, a physicist at the University of Texas. As summarized by Frederic B. Jueneman in his "Innovative Notebook" column in *Industrial Research/Development* (August 1978), Barnes found that the earth's magnetic field is diminishing at an exponential rate of decay. Back-calculating from this rate, Barnes's results showed that the earth had an implausibly large magnetic field just a few thousand years ago, "something on the order of 20 gauss at about 5,000 B.C. and approaching the magnitude of a sunspot just beyond 10,000 B.C., with all the makings of a pulsar by 50,000 B.C." Comments Jueneman:

> Obviously such a decay phenomenon for such a tiny planet as Earth could not have been continuing throughout its postulated long and varied geologic history, and since Maxwell's equations predict a decay this means that some event in the relatively recent past has injected sufficient energy into Earth's field for it to decay to its present level. So if the Earth wasn't

something of a magnetic star some 10 millennia ago, an extraordinary and catastrophic change must have taken place in the experience and memory of mankind.

Curiously enough, the mythological history of all races from around the globe is rich in the legends of a *Götterdämmerung*. But these same myths are silent about any reenergizing future event, which could start the process all over again.

Jueneman, a Velikovskyan catastrophist, has been raising the question of magnetic field effects in his column and articles. His May 1976 "Innovative Notebook" column concerns *Klimasturze*, or climatic plunges that have brought on great and minor ice ages. The last great ice age, Jueneman notes, occurred in both hemispheres, with the northern ice sheet extending down as far south as 38° N. in the Americas, ending near St. Louis, but only about 48° N. in Europe, somewhat above Moscow, and generally above 60° N. in Siberia. The ice cap was not centered on the geographical North Pole, being displaced about 15° southward, at a point below Thule, Greenland. "It might be of subtle significance to note that this center of the ice mass was at about the same latitude as the present north magnetic pole. And, moreover, the westward wandering of the magnetic pole may have moved the 950 kilometers from this position over the last 11,000 years or so" (p. 13).

What caused the ice ages? Jueneman speculates on the possibility of solar variability giving rise to anomalous magnetic activity that could create "raging geomagnetic storms around the earth, lasting tens of thousands of years or merely an occasional year-long short hiccup."

Our residual polar icecaps may be fossils of such cataclysms, where the menagerie of summer snows, brilliant auroras, worldwide myths, and reversed polarity were the result of magnetic-coupling between the Earth and the sun in a kind of super-Peltier field effect. Perhaps the woolly mammoths were quick-frozen where they grazed by what is called in physics the Ettingshausen effect, as the current of charged particles interacted with the magnetic fields; or by an even more dramatic mechanism known in cryogenics as the Giauque-Debye adiabatic demagnetization, where an anomalous cooling can be nearly instantaneous. But the Giauque-Debye effect requires a nearly complete, though temporary, loss of the Earth's magnetic field.

If these *Klimasturze* occurred before, they can happen again. But how soon, who's to say (p. 13)?

Who's to say? Actually, two people recently ventured statements bearing directly on the answer to Jueneman's question. The Canadian journal *Phenomena* (May–August 1978) reported it in a section entitled "Climate Discoveries Link Sun, Earth, Cosmos." Consider this summary of a new statistical study by Christopher Doake, of the British Antarctic Survey in Cambridge, who found data that further increased evidence for a correlation between changes in climate and the planet's magnetic field. According to *Phenomena*,

> climatic data from 4.3 to 1.5 million years ago show "significant correlations with reversals of the Earth's magnetic field." By examining five climatic indicators, Doake located their specific turning points.... "The significance of the climatic turning point, and not, for example, times of maximum rate of change of climate being apparently correlated with polarity transitions, is not clear," Doake admits, "but turning points ... provide an objective means of selecting datable events."
>
> Doake's hypothesis is that as the volume of global ice changes, it affects the Earth's rotation. This, in turn, might interfere with the interface between the Earth's crust and its liquid core, where the geomagnetic field may originate. Doake believes the interface could from time to time be sufficient to cause a reversal of field (pp. 11–12).

Immediately following this summary was another, which reported even more positive evidence linking polar shifts with climatic changes. An article by Edward M. Weyer, appearing in *Nature* (4 May 1978) warns that an ice age could be the trigger mechanism for some degree of pole slippage. Weyer's abstract preceding the article states, "If the centrifugal force generated by a surface load caused the Pole to shift, sea levels would fluctuate differentially around the world. Application of the 'hydrodynamic' formula to seemingly incongruous shoreline samples dated 14,700 to 28,000 BP suggests rhythmic polar oscillations on a 5,600-year cycle, synchronised with two glacial episodes."

Weyer's four-page article is quite technical. The *Phenomena* summary is therefore worth quoting in full:

> Rhythmic oscillations of the Earth's poles, synchronised with periods of glaciers, have been found in fossil shoreline organisms located above or below the modern water line. The discovery lends support to the idea that if a part of the land

mass of the Earth moves, it can generate a force which makes the poles shift, causing sea levels to fluctuate around the world. "The potential consequences of polar slippage is substantial," writes Edward Weyer in *Nature*. . . . "Theoretically, a slippage of only one degree would raise or lower sea levels relative to land as much as 373 metres."

In 1955, it was calculated [by Thomas Gold] that if South America rose 30 metres, it would cause the Earth to topple over at the rate of 1° per 1000 years. By examining samples from ancient shorelines, Weyer was able to detect a rhythmic rise and fall of sea levels up to 28,000 years ago. By applying various formulas to the data, it was shown that the polar curve starts downward each time glaciers begin to grow. "This is consistent with the theory that the centrifugal force generated by the ice masses would cause the Earth to slip its pole in this fashion," says Weyer (p. 12).

Pole-watching supporters of Hugh Auchincloss Brown and The Stelle Group will probably smile with delight at the Weyer quote, but it is not such dramatic confirmation of the predictions as it may seem. The crucial question is: How much might the pole slip? "The actual polar movement would not be less than the scale figure in tenths of a degree," Weyer writes. "It could be considerably greater, depending on the extent of the Earth's response to the stress" (p. 20). Since the scale of Weyer's figure 2 (bottom portion) did not exceed one degree, the slippage, though considerable, would not begin to approach the magnitude of change predicted by Brown and The Stelle Group.

Nevertheless, I sought through personal correspondence to learn if actual distances of pole slippage might be determined from the calculations. Weyer replied that his analysis didn't permit conversion of the findings into a scale that would measure distance. "A lot of intricate work is being done on the physical properties of the crust and on its behavior in relation to the underlying layers," he remarked. "However, I doubt that enough is known even to make a stab at the conversion you hoped I could give you."

Might *convection currents* trigger a pole shift? Edgar Cayce spoke of "upheavals in the interior of the earth," which the anonymous geologist who examined his predictions interpreted as possibly being convection currents. As we saw in the Introduction to Part II, convection currents are proposed by some scientists as the mechanism by which tectonic plates are moved.

They may also account for the puzzling phenomenon of variations in the earth's spin rate. According to Don L. Anderson, director of the Seismology Laboratory at California Institute of Technology, the rate of rotation has been noticeably slowing since 1965, by about nine milliseconds a day (about three seconds a year). In a *Los Angeles Times* article (29 August 1976) entitled "Giant Quakes Increase: Seismic Storm Lull Ends," Anderson speculates that earthquakes, volcanoes and planetary rotation rate may be connected via the global mechanism of convection currents. According to the *Times:*

> The Caltech seismologist believes that the process begins with small changes in the fluid outer core, a layer of superhot, superdense molten material that surrounds the solid inner core like an immersing bath of oil around a ball bearing.
>
> That fluid layer is in constant motion, Anderson explained, rising up against the overlaying mantle and then sinking back down again. Far from being uniform, that motion is turbulent and small changes either in the speed of the many small convection cells that make up the layer or in their orientation could easily develop.
>
> According to Anderson's model, such changes would affect the mantle and retard the motion of that intermediate layer of viscous matter. "There is an immense reservoir of energy stored in the earth's rotation," he said, "so any changes in the rotation rate could dump an enormous amount of energy into the crust and conceivably trigger earthquakes in regions where the crust has already been stressed to its breaking point."
>
> That would explain the increased frequency of earthquakes, Anderson hypothesizes, as well as variations in the length of day (the earth's spin rate would be lessened or increased, a variation which scientists have long puzzled over.)
>
> As for volcanoes, Anderson regards them as nature's own pressure gauges. "They sense pressure changes below the lithosphere," he said. "When the pressure increases, so does their venting. In places like Japan and Chile, we know that earthquakes and volcanoes are intimately connected." . . .
>
> (Asked to comment on Anderson's admittedly speculative model, another earth scientist—who asked not to be identified—said: "It's very imaginative. It's also very tenuous. When you link the fluid inner core to the mantle and then the mantle to shallow earthquakes in the lithosphere, it becomes a series of interdependent links. And each of those links has yet to be shown to exist, although many of us do believe there is a connection between slowdowns in the earth's rate of rotation and seismicity.") (p. 3)

While Anderson does not comment on pole shift, it is interesting to note his suggestion that convection currents might cause a differential change in spin rates of the earth's interior layers. Energy could be released only if a potential built up between layers. Thus, if there were a slowdown of the core's rotation, it might trigger a ripple effect outward to the crust, where the disruption could take the form of a slippage. What might cause that change in spin? To answer that would be—for me, at least—to speculate upon a speculation, and so I merely raise the question here. It is interesting to recall, however, that Aron Abrahamsen's prediction specifies a distinct difference between the amount of shift that the core and the crust undergo.

Earthquakes and volcanic eruptions have also been identified by some psychics as a natural trigger mechanism for a future pole shift. In the past decade, seismologists recognized a relationship between earthquakes and the Chandler wobble that may be important in this regard. For example, in *Earthquake Information Bulletin* (Sept.–Oct. 1971), the chief geodesist of NOAA's National Ocean Survey, Charles A. Whitten, reported that "the correlation between the total energy release and the daily movement of the pole during the last 20 years is so strong that I intend to keep 'tuned in' for the next year or two inasmuch as the daily movement of the pole is increasing, and during the past few months, the number of fairly large earthquakes has increased." Later that year Whitten confirmed that a definite correlation exists between polar motion and earthquakes.

More recent reports strengthen the case. In a *Science* (4 October 1974) report, "Earthquakes and the Rotation of the Earth," Don L. Anderson noted "an interesting correlation between the length of the day, Chandler wobble amplitudes and the incidence of great earthquakes . . . in particular, the large deviation in the length of the day around the turn of the century with the worldwide increase in global seismic activity at the same time" (p. 49).

The strongest confirmation so far comes from geologists Richard J. O'Connell and Adam M. Dziewonski in their *Nature* (22 July 1976) report, "Excitation of the Chandler Wobble by Large Earthquakes," which concludes that earthquakes represent the major factor in the wobble excitation. Their research showed that "the cumulative effects of major earthquakes can account for at least a large part of the excitation of the Chandler wobble, and in particular for the long-term variations of the amplitude of the wobble. This, together with recent estimates of the atmospheric excitation of the wobble, indicates that these two mechanisms may well account for the entire excitation" (p. 262).

This discussion of an earthquake/axis wobble relationship should not be taken to mean that these scientists predict a pole shift or even support the concept. Rather, I have tried to indicate that there is scientific support for psychic predictions such as those reported by Goodman in *We Are the Earthquake Generation*. The psychics he investigated—Edgar Cayce, Aron Abrahamsen, Bella Karish, Ray Elkins, Beverly Jaegers and Clarisa Bernhardt—uniformly foresee two decades of increasingly severe seismic activity that will repeatedly jolt the globe with earthquakes greater than any ever recorded.

The Richter scale measures energy released. It ends at 10, and each whole number represents a tremor ten times more severe than the preceding whole number. Goodman's psychics predict quakes of 11 and 12—right off the scale! Would this be enough to initiate a pole shift? The psychics say so, and Goodman adds that the magnitude of such seismic activity repeatedly shocking the axis might indeed do the job. A 12.0 quake, he notes, would be almost half a million times stronger* than the 1964 8.5 Anchorage, Alaska, quake, which lifted areas of the ocean floor more than fifty feet.

All this remains to be seen, of course. But it is a fact that we are entering a period of increased global seismic activity. The year 1976 demonstrated this with a horrifying death toll from earthquakes: twenty-three thousand fatalities in Guatemala, nine thousand in New Guinea, eight hundred in Italy, three thousand in the Philippines, six thousand in Bali, five thousand in Turkey, many thousands in the Soviet Union (official figures are lacking) and an incredible 750,000 people killed in China in a single quake at Tangshan. This upsurge, oddly enough, appears to be normal, rather than abnormal. In the *Los Angeles Times* article just noted, Anderson remarked, ". . . I would say that rather than entering a time of above-average seismicity, we're getting back up to the long-term average of earthquakes, both in magnitude and frequency." He added, "We're nowhere near the peak of seismic activity that we saw at the beginning of the century." In other

*This is apparently a mathematical error. The increase in strength would be only 5,000 times greater.

words, the worst is yet to come. Will it arrive just as other factors reach a critical state?

Trigger Mechanisms—Human

Although the psychics foresee what Goodman calls "a Decade of Cataclysm" from 1990 to 2000, in which the face of the earth changes and which ends with the planet tumbling in a pole shift, we have also heard from them that their previsions of the worst are modifiable by human consciousness. Cayce, Abrahamsen and Solomon are quite explicit about this. In particular, the psychokinetic effort of human thought energy can have a pacifying or aggravating influence upon geophysical processes. At this point, therefore, we must consider the second group of possible trigger mechanisms: the human ones.

SOCIOECONOMIC.

The effects of *atmospheric pollution* from industry and worldwide deforestation are well understood. Chemicals in the air produce "acid rain," and industrial particulate matter released through smokestacks blocks sunlight, lowering temperatures in some areas. Shrinking forest cover, due as much to fuel needs and primitive farming methods in Third World nations as to urbanization in the West, allows carbon dioxide to build up. This in turn causes the *greenhouse effect*, which raises atmospheric temperature by trapping radiant heat below the CO_2 layer. Stripping forest cover from the earth also allows topsoil to be eroded by wind. This increases rainwater runoff to flood conditions in some areas and puts more particulate matter in the atmosphere. The balance of nature is upset in myriad ways, setting the stage for an accelerated period of glaciation.

Mining, drilling and damming have also been named as factors contributing to increased crustal instability. The amount of influence they might have is probably very small but should not be discounted, because there are clear indications that these activities are associated with earthquakes. Goodman cites the case in which the U. S. Army, in the mid-1960s, disposed of lethal waste by dumping it down deep wells outside Denver. As a result of all the water used to flush the waste below ground, inactive faults began to activate. Hundreds of minor quakes were recorded, some large enough to cause property damage. Local protest caused the Army to halt its dumping, and the earthquakes ceased as suddenly as they had begun. In similar fashion, oil drilling in the Santa Barbara (California) Channel is thought by some seismologists to be the cause, or at least a contributing factor, of quakes recently experienced there.

Pumping liquid into or out of the ground contributes to crustal instability in other ways. The groundwater situation in Arizona was mentioned earlier. Goodman notes the subsidence, or slumping, of the ground in oil fields around Los Angeles—as much as twenty feet in some places. This being so, we should ask what the effect would be of creating a truly great void below the surface— in Alaska, for example, as the great pipeline to Prudhoe Bay removes five billion barrels of oil.

Damming also contributes to a higher incidence of earthquakes, Goodman points out. He cites two USGS geologists who reported evidence that large dams back up enough water to trigger quakes, presumably through a change in surface mass loading, as well as whatever "pumping" there might be into faults in the area. "There is no telling what the ultimate effects of man's increasing activity in modifying the natural environment will be," Goodman writes. "Man himself may be cocking the hammer for the earth's next cataclysmic blast" (p. 125).

This image of weaponry is appropriate for introducing consideration of man's most recent and most fearsome means of modifying the environment: *nuclear war*. The possibility of atomic conflagration is so near and so familiar that it has become almost an accepted part of our lives—an irritant that one learns to live with, like chronic pain or a missing limb. Yet our lulled sense of danger does not diminish the reality of what a nuclear war would entail. Albert Einstein said that if World War III were fought with atomic bombs, World War IV would be fought with stones. My own experience as a nuclear weapons officer in the Navy—during which time I was personally responsible for the operational readiness of several dozen Mark 101 "Lulu" atomic depth bombs aboard my ship, U.S.S. *Pine Island,* in Danang Harbor, Vietnam—led me to reject the use of nuclear weapons as acceptable in war in any form. Now the spread of nuclear technology in supposedly peaceful forms has led to the possibility that terrorist groups and small nations might construct their own atomic bombs. A 1977 CIA study, reported widely by wire services, concluded that nuclear war was inevitable by 1984, because thirty-five nations would have developed nuclear arms.* In addition, terrorist groups would have them by that time, either through theft, black market purchase or building their own. The probability of their use in such circumstances is absolutely certain. Nonnegotiable demands would be made under threat of nuclear destruction of a city. Whether initiated by terrorists, small nations or superpowers, the results would be the same.

What would result? That depends on the extent to which they are used, of course. But even if the bulk of humanity survived the immediate effects of a major nuclear strike and counterstrike, the long-term effects are almost impossible to calculate. Radiation effects have been considered by the experts, but has anyone given thought to the effect upon the earth's rotational axis or crustal stability?

The answer is: yes, but barely. The first reference I've come across is, surprisingly, a newspaper article in the now-defunct *Washington Daily News* (19 October 1956) in which the 1956 Democratic candidate for Vice President, Estes Kefauver, called for a ban on H-bomb tests. Kefauver asserted that the earth could be "knocked 16 degrees off its axis." Where he got his information isn't stated, but reporter Milton Britten commented, "The Atomic Energy Commission has greeted his assertion with a deafening hush."

An astrophysicist, Dr. Homer Newell, Jr., head of the Naval Research Laboratory's astrophysics division, told Britten that he didn't know whether an H-bomb could induce a pole shift, but, assuming that it could, there were three possible effects. First, the axis might be knocked forward along its path of precession. This would cause a "sliding" of seasons.

Spring might be skipped, or winter, or maybe a whole cycle of seasons, but the length of the seasons wouldn't change. And

*This would inevitably require an increase in *nuclear testing,* most probably underground. This, too, could contribute a measure of destabilization to the planet. If the underground testing were done primarily in the north polar region, conceivably it would have the greatest destabilizing effect. An Associated Press release in late September 1978 raised the question in my mind. It reported, "U.S. officials said...that seismic signals, presumably from a Soviet underground nuclear explosion, were recorded by this country's atomic energy detection system.... The signals originated...in the Novayazemlia test site in the Arctic, according to...a spokesman for the Energy Department."

the temperature zones wouldn't change either.

Even this could play hob. Leaping from one winter into the next, say with no intervening seasons, could cause food shortages and otherwise disrupt world economy.

Also, navigators would have to learn to read a new map of the heavens. The North Star, for instance, which is now Polaris, would be replaced by another.

Dr. Newell compared this first possibility to suddenly pushing the axis of a spinning top a little farther around the periphery of its normal wobble. The second possibility would be more serious, like pushing the spinning top 16 degrees toward the floor.

This would change the angle between the earth's axis and the plane of its orbit around the sun. Which is a little complicated, but to you and me, it means temperature regions—polar, temperate, tropical—would be shifted around.

As a result we might be shivering to death in the "sunny South" and sunbathing in Labrador.

Arctic cold might grip farmlands. Tropical zones might become temperate, or vice versa. This would bring population shifts, cause disruption of farm or industrial enterprise, and otherwise disjoint the world picture.

The third possibility cited by Newell is the least drastic: a tilting that "would tend to make the climates in all the zones more uniform. . . . In this case, it would be like straightening up, rather than pushing it toward the floor" (p. 18).

Except for Kefauver's lone voice in the wilderness almost two decades ago, only two scientists, so far as I know, have considered the possibility of polar shifts resulting from atomic explosions. Shigeyoshi Matsumae, president of Tokai University, and Yoshio Kato, head of the Department of Aerospace Science there, report in a privately published paper, "Recent Abnormal Phenomena on Earth and Atomic Power Tests" (30 September 1976) that abnormal meteorological phenomena, earthquakes and fluctuation of the earth's axis are related in a direct cause-and-effect manner to atmospheric and underground testing of nuclear devices. Nuclear testing, they say, is "the cause of abnormal polar motion of the earth" (p. 3).

Using data from standard international sources of information, Kato plotted north polar motion during 1975. The result was "very unusual shock-wave-like sudden changes of considerably acute angle" (p. 5). By applying the dates of nuclear tests with a force of more than 150 kilotons, Kato reports, "I found it obvious that the position of the pole slid radically at the time of nuclear explosion." Some of the sudden changes measured up to one meter in distance.

What would be the effect of megaton jolt after megaton jolt to the earth's axis? We have already seen that large earthquakes are thought to have a dampening or accelerating effect on Chandler wobble. A series of great explosions in an all-out war might very well do the same—to a degree that seems impossible to assess but that would certainly be significant, assuming, of course, that there is anything more than an asteroid belt left of our planet.

BIORELATIVE.

As described in the Introduction to Part III, biorelativity is a term coined by Goodman to denote the psychokinetic interaction of people with their environment via psychic or mind energy—the energy of thought. From the psychic point of view, the energy upon which thoughts are impressed gives rise to *thought forms*. Thought forms are produced constantly, whether or not we are aware of it, the psychics say, and they constantly impress themselves upon the energy matrix sustaining the physical environment, including the planet itself. "The psychics [say]," Goodman reports, "that the thought forms given off and created by man interact with the factors behind earthquakes, volcanoes, and geological activities, as well as the factors behind climatic change" (pp. 197–98). The effect of humans is there all the time, inescapably. The only question, therefore, is whether we are to have our thoughts affect the total process of the world's energy activity in a positive or a negative way.

The traditionally disapproved character traits of anger, greed, hatred, fear, self-aggrandizement, aggression, lust for power and so forth are powerfully negative influences on the energy processes of the earth. On the other hand, virtuous thought and behavior act to maintain harmony and balance. Most important of all is to maintain a loving sense of relatedness to the planet and its life forms as a single living organism—a senior member in the community of life that extends upward in a great chain of being to the Creator. This is what Native Americans call "walking in balance on the Earth Mother." Violation of this biological-moral principle, the psychic sources say, will surely bring on our destruction. It has happened before, with Atlantis, Lemuria and other high civilizations before ours, they claim, and it can happen again. If there is atomic conflict and the human race survives it, those detonations could start chain reactions in the subsurface geology that build up just as other naturally occurring factors, including thought form influence, reach a critical state. In that case, we will have directly brought on a pole shift and will have no one to blame but ourselves.

But it need not happen. From the psychic point of view, the choice is ours. The quality of our living can change at any time, and with that change will go all the positive effects upon the energy matrix of the earth. Consciousness is the key to intelligently controlling and directing psychic energy and thought forms.

This brings us to another aspect of biorelativity—specifically, one of the new topics that I said earlier I would introduce here. *Etherian physics,* or "the physics of the ethers," denotes an occult science both theoretical and applied. I am using the term occult to mean secret, hidden, beyond the bounds of ordinary knowledge for both the general public and the scientific community. It might be better understood as metaphysics—that is, dealing with processes of nature operating at a level beyond that which official science can presently observe. The aim of etherian physics is to control these processes, which involve certain forces—the ethers—said to be more fundamental than the four "basic" forces recognized by official science: electromagnetism, gravity, and the weak and the strong nuclear forces.

Etherian physics owes its modern formulation to Dr. Rudolf Steiner, founder of the spiritual science called Anthroposophy. However, the principles of etherian physics appear to have been grasped to some degree intuitively by many ancient prescientific traditions and also by some modern scientists researching the physics of paranormal phenomena, such as Dr. Wilhelm Reich. Dozens of terms exist for an all-pervasive life force, or vitalizing principle, in nature. They come from cultures around the world, ranging from *ch'i* (Chinese) and *prana* (yogic) to the Holy Spirit (Christian). I have listed more than one hundred of these terms in an appendix to *Future Science.* According to Trevor James Constable, whose book *The Cosmic Pulse of Life** offers the best introduction to etherian physics, "Nobody can speak with precision or accuracy about polar shifts without a knowledge of etherian physics."

---

*Available through Multimedia Publishing Corp., 72 Fifth Ave., New York, NY 10011. A condensed version, *Sky Creatures: Living UFOs,* is available in mass market paperback through Pocket Books.

Because the subject is vast and complex, I can do no more than give the barest summary here as an indicator to further research. I am indebted to Constable for this brief explanation of the Steinerian doctrines, and he refers interested readers to the following basic texts, from which he has drawn: *The Etheric Formative Forces in the Cosmos, Earth and Man,* by Guenther Wachsmuth (Anthroposophic Press, 1932); *Man or Matter,* by Ernst Lehrs (Harper & Row, 1957); and many works by Rudolf Steiner, primarily *Occult Science* (Anthroposophical Press, 1972). I would add the recent book *The Loom of Creation,* by Dennis Milner and Edward Smart (Harper & Row, 1976).

According to Steiner, the physical universe and its space-time processes are unfolded or manifested by the action of etheric formative forces. They are the warmth ether, the light ether, the chemical ether and the life ether, each of which successively evolves out of the previous one (except for the warmth ether, which is fundamental). These ethers are not the stationary ether of classical physics. Rather, they are active—the motive power of nature, if you will—and their properties give rise to the observable universe, including life in its various forms.

With regard to pole shifts, from the Steinerian perspective they occur when the life ether, coming principally from the sun, becomes overconcentrated at the North Pole. Life ether radiates from the sun to the earth, concentrating most strongly at the poles and exerting a steady attraction upon the North Pole. Over a period of millennia, the North Pole is drawn toward the sun. Pole shift occurs when alignment nears and the attraction of the sun becomes irresistible. The North Pole comes around sharply to align with the sun. Ice caps are then shifted to tropical zones, cultures are drowned in great floods, mammoths are flash frozen, etc., and a new North Pole, 90° or so away, is established.

Experimental evidence to support these assertions is relatively sparse, and, in fact, etherian or Steinerian physics has been most often criticized as being without empirical data or demonstrable results. I make no defense of it here, but simply offer information that may bear on the critical question before us. However, it should be of great interest to the scientific community to know that one of the technological applications of etherian physics is *orgone weather engineering.* Trevor Constable claims to have developed this to an advanced state, using nonchemical, nonpolluting means based upon the work of Steiner and Reich.

According to Constable, Reich's discovery of orgone energy was actually a rediscovery—quite independent of Steiner and Anthroposophy—of the chemical ether. Reich's orgone technology is predicated upon the existence of a nonelectric, primary, pulsatory, massless energy. This energy can be directed to modify and control weather on a global scale. The means for accomplishing this was invented by Reich and called a "cloudbuster" because of the effect he first noticed: its ability to dissipate clouds. Constable has demonstrated this tubular gunlike device repeatedly and reported it in professional literature (see Appendix II of *Future Science*). His most recent weather engineering feat, he says in *Energy Unlimited* (No. 3),* was to reverse the wind patterns during the 1977 Santa Barbara forest fire, drawing offshore air over the city and thereby slowing the advance of the fire sufficiently for firefighters to extinguish it.

The claims for orgone weather engineering are doubly important. First, the practice of orgone weather engineering apparently offers a means to modify world climate in a manner that works with the laws of nature and is therefore ecologically acceptable. This could allow us to mitigate climate changes as a factor influencing a pole shift. Second, the theory of orgone weather engineering purports to offer direct insight into major mysteries of geophysics, including why the planet rotates.

Reich claimed to have discovered that orgone energy forms an envelope around the earth, moving from west to east slightly faster than the planet revolves. Streaming in from space to "pool" around the earth (and other planets), orgone imparts spin to the planet. It, not inertia, is the bioenergetic torque driving the earth, and it would continue to operate after a pole shift. Thus orgone theory and etherian physics provide a reply to the question raised by pole shift opponents who ask, "How will the planet start spinning again? What could provide the tremendous energy needed to 'rev it up'?" Admittedly, from the perspective of conventional science, etherian physics and orgone weather engineering appears to be mere fantasy and magical thinking. But the results of Constable's weather-engineering projects are there for all to examine.†

---

*Available from the publisher at Box 288, Los Lunas, NM 87031.

†Constable can be reached at TJC-Atmos, Inc., 1605 East Charleston Blvd., Las Vegas, NV 89104. He conducts his weather engineering projects on a "no results, no pay" basis.

If orgone weather engineering appears so beneficial, why have I listed it as a possible contributor to a pole shift? Again, the answer is the potential of human consciousness. Orgone weather engineering experiments conducted by unthinking or irresponsible people could bring disastrous results, and Constable says that he has already seen this from crude devices operated by uninformed, inexperienced dilettantes. In a recent letter, he raised the question of our own government's disruptive chemical-based weather modification experiments in Vietnam and the likelihood that a military or intelligence agency would soon begin to try the same thing with cloudbusters. "Suppose the USAF," he writes, "decides that it wants to build cloudbusters so that the Soviet Union may be drowned. They build a unit in greatest secrecy that has a thousand tubes 9 feet in diameter. They point it at the sun, during evaluation tests. Right away, their local area for miles around is drained of life ether, which goes tearing out through the tubes to its high potential home [the sun]. The huge area of lowered potential on the earth's surface now begins in turn to drain into the poles. Pole potentials, especially the north pole, rise rapidly. The process becomes self-sustaining and you have created the *physical* conditions for imbalance."

The specter of governmental and military applications of etherian physics is already a reality, according to persistent rumors circulating among researchers of the paranormal. The stories indicate that Soviet researchers have advanced far beyond their Western counterparts in developing technology that functions on psychic or etheric energy. Their creations are formally named *psychotronic weapons*, although these devices are unlike any weapons seen before because their operation is such a radical breakthrough in parascience.

One class of psychotronic, or PT, weapons is based on the pioneering work of Nikola Tesla, the Yugoslavian-born genius who revolutionized the field of electrical technology with his inventions. Early this century, Tesla demonstrated wireless transmission of electricity over a twenty-six-mile distance by sending it *through the ground*. Apparently, Soviet scientists have progressed from Tesla's discoveries to the point where electromagnetic signals can be broadcast through the earth to form standing waves in the earth itself. By triangulating signals from transmitting stations (Riga, Gomel, Semipalatinsk and Novosibirsk), coherent patterns can be set up that, through an effect known as "kindling," are amplified by drawing energy from the core of the planet. The amplified energy in the standing wave can in turn be directed and focused to induce a variety of effects, including earthquakes that appear to naïve observers as natural (rather than man-made) phenomena. It is suspected by one researcher that the Iranian earthquake of 1978 may have been created by this means.

The researcher, retired U. S. Army Lieutenant Colonel Thomas A. Bearden, a nuclear engineer and former intelligence officer, writes in the journal he edits, *Specula* (July-September 1978),* "It is the author's thesis that the foregoing . . . [accounts] for some of the drastic weather effects that have occurred in the past two years, as well as *some* of the major earthquakes that have occurred throughout the earth in the last several years. The Soviets, I believe, have been orienting and aligning—and operationally testing—actual weapon systems to be used as precursors to war or during war. The recent Iranian earthquake, e.g., may well have been Soviet-induced." American and Canadian intelligence groups are already studying the possibility that Soviet use of Tesla's work has disrupted weather patterns over North America, causing or contributing to the unusual conditions of the 1977–78 winter.

The second class of psychotronic weapons, even more sinister, that the Soviet Union is suspected as having operationally tested involves use of hyperspace transmissions. The content of the transmissions, according to Bearden in his report, "Soviet Psychotronic Weapons," in *Energy Unlimited* (No. 3), can vary from disease effects and extinguishment of missile, aircraft or submarine electrical systems to kiloton-range nuclear explosions! The details of this new military threat are given in Bearden's recently published book *The Excalibur Briefing* (San Francisco: Strawberry Hill Press, 1980). Another useful summary of Soviet psychic/psychotronic research is given by Michael Rossman in the excellent, final essay of his *New Age Blues* (New York: E. P. Dutton, 1979).

Bearden's thesis should be carefully considered because, according to *New Scientist* (12 October 1978), geophysicists in Germany and England believe the earthquake at Tabas, Iran, in which at least twenty-five thousand people were killed, may have been triggered by an underground nuclear explosion. Several seismological characteristics of the quake make it unusual and, in light

*Available from the American Association of Meta-Science, P. O. Box 1182, Huntsville, AL 35801.

of Bearden's comments, suspicious. British seismologists believe "the Tabas earthquake implies a nuclear test that had gone awry" (p. 91). Moreover, a seismic laboratory in Uppsala, Sweden, recorded a Soviet nuclear test of unusual size—ten megatons—at Semipalatinsk only thirty-six hours before. One German scientist, the article states, specifically implicated this test in the origin of the Tabas disaster. Was the quake induced by Soviet psychotronic technology? Was the "awry" test actually a new sort of nuclear event in which explosive energy was transmitted artificially to induce an earthquake? At the very least, the Tabas event offers striking evidence that nuclear explosions are associated with earthquakes through natural means of underground transmission.

Etherian physics is essentially the ability to go "through the vacuum" of space into hyperspace and tap the "zero-point energy" of the prematerial void from which subatomic particles form to yield the physical universe. The energy of hyperspace is enormous—witness the ability of black holes to trap light itself and draw it into the void. As described in my *Future Science,* the technology of hyperspace travel would quite literally open up whole new universes. These universes, or sets of dimensions, interpenetrate our familiar space-time continuum in a normally unperceived fashion without noticeable effect. If the Soviet military has achieved the ability to enter and return signals from hyperspace, modulating onto them biological, electrical and nuclear effects, then conquest of non-Communist lands could be achieved without firing a shot. All this is highly speculative, of course. But, again, I prefer to report the information and suggest that an investigation be made because of its possible importance to our theme, rather than ignore what might be useful data.

Intervention by higher life forms has been suggested by some people in various psychic and New Age groups as the means whereby a pole shift will occur. I mention it here in the interest of completeness but, speaking personally, I discount the suggestion. Without debating the question of malevolent supernatural entities, I'll merely comment: first, human beings have repeatedly shown themselves quite capable of performing enormous evil without any outside help and, second, if such entities exist, then surely there must also be benevolent entities who would be equally eager to influence planetary affairs. Thus, the forces of light and darkness may be waging their eternal war for control of the earth, but human consciousness has the potential to rise or fall unaided.

* * *

Our overview of the forces and factors that might trigger a pole shift has necessarily been brief. Nevertheless, it should be apparent that the cumulative effect of some or all of these could result in a massive catastrophe unforeseen by establishment science, governmental agencies and military strategists alike. A number of natural influences with varying rates of intensity seem to be converging in their effect and will reach maximum influence about the end of the century.

Imagine that the planet's "burners" are being "turned up" internally, causing greater activity. Convection currents move with greater force or speed, causing earthquakes and volcanic eruptions, speeding up tectonic plate movement and perhaps increasing pressure on the entire crust. At the same time, magnetic field strength is dropping to zero, radically altering the magnetohydrodynamic energy of the mantle, allowing the crust to slip much more easily. Polar ice caps are building, or perhaps one expands while one decreases. This shifts surface mass loading and at the same time builds an off-center "throw" of centrifugal force that wobbles the axis more than usual. The increasing earthquakes also jolt the axis and accentuate the wobble. Meanwhile, peaceful human activity such as drilling adds to crustal instability. All is in readiness for the critical trigger. What might it be? A planetary alignment, perhaps, or a gigantic meteorite that slams into the earth? A comet that streams past, electromagnetically tilting the axis? Or a cloud of nuclear missiles that jolt the crust loose?

Might the year 2000 have more than mere numerological significance? Might the cumulative effect of the various natural trigger mechanisms be sufficient to topple the earth or shift the crust? And if not, might they be sufficient to create a critical condition in which man himself might unwittingly trigger a pole shift? *I don't know and, to be completely honest about it, I don't think anyone else knows either—not the ancient prophets, the modern psychics or the scientifically oriented researchers.*

We have seen that despite their common theme, great differences in perspective on the question exist. Consequently, there is disagreement on major aspects of the pole shift—how it will happen, exactly when, how rapidly, and so forth. Moreover, the fact that something has happened regularly before does not guarantee that it will happen again, just as you may wake up every

morning for sixty years until the night you die. There may have
been pole shifts in the past, but logically speaking, another one
is not inevitable.

The situation seems to be like the ancient fable in which some
blind men try to describe an elephant by feeling parts of its body.
One feels a leg and says, "An elephant is like a log." Another feels
the trunk and says, "No, an elephant is like a thick rope." Still
another, grasping an ear, contradicts them by saying, "You're both
wrong. An elephant is like a palm leaf." And so on and so on—
each having a piece of the truth but not, because of their blindness,
seeing the situation completely.

In the case of pole shifts, we have an even more difficult
question. The blind men, at least, know that the elephant is real.
We, however, are not even sure the elephant exists—that is, that
pole shifts have indeed occurred. Obviously there is a case to be
made for them; that is what this book is all about. But *presenting*
a case is not the same as *proving* a case. For that, the court of
scientific judgment is needed to lend its full wisdom and expertise
in weighing the critical issues.

Let us remember that there is also a case *against* pole shift.
Although I have not examined it extensively here, recall, for example, the data about the age of the Antarctic ice sheet. Recall
also the comment by Dr. William Ryan about the 60-million-year
continuity of tropical sea life in the cores brought up from the sea
floor. Are these indisputable facts or might they be subject to
reinterpretation? Is there an alternative explanation to the *Glomar
Challenger* data? Might the cores have been misread due to a
nonconscious bias?

The possibility must be considered. Nonconscious bias and
even premeditated fraud have been seen before operating in the
court of scientific judgment—witness the Velikovsky Affair.
Since I am presenting a case for catastrophism, the scientific establishment may be quick to dismiss it from consideration simply
out of prejudice, prejudgment. But again I say: premature closure
of the pole shift question could be fatal to the entire biosphere.

It seems appropriate, in support of that position, to quote from
an article by Dr. Roger W. Wescott of Drew University. In a
Kronos (Fall 1978) article entitled "Polymathics and Catastrophism," Wescott wisely and succinctly makes two points I wish to
emphasize here. They ought to give considerable pause to those
who would dismiss the question of pole shift without a full hearing.

Despite the patently sensational—and hence implicitly extremist—aura of the term "catastrophism," it is undeniable that catastrophists take a broader view of Earth history than do uniformitarians. For uniformitarians assert that our planet's evolution has been uniformly gradual and that no genuinely global catastrophes have occurred. Catastrophists, however, do not merely reverse this assertion. On the contrary, they agree that most of the terrestrial past has been a record of undisturbed gradualism. But they maintain that this record has been punctuated periodically by violent and worldwide episodes of what geologists call diastrophism—that is, major rearrangements of the Earth's surface (p. 5).

One of the charges most frequently leveled against catastrophists is that they ignore or defy "the laws of nature." In making such charges, uniformitarians overlook the shrewd polemic analysis proffered by Lawrence Dennis, who defined wars as disputes, not under laws, but over laws. In political conflicts, the victors make the laws; in scientific conflicts, they "discover" the laws. In both cases, however, the laws are labile.

In short, many of the so-called "laws of nature" are simply those formulations which happen to be fashionable among scholars of a given period, region, or school of thought. Those theorists who complacently identify their own views with natural law might then quite justifiably be said to be "playing nature" in the same sense in which people who arrogate excessive political or religious authority to themselves are often described as "playing God" (p. 4).

Although we must consider the possibilities of nonconscious bias and deliberate fraud, it seems most likely to me that the truth will be found not somewhere between the cases for and against polar shifting but *above* them. That is, I foresee a higher-level synthesis developing out of the present thesis-antithesis situation. I will conclude this summation, therefore, with a brief outline of what, in my judgment, will emerge.

On one hand, we have the data arguing against polar shifts—or at least the *interpretation* of those data. First, the Antarctic ice cap may be more than 20 million years old. Second, the equatorial region has been approximately in its present location for more than 60 million years. The third major element in the case against polar shifting—paleoclimatic research—indicates, as Munk and MacDonald put it, that there is "little positive evidence" for the concept. According to a *New Scientist* article on ice ages and continental drift (7 December 1978), "patterns of atmospheric circulation, and therefore the locations of the general latitudinal climatic zones, have remained more or less the same throughout most of the past 500 million years" (p. 777). Last of all, the celebrated "flash-frozen" mammoths did not die in a single event or even several clearly dated events.

On the other hand, we have seen, thanks to the scholarship of Velikovsky, Brown, Blavatsky and others, that ancient records speak unequivocally of the earth flipping over. We have also seen, in the work of Michelson, Warlow and others, that it is theoretically possible for such a 180° pole shift to occur. Moreover, Goodman, Warlow and a number of other scientists acknowledge that such events can sensibly explain in unified fashion a host of problems and anomalous data in the earth sciences.

A simple resolution would therefore incorporate *op*posite views into a *com*posite explanation. That is just what I propose here. Is it naïve to say that if the planet were to flip a full 180° in space, the equatorial region would retain its location? Is it too simple to observe that climatic zones would, except for loss of a single season, remain as established? Is it unacceptable to note that polar ice caps, except for changing places, would still be positioned in frigid regions?

I do not ask these questions rhetorically. The answers are, of course, obvious—within the limits of the problem as stated. But does the problem as stated adequately represent the situation regarding polar shifts? Might there also be 80–90° pole shifts, as proposed by Brown, Thomas, Barber, Sepic and some of the psychics? And what about smaller pole shifts of longer duration, as Hapgood indicates?

Here is a subject worthy, truly worthy, of the most concerted and full-scale investigation by the entire scientific and scholarly community. It represents a potential revolution in the earth sciences surpassing that which occurred through the concept of plate tectonics. It represents equally well a revolution in the biological sciences, in archaeology and in intellectual history. Last of all, it represents what may be the key to humanity's survival in the face of imminent cataclysm. For if there is the slightest bit of truth to it, we may be standing at the edge of the ultimate disaster. If we avoid or ignore the possibility of a pole shift, we will have only ourselves to blame should there be a cataclysmic repeat of the "myth" of Atlantis.

Afterword

How to Prepare for a Pole Shift

> The magnitude of disasters decreases to the extent that people believe that they are possible, and plan to prevent them, or to minimize their effect.
>
> KENNETH E. F. WATT, *The Titanic Effect*

> "Future shock" is the "aftershock" of unpreparedness. It can be avoided or greatly diminished by gathering holistic and high-probability information about the future, and by planning to sidestep its unwelcome prospects.
>
> PAUL JAMES, *California Superquake 1975–77?*

While pondering the awesome possibilities explored in this book, and saying to myself, "It just can't happen," an image of the ship *Titanic* frequently came to mind. No one thought *that* could happen either.

While I am not yet convinced that a pole shift will occur, it should be obvious that there is significant evidence indicating one *might* occur. Moreover, it might occur within a relatively short time, quite in accordance with the many prophecies and predictions that place it within the next two decades. Simple prudence therefore requires that certain basic measures be taken in order to prepare for possible disaster. No alarms need be sounded, but clearly the time may be short—a mere twenty years or so.

How should we act? What should we do?

In the broadest manner of speaking, we should avoid panic and the Chicken Little syndrome. Instead, we should proceed non-anxiously to apply the full range of our physical, mental and spiritual capabilities in preparing for the possibility of a pole shift. We must think individually and societally about the future, bearing in mind that, for better or for worse, we are all in this together.

As a first step, I suggest that a concerted international program of scientific research and civil preparedness planning be mounted. This program will delineate areas of responsibility wherein *prediction*—the when and where of a pole shift—is recognized as the domain of scientists, and *warning*—what is to be done about it—is recognized as belonging to the authority of civic officials. The purpose of the program should be:

1. To determine the probability of a pole shift being initiated by any single trigger mechanism or combination of them.

2. To make recommendations to world governments and the United Nations for steps to modify the influence of possible trigger mechanisms—i.e., to stop nature's "finger" (if it be hers, not ours) from squeezing hard enough to actually "fire."

3. To monitor ongoing conditions so that if a critical situation should develop, adequate warning could be given to world society.

4. To prepare long-range plans for the survival of world society if a pole shift should occur. The plans should cover both the health and safety of people during a cataclysm and the orderly restoration of living conditions afterward.

The heart of any long-range warning system will be geostationary satellites such as those now mapping the weather, land resources, lightning, ocean currents and other aspects of the earth's biosphere.

Other satellites, I suggest, could be used for infrared mapping of heat flow around the globe on the assumption that pressure buildup along earthquake faults would be accompanied by an increase of temperature. Since the pole shift concept as presented by psychic sources includes a prelude of increasingly severe earthquake and volcanic activity and since seismologists have clearly indicated that such activity is in fact happening, it would seem to make good sense for us to apply present technology this way.

That this might be viable is suggested by the 1978 government approval for Daedalus Enterprises of Ann Arbor, Michigan, to sell a sophisticated electronic monitoring device to the People's Republic of China for use in predicting earthquakes. Called a ground scanner, the instrument can detect minute differences in the geologic structure of the earth by use of advanced infrared and television technology.

Interestingly, *Earthquake Information Bulletin* (January-February 1979) notes in an article entitled "Space Techniques for Measuring Crustal Deformation" that NASA proposes to get underway with a program to make very accurate measurements of the Earth's rotation rate (length of the day) and tipping of the Earth with respect to its spin axis (polar motion).

This is a welcome development.

By combining space technology and seismic detectors for the purpose of saving lives—a marriage of heaven and earth—we may be able to make the most humane possible use of machines.

We should also make the best use of psychics. In the May-June 1978 issue of *Earthquake Information Bulletin*, two U. S. Geodetic Survey scientists reported the results of a study they made of earthquake predictions registered with them by private citizens, many of whom claimed to be psychic. The study showed that none of the ninety-one people were more accurate than chance, with one notable (but unnamed) exception.

This is a useful step, but to end there—as the USGS scientists did—is regrettable. It's like trying to make a gold necklace by sifting nuggets from a stream bed, as if nothing else were required.

If the USGS were to work with psychics who have a proven "track record," in addition to depending on whatever may come their way by chance, the accuracy of this mode of earthquake prediction would probably increase dramatically.

One way to begin would be to make a controlled test of psychics who have proved themselves capable of prediction beyond chance in one field or another. If many such psychics can be found, and if they could be trained to a good degree of accuracy and reliability, they could then be used as a "distant early warning" instrument. Earth scientists would have a most valuable addition to their forecast system.

Long-term civil defense preparations should also be considered. Since it may not be possible to determine with precision where safe geographical areas are—if any—in a pole shift, government officials and civilian task forces should begin to plan large-scale actions to preserve lives and property. Individual preparations by survival-minded citizens may reduce the number of fatalities and injuries and the degree of social chaos but will not eliminate them. Coordinated social action is necessary, especially since it is now widely acknowledged that our civil defense program is at best antiquated and inadequate. In fact, the head of the Pentagon's Civil Defense Preparedness Agency, Bardyl Tirana, recently admitted to *Parade* magazine (21 May 1978): "We have no civil defense program, merely the apparatus to start one. When you look at civil defense in the United States, you find the emperor has no clothes."

In addition to scientific research, advanced technology and civil defense measures, one more societal means is available to us in preparing for a pole shift: legislation. Not that we'd outlaw pole shifts! Rather, thoughtful political measures could be taken to allocate research funding, to expand civil defense efforts and to mandate educational programs for dissemination through channels such as public schools, television, the armed forces and various government agencies.

Another possibility for legislative action would be to require certain safety measures in public buildings. The principle is already in effect in many areas such as transportation, public housing, occupational safety and health, patent medicines, vaccination, etc., where public safety requirements are set by government fiat. As with solar energy technology, there might be a tax incentive for certain structural changes or additions made to public buildings.

Altogether, precedents exist for a vast legislative effort to be mounted in support of science, civil defense, private industrial research and various consumer/ecological movements concerned with pole shift preparations. Coordinating these activities and educating the public about them would be a massive task. But I can think of no more noble one.

Epilogue

POLE SHIFT UPDATE

It has been ten years since I published *Pole Shift*. During that time, many new data have come to light which bear on the two principal questions I raised: Have there been previous pole shifts? Might there be one in the near future? I will briefly survey the data and offer my assessment of its significance. First, however, I will anticipate my conclusion. I offer this at the outset to alleviate the anxiety many people have expressed over humanity's future because of the pole shift predictions and prophecies I reported.

On the basis of a decade's hindsight, I think that the possibility of a catastrophic pole shift at the end of this century is highly unlikely. To be more precise, *I do not think a pole shift will occur as predicted.*

That may seem strange to you after you've just finished reading the case for pole shifts. But remember that I said I neither believed nor disbelieved in pole shifts. I felt that a strong case for pole shifts could be made, but I recognized that presenting the case is not the same as proving the case. Therefore, I said at the end of the main text that there were still too many unanswered questions for me to draw a firm conclusion, and more research was needed.

That further research led me to a conclusion which is both startling and gratifying—scientifically startling and spiritually gratifying. I present it at the end of this Epilogue. Suffice it to say for the moment that I regard my time and experience invested in researching and writing about pole shifts to be immensely valuable, and I trust that my effort will prove equally useful for you without your having to go through the large amount of writings and data I wrestled with in order to gain insight about "the ultimate catastrophe."

Here are the principal categories of evidence supporting the pole shift concept, a summary of why theorists say they do, and a review of new developments. Some of the questions which prompted me to present the case for pole shifts are still unanswered. But enough has come to light to refute the pole shift theorists in two of these categories (the frozen mammoths and the ancient maps of Antarctica). In a third category, new theoretical support for sudden axis reversals has also come forth; at the same time, however, counterarguments have refuted one pole shift theoretician.

1. **Anomalous Glacial Striations.** Continental drift does not explain all anomalous glacial striations, pole shift theorists say. Those in South Africa, for example, show a direction of movement toward, not away from, the South Pole. Tectonic plate theory indicates nothing to explain them. However, no further evidence or commentary on this subject has come to my attention. These data continue to defy the conventional notion that ice sheets have always spread out from the present polar locations. (For a relevant datum to the contrary, however, see the end of the next section.)

2. **Ice Ages.** Pole shift theorists ask the following question: If slow and regular changes in the orbital geometry of Earth, reinforced by changes in the climate cycle, are the cause of ice ages, as conventionally claimed, what explains the following: first, the extremely rapid appearance and disappearance of continental-sized ice sheets; second, the vast epochs—each several hundred million years long, far exceeding the alleged periodicity of the ice ages—in which the planet was free of polar ice sheets; and, third, the fact that the North American ice sheet during the last ice age was centered in Hudson Bay while the north polar area—as presently located—was virtually ice free?

In 1982, Peter Warlow presented his position at length in a book, *The Reversing Earth*. As I presented here, Warlow proposes an extraterrestrial source as the trigger mechanism for pole shifts. He also proposes that planets are born as ejected cores of some larger bodies which lead to a sequence of disturbances in a solar system. He states that it is possible for a celestial body "to exert a torque on the Earth even though it does not come into direct contact. We thus have the means of turning the Earth over ... "

After discussing the characteristics of such a body and the forces involved, Warlow calculates that a pole shift could occur "on a time scale of days rather than weeks, months, or years." But he does not predict that such an event will occur. In fact, he specifically rejects predictions and prophecies of "the Earth's axis tilting." Asserting that there is nothing special about the year 2000, Warlow concludes:

I do not know when the next event is likely to occur. If the idea of planet birth is correct, it may well be a long time before another is born to disturb the Solar System ... it may be many millions, or tens of millions of years ... In any case, it is likely that we will have as much time to build our arks and our stone circles, or their equivalent, as our ancestors had. In all probability, we will have plenty of forewarning—but from astronomical sightings rather than from any psychic insights.

Warlow's presentation resulted in a number of papers examining his assumptions and mathematics. Dr. Victor J. Slabinski, an astronomer in the Intelsat Physics Department in Washington, D.C., challenged Warlow in a *Journal of Physics* article entitled "A Dynamical Objection to the Inversion of the Earth on Its Spin Axis" (September 1981) in which he demonstrated that Warlow's explanation was mathematically unsound. In a further comment in *Kronos* (Vol. VII, No. 2, Winter 1982), he noted that Warlow had miscalculated the necessary torque for flipping the Earth, adding, "Cosmic bodies large enough to invert the Earth act for a period much less than the 24 hours assumed by Warlow ... Not even confining the inversion to a thin crustal shell yields a practical solution."

Further refutation of Warlow's position was offered in the same issue of *Kronos* by C. Leroy Ellenberger, a longtime investigator of the pole shift question. He began as an advocate of Velikovsky's work. However, the weight of evidence in his continuing research has led him to conclude that Velikovsky was thoroughly wrong. Greenland ice cores, bristlecone pine rings, and ocean sediments show that Velikovsky's version of our solar system history did not happen, he asserts. (See the section below, "The Case Against Velikovsky.") Ellenberger reviewed

Warlow's argument and mathematics, noting various errors, and concluded, "For the entire Earth or its crust to flip over solely under the gravitational influence of a passing cosmic body of any size seems impossible."

In a long technical letter submitted for publication by the Society for Interdisciplinary Study's *Workshop* (1988:2), Ellenberger wrote, "There simply is no physical evidence for a geographical inversion ever having happened."

Warlow's book discusses four lines of physical evidence that supposedly support their occurrence: geomagnetic reversals, ice ages, sea-level changes, and mass extinctions. Perhaps Warlow ... now realizes that these events are inadequate as evidence since they can be readily explained by other less extravagant processes ... I believe the best physical evidence would be uniquely related to an inversion, e.g., formations on Earth reminiscent of the tidally induced bulge or some other mass concentration that enabled a torque to flip Earth or systematic worldwide coastal destruction caused by the accompanying ocean floods. I am unaware of any evidence for the former presence of such a bulge or mass concentration.

(For an elaboration of the "less extravagant processes," see the Velikovsky section below.)

About the same time as Warlow's book appeared, a Swedish theoretical physicist, Dr. Stig Flodmark, presented a long paper, "The Earth's Rotation," at the European Geophysical Society's annual meeting in Uppsala in which he proposed a mechanism for explaining pole shifts. Flodmark is associated with the Institute of Theoretical Physics at the University of Stockholm. In August 1981, he offered a new model of polar shifts which also accounts for ice ages and their anomalous (rather than regular) appearance in Earth's history. He proposed a "double-top" model of the planet in which the solid inner core is separable from the solid mantle and viscous part of the core. Frictional forces normally keep the "tops" rotating in unison or nearly so, but there is a slight differential which can explain observed small polar motions known as the annual wobble and the Chandler wobble. The double-top model, Flodmark asserts, also can explain glacial ages, magnetic role reversals, faunal extinctions, and other enigmatic topics in Earth's history and geology. Unlike Warlow, Flodmark predicts that Earth *is* nearing the moment when another pole shift will occur within the space of a single day and that "some perturbance of the smooth rotation of the earth could be expected shortly."

In subsequent papers, Flodmark modified his model to a triple-top.

With it, he also proposes explanations for the end of ice ages through frictional heat from mantle wobbling, the necessary power for geomagnetism, and the cause of earthquakes, volcanism, and mountain building.

The only commentary on Flodmark I've found was offered by Ellenberger in the same *Kronos* article noted above, in which he wrote, "In light of the unsolved and apparently unsolvable problems attending the solid body tippe top [gyroscope] model, Flodmark's double-top model, at this same, seems a viable replacement ... however, a high priority should be placed on validating the double-top model." As already noted, Flodmark himself revised the model. I am not aware of any further commentary or criticism of Flodmark's work.

That work may provide the mechanism sought by Dr. Michael Herron, research assistant professor at State University of New York at Buffalo, who, at about the time of Warlow and Flodmark's pronouncements, offered the results of a ten-year drilling project completed in Greenland. A 7,000-foot core from the Greenland ice sheet revealed climatic data on a year-by-year basis as far back as 10,000 years ago, when the last great glaciation—the Wisconsin glaciation—ended.

According to Herron, the data show that the change between "normal" and ice age conditions on Earth has been surprisingly and dramatically abrupt. In fact, Herron said, climatic change may have been so sudden at the end of the last glaciation that he has no idea what mechanism might account for such a quick change.

In a related matter—the fact that the last north polar ice sheet was centered in Hudson Bay—an article, "The Earth's Orbit and the Ice Ages," by C. Covey in *Scientific American* (February 1984) points out that recent advances in climatic modeling (the Milankovitch theory connecting ice ages and Earth's orbit) show that no pole shift or axial tilt is required to explain that fact. Rather, Covey says, the eccentric location with respect to the pole is a natural consequence of land-water distribution.

Altogether, then, the latest evidence in this category weighs against pole shifts, although (1) the onset of deglaciation is unexplained and (2) the theoretical work of Flodmark warrants consideration.

3. **The Frozen Mammoths.** The evidence in this category, as interpreted by pole shift proponents, suggests that the famed Berezovka mammoth died suddenly by asphyxiation in late summer in a temperate climate and that it was frozen by the imposition of temperatures in excess of -150° F. in ten hours or less. Contrary to popular belief, they claim, the mammoth was not an Arctic mammal because it lacked the sebaceous oil glands which cold-adapted land animals have to lubricate their skin and thereby avoid death by dehydration. Moreover, the Arctic

could not possibly supply enough vegetation to support vast herds of these herbivores, who required several hundred pounds of vegetation daily for each member. Yet their skeletons litter the tundra by the hundreds of thousands.

This has been a topic of lively debate among some of my correspondents. William White of England, an opponent of the flash-freeze school of thinking, wrote a critique in which he makes these points. First, the findings of the new science of taphonomy, which is the study of all the processes an animal goes through from the time it starts to die until its remains are finally embedded in a geological stratum, demonstrates that mammoths died not as a result of disastrous temperature change but from asphyxia (e.g., drowning in an icy stream, suffocating from a landslide, etc.). Second, other studies demonstrate that mammoth flesh is not so well-preserved as has been claimed, but rather that the flesh had begun to putrefy *before* being frozen in permafrost. Third, sebaceous glands, which are said by pole shift theorists to be necessary in all Arctic mammals to tolerate extreme cold, are missing in mammoths but present in wooly rhinoceroses. Since rhinoceroses are often found frozen in company with mammoths, how can the two—supposedly living in widely separated locations at the time of a catastrophic pole shift—be accounted for being interred together?

Dwardu Cardona of Canada, a catastrophist who defends a sudden freezing of mammoths, rebutted White's position in comments entitled "The Mammoth Controversy" in *Kronos* (VII:4):

Despite White's statement to the contrary … the commencement of putrefaction *prior* to freezing has never been satisfactorily proven. Eyeballs are among the first parts of the body to rot after death, yet some of the mammoths discovered in Siberia had their eyeballs intact. Dima [a baby mammoth found in Siberia in 1978] is the only specimen so far to have been discovered in an unthawed condition. I might be wrong but, to my knowledge, it showed no signs of putrefaction. Yet even if it did it would not much matter, for there is nothing in prevalent catastrophic theories which excludes interim, even if minimal, thawing between catastrophes.

The fact remains that where mammoth carcasses, in whole or in part, have been discovered, decomposition has been minimal. They did not decompose away … [If] climatic conditions have not changed since the mammoths roamed, why is it that only extinct species are ever discovered entombed in ice?

What I do grant William White is that the direct cause of the Berezovka mammoth's death was asphyxia *before* freezing. That has always been known and admitted by catastrophists. Suffocation, however, is not necessarily the result of drowning and/or landslide burial as White and others would have us believe. Ivan Sanderson, Immanuel Velikovsky, and Charles Hapgood have all described extraordinary, but *possible,* atmospheric conditions which *could* have asphyxiated the mammoths just prior to freezing. While not necessarily correct in the details these investigators have supplied, it is a fact that, both in Alaska and Siberia, mammoth remains are associated with evidence of atmospheric tempests of unprecedented dimensions. *And it is this overall picture, not the hair-splitting issues we have been debating, that is the crux of the matter.*

The "young but powerful science of taphonomy," upon which White relies, has shown that the carcass of an African elephant decays in about three weeks, leaving nothing but the tough skin covering the bones. In temperatures which, according to Farrand, were higher than the present 90-100° F. of the Siberian summer, the Berezovka mammoth should have likewise decomposed. The position in which this beast was found clearly indicates that it could neither have been drowned nor crushed beneath a slide. Its stance suggests that it was felled on its haunches, that it attempted to regain its feet, that it was then somehow asphyxiated, and that it froze in this animated position. It did not even keel over.

The latest word on the mammoth controversy comes from White's rejoinder to Cardona and Ellenberger in a three-part *Kronos* statement (XI:1, Fall 1985 to XI:3, Summer 1986). White effectively countered nearly all their arguments, convincing Ellenberger, at least, that processes and mechanisms less extravagant than a pole shift can account for the frozen mammoths.

White shows, first, that prevailing Arctic conditions—namely, blizzards whose low temperatures are compounded by high winds, producing wind-chill factors equivalent to -150° F.—are sufficient to freeze a mammoth, including its stomach contents, into the state in which they are presently found. "It is therefore glaringly apparent that the allegedly compulsory rapid chilling of a dead mammoth, including its stomach, would have been readily accessible under known conditions."

The Berezovka mammoth died in late summer, according to the analysis of Pfitzenmeyer, who dissected the mammoth in 1901 and noted the vegetable contents of its stomach and mouth. White points out that it would have been possible to preserve the animal even in a Siberian August because sufficient meteorological conditions occasionally prevail.

What about the amount of vegetable biomass needed to support "truly vast herds" of mammoths? White writes:

> Where is the evidence for such a large population and herd size? Neither of the living species of elephant moves in herds more than a fraction of this number. He is on firmer ground when he speaks of "untold numbers of mammoths *in the past.*" Few would care to dispute this, given that their skeletal remains runs into tens of thousands of individuals! Yet, such numbers are unremarkable considering the ideal conditions for the preservation of skeletons in the cold ground plus the fact that the wooly mammoth flourished from the penultimate glaciation of Europe (240,000 years before the present?) until circa 10,000 years B.P.—or later still, according to my critics. Thus, in 200,000 years or so, the death and preservation of only *one mammoth per year* would be adequate to account for all those tusks found. This makes it obvious that the Mammuthus primigenius population was much smaller than certain writers have chosen to presume and/or that only a small proportion of their remains have been preserved (in any form).

Another argument in favor of flash-freezing of mammoths has been their relatively good state of preservation. White states that "putrefactive degradation [of tissues] is solely under the control of bacterial enzymes and, *when these are absent, decomposition does not ensue* ... the available evidence dictates that decay upon thawing was *very slow* indeed. So low was the ambient temperature and bacterial population that some of the mammoths in question were substantially complete when examined years after the initial re-exposure of the carcasses."

He goes on to point out that rare photographic views of Dima's right profile reveal a large hole, with exposed ribs and thoracic cavity. "Moreover, decomposition was much more extensive than was claimed initially, such that microstructural studies of the brain and muscles have had to be abandoned while the analysis of albumin was marred by the general decay of tissue proteins which was evident." Thus, Dima was not "perfectly preserved" as some claimed and I reported.

Even the survival of eyes, while remarkable, is not miraculous, White notes. Eyes are often present in Peruvian and Egyptian mummies. Hence, preserved mammoth eyeballs are not *prima facie* evidence of a pole shift. Moreover, if the wooly mammoths were the victims of a widespread calamity, why are so few found? "It is the paucity of their complete bodies in a frozen state that is remarkable ... Human agents were, in fact, responsible for many of the vast caches of mammoth bones which seem to cause so much consternation."

As for the oil-gland dispute, White demolishes it with a number of morphological arguments. Mammoths were adapted to cold conditions, he demonstrates, not only because of the morphological evidence but also because of evidence from paleontology, which shows mammoth bones in association with cold-loving species, and archaeology, where paleolithic cave art shows the same thing.

White's final argument is radiocarbon dating of frozen mammoth remains. The dates alloted to mammoth carcasses "have too wide a spread to be consistent with the theory of mammoth extinction in a single cataclysm."

In concluding his statement, White emphasizes that the known complete mammoth remains have been found in "fossil traps, which were responsible for both the demise and the preservation of the individual beasts." He acknowledges that no categorical solution to the problem of the frozen mammoths has yet been given because the evidence presented by the preservation of the animals is equivocal in that it can be used in support of a wide range of hypotheses, some even diametrically opposed. "My chief concern," he writes, "has been that frozen mammoths alone provide but a flimsy basis upon which to build a theory."

4. The Ancient Maps of Antarctica. Charles Hapgood, who died at age 78 in 1982, summarized this subject in his 1979 revision of *Maps of the Ancient Sea Kings*. He wrote, "The maps in this book show that an ancient advanced culture mapped virtually the whole earth [about 15,000 years ago or more, and] that its cartographers mapped a mostly deglacial Antarctica …" (p. 239). And in a 1980 article in *Catastrophism and Ancient History*, he wrote, "Our best indication of a warm Antarctica is an authentic map [the Oronteus Finaeus map of 1531] showing Antarctica free of ice …" At the time of *Pole Shift'* s publication, this research had never been refuted. It had only been ignored.

Since then, Hapgood's work has been challenged by researchers who claim that Hapgood's interpretation of the ancient maps was wrong. In the words of one critic, David C. Jolly, "I do not think that Hapgood was intellectually dishonest—merely that he uncritically accepted any evidence supporting his views and did not try very hard to come up with alternative explanations. Ultimately, he became a victim of his own enthusiasm."

In 1980, a journalist-scholar team concluded that Hapgood's work was open to serious question, although certain moot factors could weigh in his favor. Paul F. Hoye, editor of the Aramco Oil Company's magazine, and Paul Lunde, a graduate of London's School of Oriental and African Studies working on Arabic manuscripts in the Vatican Library in Rome, examined Hapgood's work in an Aramco World

Magazine article entitled "Piri Re'is and the Hapgood Hypotheses" (31:1, January-February 1980). They pointed to "serious weaknesses in Hapgood's case."

First, they say, Hapgood's theses depend entirely on mathematical projections and logic; there are no examples of "advanced" source maps such as he postulates, nor can he display a single artifact from the "lost" civilization which supposedly mapped the Americas and Antarctica. Also, his postulate of an ice-free Antarctica conflicts totally with accepted geological theory which says the Antarctic ice cap has been in place for millions of years. Among other objections to Hapgood's work they note, the most compelling arguments, in their opinion, concern the Andes mountains and Antarctica. "Is the chain of mountains to the left of the map really the Andes? Is the coastline at the bottom really Antarctica? Are there any mountains shown there? And is Antarctica free of ice?"

To put it more simply, Piri Re'is, or the scribe who copied his work, may have realized, as he came to the Rio de la Plata, that he was going to run off the edge of his valuable parchment if he continued south. So he did the logical thing and turned the coastline to the east, marking the turn with a semicircle of crenelations, so that he could fit the entire coastline on his page. If that was the case, then the elaborate Hapgood hypotheses—or at least those based entirely on the Piri Re'is map— would have no foundation whatsoever.

Hoye and Lunde find that the Oronteus Finaeus map, on which Hapgood puts so much store, is more reliable but also open to question. After discussing various possibilities, they conclude that simpler—and more prosaic—explanations are possible. Nevertheless, they acknowledge that Hapgood's belief that an advanced source map will be found might one day be fulfilled, since there are massive collections of documents crammed into museums and archives in Istanbul, many still unexamined. They also acknowledge that their criticism of Hapgood's work might itself be questioned. "Major historical and cartographical problems ... remain unsolved. The mystery is still there."

David Jolly disagrees. He feels Hapgood's work is completely flawed and unacceptable, and there is no mystery. Jolly, who publishes an annual handbook for the rare map trade, asked in *The Skeptical Inquirer* (Fall 1986), "Was Antarctica Mapped by the Ancients?" He examined Hapgood's sources and reasoning. On both grounds, he faulted Hapgood. With regard to the Piri Re'is map, Hapgood assumed that a portion of the South American coast was really a misplaced section of Antarctica. Jolly shows that to be most unlikely, based on comparison with other ancient maps and mapmaking techniques of the time. Likewise, Jolly argues, Hapgood's assumptions about the Oronteus Finaeus map, so important to his position, were not well founded. Again, Jolly surveys other ancient maps and concludes that "the Finaeus map may be a blend of imagination and fact." The simplest theory, he writes,

> ... is that Finaeus drew an asymmetrical, bilobate blob of no special shape to conform to ancient belief in a southern continent and to Magellan's discovery. The blob theory is consistent with the varying shape of Terra Australis on maps both before and after the Finaeus map. If there were actual observations or maps of the coastline, one would expect a more consistent representation.

Jolly also cites M. M. du Jourdin and M. de La Ronciere's recent book *Sea Charts of the Early Explorers* (Thames and Hudson, 1984), which shows that "the so-called portolan charts of the Mediterranean evolved from crude prototypes and were not derived from ancient sources as Hapgood claimed." He concludes his examination of Hapgood's work by saying that Hapgood's ideas were rightly rejected in scholarly circles "not because of animus but because he had not proved his case. Too many leaps of faith were needed to establish his thesis."

It therefore appears that Hapgood's thesis—the Earth was mapped by a prehistoric civilization thousands of years ago—has been refuted.

The Case Against Velikovsky. Although the work of Immanuel Velikovsky is contained by three of the four categories of evidence for pole shifts—he does not refer to ancient maps of Antarctica—it is so well known that I feel it deserves special attention. I single it out here because of Velikovsky's prominence in the pole shift debate.

A major critique of Velikovsky appeared in 1984 entitled *Beyond Velikovsky.* The author, Henry Bauer, gives an account of the Velikovsky controversy and reviews literature through the early 1980s.

The most comprehensive, although succinct, refutation of Velikovsky's position was offered by C. Leroy Ellenberger. In a two-part *Kronos* article entitled "Still Facing Many Problems" (X:1, November 1984 and X:3, July 1985), he commented on a wide range of topics bearing on Velikovsky's hypothesis, ranging from orbital changes, geomagnetism, and the cooling of Venus to plate tectonics, ice age dynamics, and sea level changes. For each topic he shows that Velikovsky's data were wrong or that his use of that data was wrong—and sometimes both.

In a letter to *The Skeptical Inquirer* (Summer 1986), Ellenberger further sealed the case against Velikovsky. " ... the Terminal Cretaceous Event 65 million years ago, whatever it was, left unambiguous worldwide signatures of iridium and soot. The catastrophes Velikovsky conjectured within the past 3,500 years left no similar signatures according to Greenland ice cores, bristlecone pine rings, Swedish clay varves, and ocean sediments. All provide accurately datable sequences covering the relevant period and preserve no signs of having experienced a Velikovskian catastrophe. Although Velikovsky believes *Earth in Upheaval* proved his scenario happened, his evidence can be explained without invoking cosmic catastrophes."

Two letters by Ellenberger in the *New York Times* complete this summary of the case against Velikovsky. On May 16, 1987, his letter to the editor stated:

Velikovsky ... does not deserve to be taken seriously for two reasons: (1) his use of mythohistorical sources was discredited by Sean Mewhinney in the winter 1986 issue of *Kronos* ... and by Bob Forrest in the winter 1983-'84 issue of *The Skeptical Inquirer* ... and (2) no physical evidence exists to support recent catastrophes of the magnitude described in "Worlds in Collision" ... That Velikovsky's catastrophes did not happen is proved by the Greenland ice cores. Their annual layers can be counted back 7,000 years. The ice contains no trace of the cometary debris that Velikovsky said caused 40 years of darkness following the Biblical Exodus.

And on August 29, 1987, the *Times* published this statement by Ellenberger:

... your August 6 article about the redating of the Minoan eruption of [the Agean volcanic island] Thera from 1500 B.C. to 1645 B.C., using Greenland's Dye 3 ice core ... should be the final evidence needed to end the Velikovsky controversy.

According to the revised chronology of Velikovsky's "Ages in Chaos" books, the date for Thera's eruption [whose remains are now called Santorini] is required to be about 950 B.C. With this keystone removed, Velikovsky's revised chronology collapses utterly. The absence of copious cometary debris in the cores further proves that the catastrophes ... never happened.

The ice cores have thus repudiated both Velikovsky's catastrophes and his chronology.

So where are we in the debate about pole shifts? From the *scientific* perspective, the question of previous pole shifts appears resolved in favor of a noncapsizing planet with stationary poles. Velikovsky's case has been demolished; the evidence he offers for recent catastrophes can be explained by noncatastrophic processes. The theoretical support for pole shifts offered by the work of Warlow appears disproven, although Flodmark's remains in need of thorough examination. When disproof of Hapgood's cartographic contentions is added to the mundane explanation for frozen mammoths, the case for pole shifts is seemingly beyond recovery. In addition, although the question of ice ages remains open—insofar as the absence of ice sheets over long periods is not explained by the conventional explanation—those absences in and of themselves are not evidence of pole shifts. So with these refutations of the major categories of evidence for the pole shift concept, the case for pole shifts is now virtually nonexistent.

THE PREDICTIONS WERE WRONG

If the data from science regarding the existence of previous pole shifts and the theoretical possibility of such an event remain ambiguous to any degree, the predictions from psychics regarding future pole shifts do not. A number of signposts or precursors of a pole shift were predicted which never came to pass. The period from 1982 to 1984, for example, was to be a time of major seismic activity around the globe. Some psychics went on record as saying that the "Jupiter Effect" would be devastating.

The 1982-'84 period passed without significant seismic events— at least not significant enough to fulfill even remotely the various psychic predictions for that time.

Paul Solomon had stated that Japan would disappear beneath the sea, the Great Lakes would empty into the Gulf of Mexico, new continents would begin to rise in the Atlantic and Pacific oceans, much of the West Coast of North America would disappear, the North American continent would be split in half down the center, and other disasters of similar magnitude would occur in a general scenario of global seismic upheaval.

Likewise, Aron Abrahamsen stated that by 1981 the point of no return would have been reached. At that point, he said, "people will have improved or the momentum will be so great that it will be difficult to induce a change in the events which are to follow." I've noted that some of Abrahamsen's predictions for previous years were not accurate. In light of world events during the years following 1981, which are notably lacking in geophysical events of the sort he forecast, I conclude either that human consciousness and behavior have radically improved, thereby offsetting geophysical disaster through the mind-matter linkage psychics identify, or Abrahamsen's earth change predictions were incorrect. The latter seems more likely, although the dizzying pace of change in civilization's sociopolitical character in the last few years— as monolithic communism moves toward democracy and a free-market economy—does indeed indicate a profound shift in human consciousness for the better.

What about Edgar Cayce, whose readings had "set the baseline" in the psychic community regarding earth changes? I've pointed out that Cayce's readings about earthquakes striking the West Coast of America, even when interpreted in the broadest possible fashion, would have reached their limit of reliability by the end of 1982. That, in turn, would bear directly on the accuracy of his pole shift predictions.

The old saying, "Time will tell," has proven true. Time *has* told us that the psychic predictions for events to date (1991) were wrong. They were wrong with regard to both type and magnitude of events. For example, while it is true that there have been volcanic eruptions in the Cascade Range along the West Coast of North America—notably, Mt. Saint Helens in Washington—there has been nothing even remotely resembling the loss of the West Coast, with the Pacific Ocean flowing inland through many states. In fact, as geologist Dr. John Peterson says in an unpublished report, "There has been no significant change from the normal and anticipated geological pattern."

Peterson is a registered professional geologist in San Diego, California, and has been a student of the Cayce material for 25 years. He has lectured on earth changes for A.R.E., at times with Hugh Lynn Cayce. During most of that time he accepted the conventional understanding of Cayce's earth changes readings in an uncritical manner. But in 1988 he reviewed the readings from start to finish and concluded that they tell a far different story than what is commonly believed and stated. Because his work is so new and important, I am going to summarize it here. I am grateful to him for allowing me to use his research in this new edition. Arrangements are being made for it to be published in the A.R.E. members' magazine, *Venture Inward.*

Peterson finds that the Cayce earth changes readings fall into three categories: the gradual readings, the 1936 readings, and the catastrophic readings. Each group appears to have a different meaning or intent. The gradual readings, of which there are seven, generally say that earth changes will be gradual, not catastrophic, and the time of them cannot be foretold. The 1936 readings predict major and catastrophic earth changes. They focus on physical events which were predicted for the year 1936. Peterson notes that 1936 was not a particularly active year, seismically speaking. In fact, seismic activity was below normal for that year. The last group of readings—the catastrophic readings—predicted widespread earth changes for periods after 1936. This group has only three readings in it. In general, Peterson says, these predicted changes have not occurred. Moreover, the current geographic and geologic record does not support the readings, and one of the readings was not from Cayce's usual source but identified itself as different, leading the Cayce group to reject its authority, although it remains recorded as an earth changes reading. So in general, he concludes, the Cayce material does not paint a clear picture of future physical earth changes and the evidence from history shows that many Cayce predictions have not occurred.

So the psychics were wrong. Why? A plausible explanation can be found in the near-death experience (NDE) research of Dr. Kenneth Ring, professor of psychology at the University of Connecticut and author of the 1984 book, *Heading Toward Omega*, which is a study of the meaning of NDEs. One chapter surveys what Ring calls prophetic visions (PVs), which are particular sequences of imagery perceived during an NDE which have to do with earth changes and a pole shift.

Ring found about two dozen persons who report PVs as part of their close approach to death. Several other investigators, both American and British, have likewise noted the feature as part of the NDE and have confirmed Ring's description of the general scenario, which he summarized with these words in an *Anabiosis* article, "Precognitive and Prophetic Visions in Near-Death Experiences" (1982:2):

There is, first of all, a sense of having total knowledge, but specifically one is aware of seeing the entirety of the earth's evolution and history, from the beginning to the end of time. The future scenario, however, is usually of short duration, seldom extending much beyond the beginning of the twenty-first century. The individual reports that in this decade there will be an increasing incidence of earthquakes, volcanic activity and generally massive geophysical changes. There will be resultant disturbances in weather patterns and food supplies. The world economic system will collapse, and the possibility of nuclear war or accident is very great (respondents are not agreed on *whether* a nuclear catastrophe will occur). All of these events are transitional rather than ultimate, however, and they will be followed by a new era in human history, marked by human brotherhood, universal love, and world peace. Though many will die, the earth will live. While agreeing that the dates for these events are not fixed, most individuals feel that they are likely to take place during the 1980s.

In "Prophetic Visions in 1988: A Critical Reappraisal," which appeared in the *Journal of Near-Death Studies* (1988, 7:1), Ring noted that the PVs' forecast of cataclysmic destruction tended to converge on a single year—1988—as the critical time. Writing toward the end of that year, Ring reviewed the six interpretations of PVs which he offered in his book. The most obvious fact, of course, is that the PVs did not come to pass in a literal sense. From that point of view, they were totally wrong.

The central lesson for us, Ring concludes, is that modern NDE-inspired PVs are "not the literal forecast of planetary doom, but rather the foreshadowing of the need for a new myth of cultural regeneration." This conclusion forced itself on Ring for two reasons. First, the critical time period passed safely. Second, he learned from others that such PVs have a long history. The first PV to come to his attention from a source outside his data base stated that planetary upheaval was imminent; that PV was recorded in 1892! In 1987 Ring read John Perry's *The Heart of History,* which argues that many such PVs can be found throughout history, particularly in sensitive individuals whose cultures were undergoing a period of deep crisis. Perry states that "the horrific vision of world destruction is part and parcel of the mythic imagery of rapid culture change and world views in transition ... Beholding the world coming to its end amid storm, earthquake, flood, and fire we have found to be a typical experience of a prophet whose psyche is registering the emotional impact of the end of an era." Ring, agreeing with Perry, concludes that such prophets arise during times of cultural crisis and they bring a messianic message of the need for cultural renewal.

What I am asserting is that whereas PVs are indeed prophetic utterances, they should not, in my opinion, be regarded as truly precognitive or taken literally. Rather they seem to be reflections of the collective psyche of our time, which is generating its own images of planetary death and regeneration for which the sensitive souls of our era serve as carriers.

Ring cautions, on the basis of his research and the lessons of history, that images of general apocalypse should not be taken literally. Rather, they should be seen metaphorically or symbolically as "harbingers of hope that point to the possibilities of human regeneration and planetary transformation."

CONCLUSION: SPIRITUAL ADVICE

I think that is the most sensible—and spiritual—note on which to conclude this update on the pole shift issue. From a scientific perspective, pole shifts are nothing to worry about because there is almost no scientific basis to the concept. But even if there were, according to the predictions and prophecies, a pole shift is not inevitable. The possibility of it is influenceable by our manner of living and thinking. Human consciousness, the predictions and prophecies say, will be of critical importance in triggering or preventing a pole shift. In essence, they say, the state of consciousness among people will determine the outcome of an approaching global crisis. Throughout the book I've delineated the process through which, according to metaphysics and esoteric psychology, human consciousness interacts with the matrix of geophysical and astrophysical factors operating to stabilize Earth in space.

Some people fearfully watch the poles to detect the slightest movement. But that is self-defeating, the psychics and prophets say. We should be soul watchers rather than pole watchers. In other words, if there are people changes for the better, earth changes for the better will follow.

Predictions of pole shift serve to point out that we must change our consciousness from ego-centered to God-centered living and recognize that there is a benevolent wisdom governing the universe, including us.

From a spiritual perspective, there are no problems—there are only situations. Problems don't exist in nature. Only situations do, only sets of circumstances. It is the human mind which projects attitudes and values onto those situations and then labels them as problems. But that label doesn't describe what nature is doing. It describes the state of mind of the human who labeled the situation.

Problems are a reflection of a state of mind—a state based on self-centeredness and fear and an unwillingness to face new experiences. But change that attitude and, suddenly, there are no problems. The set of circumstances remains, but the situation now becomes an exciting challenge in which to learn and deepen understanding and familiarity with the unknown. New values come into mind and are projected onto the situation so that what was once seen as a problem becomes a fortunate opportunity for growth and discovery.

A pole shift probably will not occur in the near future, but even if it does, we don't have to see it as a problem. We need not dwell upon it as a source of fear and destruction. Prudent, practical survival preparations can and should be made if evidence convincingly indicates the need for it, but our primary task as citizens of Earth is to attune ourselves spiritually with Life—with the processes of the planet and the cosmos—and thereby understand that, if we are at the end of a cycle, we are being given an occasion to grow, to evolve, to transform ourselves on the basis of deeper understanding and wider vision.

Spiritually speaking, we should remember the words of Jesus to be not anxious about the morrow, but rather to consider the lilies of the field which are arrayed in glory and are tended by a loving Providence who tends us every bit as well. To the awakened mind, every experience is a blessing, even situations commonly labeled misfortune or even tragedy, disaster, catastrophe, cataclysm. The attuned consciousness will receive all it needs, and more, from a loving universe whose whole purpose is to nurture the evolution of organisms such as us to a higher state of being.

This can be likened to the birth of a baby. In such a situation, we don't focus all our attention and preparation on a few brief moments of birth pains. Rather, we joyfully look ahead to the presence among ourselves of a new child. We prepare to receive another into the family. We don't cower in a corner because there will be some degree of suffering during delivery. The mother prepares for delivery to ease the birth passage, knowing that the hours spent in delivery are worth the years spent in company with the child.

I'll conclude with a brief story given to me by a wise man from India who read *Pole Shift.* He told me: One day Mr. Plague was talking to the Keeper of the Akashic Records when he remembered that he had an appointment at a distant city. So he broke off conversation and rushed to keep his appointment. Now, the Akashic Records are a kind of metaphysical account book of everything which happens to humanity, so when Mr. Plague returned from the distant city, the Keeper of the Akashic Records asked him how many people had died on his mission. Mr. Plague, recognizing the need for accuracy, replied, "Five thousand died due to me and ten thousand due to fear."

Think about that. According to the predictions and prophecies, humanity is approaching one of the most critical junctures in the history of our planet. It seems clear that the significance of those predictions and prophecies is not in offering geophysical data but psychological guidance. Will there be global destruction or planetary transformation to a new world order—a New Age—based on love and wisdom? The choice, they say, is ours. The proper choice, the wise choice, is to shift our consciousness from the pole of egotism to the pole of enlightened living. We must realize that from the standpoint of God-realization, pole shifts are merely an exotic way to die. The *real* question is: Are we ready to die? Are we living our lives in surrender to God under all circumstances, even our own inevitable demise? Since the immortality pill hasn't been invented, none of us will get out of here alive. Pondering our mortality is the beginning of wisdom. If we do that, the pole shift predictions and prophecies will have served humanity and the planet alike by proving themselves "wrong" in the best manner possible.

Spring, 1991

Who was First?
—A Note on Marshall Wheeler
and Alfred Drayson

Marshall Wheeler and his short book *The Earth—Its Third Motion* were brought to my attention by Larry E. Arnold, a parascience researcher-journalist in Harrisburg, Pennsylvania. In a 1978 letter to me, Arnold noted two passages that he feels qualify Wheeler as "the father of modern pole shift predictions." I have already quoted one. The other is as follows:

> Just when the Earth experienced its last central turning [which Wheeler thought would cause comets to spin off the Earth, and later would revisit "the scene of its birthplace" to wreak another catastrophe; he attributed the comet of 1861 to a previous axis shift], *and at what future time the next may be expected* [emphasis added] may be approximated with a certain degree of accuracy by a series of observations at the 284th degree of longitude east from Greenwich, near Quito, Ecuador, South America. In my opinion the equator will become fully

rounded out to a perfect circle before another central turning... and that it will be found upon proper investigation that the equator is rising at the point indicated. The rate of its rising ascertained, the information sought may be given by simple calculations. It can at least be approximately ascertained what time has elapsed since the Earth's last central turn. It is now about twenty-four twenty-fifths of the time, judging by the Earth's equatorial movement" (p. 36).

Arnold remarks that there is no indication of the basis on which Wheeler determined the 24/25 factor, nor exactly when he thought the last axial shift occurred. "However, he clearly senses another shift is probable and attempts a rudimentary estimate as to when. Wheeler's estimate, whether a lucky guess or based on a higher level of awareness, looks rather impressive—especially when viewed in the scholastic climate within which it was made."

Even Wheeler may not be the "father of modern pole shift predictions," however, because a brief report in the January 1979 *Pole Watchers' Newsletter* notes that a British army officer, Alfred Wilks Drayson (1827–1901) wrote several books about pole shift that preceded Wheeler's. In 1859 Drayson published *Great Britain Has Been, and Will Again Be Within the Tropics*. He followed this with several others, including his 1888 *Thirty Thousand Years of the Earth's Past History, Read by Aid of the Discovery of the Second Rotation of the Earth.*

These books, writes pole watcher Robert J. Shadewald, inspired followers of Drayson to write their own. Among them were *Warmer Winters and the Earth's Tilt Fully Explained,* by Alfred H. Barley, and *The Change in the Climate and Its Cause,* by Reginald Adam Marriott.

"I haven't seen any of the above books," Shadewald reports, "but they are all listed in the printed *Catalogue of the British Museum Library.*" Nor have I seen them. I simply offer this note on Wheeler and Drayson for the historical record.

Appendix Two

Safe Geographical Areas

Various predictions name areas, principally in North America, that will be relatively undisturbed by earth changes and a pole shift. These "safety lands," as Edgar Cayce called them, are probably desirable to be in, assuming that the predictions are valid.

Cayce's reading 1152–11, given on 13 August 1941, contained this statement:

> Then the area where the entity [1152] is now located [Virginia Beach] will be among the safety lands—as will be portions of what is now Ohio, Indiana and Illinois and much of the southern portion of Canada, and the eastern portion of Canada; while the western land, much of that is to be disturbed in this land, as, of course, much in other lands.

Another Cayce reading describes Livingston, Montana, as a major seaport after the earth changes have ceased. Likewise, a dream that Cayce had late in life—which he considered to be prophetic—showed him reborn in a twenty-second-century lifetime and living in Nebraska on the seacoast.

Paul Solomon's readings—essentially in accord with Cayce's—indicate the Shenandoah Valley as a primary safety land. (Solomon has established Carmel-in-the-Valley there as a spiritual community that is expected eventually to contain thousands of people.) Bermuda will also be safe. The northeastern United States, away from coastal areas, is another area of relative stability, as is central and southern Florida (except for the coasts).

Hugh Auchincloss Brown's research pointed to East Africa—specifically Kenya—and the mid-Pacific as the pivot points for the next pole shift and thus as being relatively safe.

Aron Abrahamsen's readings indicate that New England, except for the coastal areas, will be safe. Vermont and New Hampshire, he told me privately, will probably be quite stable and relatively unaffected. His reading that outlines the western coastline after A.D. 2000 names Houston, Texas, as a major seaport.

Sun Bear points out that Indians generally feel the Rocky Mountains are a secure zone for passage through the time of purification. The same applies to the Allegheny range. Hopi land—the Four Corners area in Arizona and New Mexico—is considered by the Hopi to be the best possible haven during the geophysical cataclysms forecast to wrack the planet.

I have put this information into chart form, allowing space for you to add data as they may come your way from new sources of earth change and pole shift prophecies.

Major Safety Lands

Hugh Auchincloss Brown	Greenland, Antarctica, East Africa, Pacific Ocean pivot area
Chan Thomas	Southeast Africa
Emil Sepic	Far inland at high altitude
Edgar Cayce	Southern Canada, eastern Canada, Norfolk (Virginia) area, Livingston (Montana), portions of Ohio, Indiana and Illinois
Aron Abrahamsen	New England (inland), Houston

Paul Solomon	Shenandoah Valley, Bermuda, southern Florida, New England (inland)
Native Americans	Rocky Mountains, Appalachian Mountains, Four Corners area of Arizona-New Mexico-Utah-Colorado
Ruth Montgomery	Egypt, Mediterranean area, Gobi Desert
Tom Valentine	Western coast of Europe, mid-Pacific pivot area (near Hawaii)

Notice that I said these areas *probably* are good to be in if a pole shift occurs. Let me explain.

One of the difficulties I've had to deal with in writing this book is the danger of creating a self-fulfilling prophecy of disaster—bringing on through "scare tactics" that which is predicted. Another is the danger of releasing information to parties who are intellectually unprepared or—worse yet—spiritually unfit to receive it. In considering the situation, I have decided to be nonjudgmental and egalitarian. I am therefore releasing the data freely, letting what I see as self-correcting tendencies in society and nature do the work of nurturing those ready for what I call "survivolution" while "recycling" the rest. That is how evolution works, and my trust in a benevolent universe is greater than my sense of duty to various occult or metaphysical maxims about who should be given access to privileged information or esoteric wisdom. When in extremis, the standard regulations no longer apply.

I have given the foregoing data about safety lands with full awareness that I may be responsible in part for a sort of "land rush" even more aggressive than, say, the opening of the Oklahoma Territory. There are two basic modes of behavior in people that can be counted upon to operate in this situation. They are based respectively on fear and greed: the desire to "save your skin" and the desire to "get ahead." The mere fact that I have named certain localities may generate some degree of population relocation, which in turn would cause land values to rise. Real estate speculation could add to the pressures of an economy already in serious difficulty. If there is a way to "make a buck" from disaster, even if it is the end of the world, you can be sure someone will try. Hence the safety lands could become areas of extreme economic and social pressure well in advance of any major earth change activity. And if a geophysical cataclysm occurs, the most populous areas—including the so-called safety lands—could well be the sites of greatest danger, simply because of the population density. There would be greater competition for the necessities of life, and greater exposure to those who would act ruthlessly to get them.

A person would do well, therefore, to think carefully before selling his house and rushing to relocate in an area where he believes he will be able to save himself. He could, like the man in the ancient folk tale who sought to avoid a meeting with Death, actually be keeping an appointment in Samarra.

About the Author

JOHN WHITE, M.A.T., is an internationally known author and editor in the fields of consciousness research, human development, and parascience. He has held positions as Director of Education for the Institute of Noetic Sciences, a research organization founded by Apollo 14 astronaut, Edgar D. Mitchell, to study human potential and planetary problems, and as president of Alpha Logics, a school for self-directed growth in body, mind, and spirit.

White is author of *Pole Shift, A Practical Guide to Death and Dying, Everything You Want to Know about TM,* and a children's story, *The Christmas Mice.* He has also edited a number of anthologies, including *The Highest State of Consciousness; Psychic Exploration; What Is Meditation?; Frontiers of Consciousness; Future Science; Relax; Other Worlds, Other Universes;* and *Kundalini, Evolution and Enlightenment.* Articles and reviews by him have appeared in popular magazines such as

Reader's Digest, Saturday Review, Esquire, Science of Mind, Omni, Science Digest, and *New Age* and in professional journals and major newspapers such as *The New York Times, San Francisco Chronicle,* and *Chicago Sun-Times.*

White holds degrees from Dartmouth College and Yale University, and has taught English and journalism on the secondary and college levels. He is on the boards of various academic and spiritual organizations, and is an editorial contributor to several national publications, including *Venture Inward.* As a lecturer and seminar leader, he has appeared at colleges and universities and before public and professional organizations throughout the U.S. and Canada. He has also made radio and television appearances across the nation.

He and his wife Barbara have four children and live in Cheshire, Connecticut.

Index

What Is A.R.E.?

The Association for Research and Enlightenment, Inc. (A.R.E.®), is the international headquarters for the work of Edgar Cayce (1877-1945), who is considered the best-documented psychic of the twentieth century. Founded in 1931, the A.R.E. consists of a community of people from all walks of life and spiritual traditions, who have found meaningful and life-transformative insights from the readings of Edgar Cayce.

Although A.R.E. headquarters is located in Virginia Beach, Virginia—where visitors are always welcome—the A.R.E. community is a global network of individuals who offer conferences, educational activities, and fellowship around the world. People of every age are invited to participate in programs that focus on such topics as holistic health, dreams, reincarnation, ESP, the power of the mind, meditation, and personal spirituality.

In addition to study groups and various activities, the A.R.E. offers membership benefits and services, a bimonthly magazine, a newsletter, extracts from the Cayce readings, conferences, international tours, a massage school curriculum, an impressive volunteer network, a retreat-type camp for children and adults, and A.R.E. contacts around the world. A.R.E. also maintains an affiliation with Atlantic University, which offers a master's degree program in Transpersonal Studies.

For additional information about A.R.E. activities hosted near you, please contact:

A.R.E.
67th St. and Atlantic Ave.
P.O. Box 595
Virginia Beach, VA 23451-0595
(804) 428-3588

A.R.E. Press

A.R.E. Press is a publisher and distributor of books, audiotapes, and videos that offer guidance for a more fulfilling life. Our products are based on, or are compatible with, the concepts in the psychic readings of Edgar Cayce.

For a free catalog, please write to A.R.E. Press at the address below or call toll free 1-800-723-1112. For any other information, please call 804-428-3588.

A.R.E. Press
Sixty-Eighth & Atlantic Avenue
P.O. Box 656
Virginia Beach, VA 23451-0656